ENVIRONMENTAL
CLEANUP AT NAVY FACILITIES

ADAPTIVE SITE MANAGEMENT

Committee on Environmental Remediation at Naval Facilities

Water Science and Technology Board

Division on Earth and Life Studies

NATIONAL RESEARCH COUNCIL
OF THE NATIONAL ACADEMIES

THE NATIONAL ACADEMIES PRESS
Washington, D.C.
www.nap.edu

THE NATIONAL ACADEMIES PRESS • 500 5th St., N.W. • Washington, D.C. 20001

Support for this project was provided by the U.S. Navy under contract #N47408-00-C-7102.

International Standard Book Number 0-309-08748-1 (Book)
International Standard Book Number 0-309-51720-6 (PDF)
Library of Congress Control Number 2003108132

Printed in the United States of America.

THE NATIONAL ACADEMIES
Advisers to the Nation on Science, Engineering, and Medicine

The **National Academy of Sciences** is a private, nonprofit, self-perpetuating society of distinguished scholars engaged in scientific and engineering research, dedicated to the furtherance of science and technology and to their use for the general welfare. Upon the authority of the charter granted to it by the Congress in 1863, the Academy has a mandate that requires it to advise the federal government on scientific and technical matters. Dr. Bruce M. Alberts is president of the National Academy of Sciences.

The **National Academy of Engineering** was established in 1964, under the charter of the National Academy of Sciences, as a parallel organization of outstanding engineers. It is autonomous in its administration and in the selection of its members, sharing with the National Academy of Sciences the responsibility for advising the federal government. The National Academy of Engineering also sponsors engineering programs aimed at meeting national needs, encourages education and research, and recognizes the superior achievement of engineers. Dr. Wm. A. Wulf is president of the National Academy of Engineering.

The **Institute of Medicine** was established in 1970 by the National Academy of Sciences to secure the services of eminent members of appropriate professions in the examination of policy matters pertaining to the health of the public. The Institute acts under the responsibility given to the National Academy of Sciences by its congressional charter to be an adviser to the federal government and, upon its own initiative, to identify issues of medical care, research, and education. Dr. Harvey V. Fineberg is president of the Institute of Medicine.

The **National Research Council** was organized by the National Academy of Sciences in 1916 to associate the broad community of science and technology with the Academy's purposes of furthering knowledge and advising the federal government. Functioning in accordance with general policies determined by the Academy, the Council has become the principal operating agency of both the National Academy of Sciences and the National Academy of Engineering in providing services to the government, the public, and the scientific and engineering communities. The Council is administered jointly by both Academies and the Institute of Medicine. Dr. Bruce M. Alberts and Dr. Wm. A. Wulf are chair and vice chair, respectively, of the National Research Council.

iii

COMMITTEE ON ENVIRONMENTAL REMEDIATION
AT NAVAL FACILITIES

EDWARD J. BOUWER, *Chair*, Johns Hopkins University, Baltimore,
   Maryland
GENE F. PARKIN, *Vice Chair*, University of Iowa, Iowa City
SIDNEY B. GARLAND, Bechtel Jacobs Co., Oak Ridge, Tennessee
PATRICK E. HAAS, Mitretek Systems, San Antonio, Texas
ROBERT JOHNSON, Argonne National Laboratory, Argonne, Illinois
MICHELLE M. LORAH, U.S. Geological Survey, Baltimore, Maryland
FREDERICK G. POHLAND, University of Pittsburgh, Pittsburgh,
   Pennsylvania
DANNY D. REIBLE, Louisiana State University, Baton Rouge
LENNY M. SIEGEL, Center for Public Environmental Oversight,
   Mountain View, California
MITCHELL J. SMALL, Carnegie Mellon University, Pittsburgh,
   Pennsylvania
RALPH G. STAHL, Jr., DuPont Company, Wilmington, Delaware
ALICE D. STARK, New York State Department of Health, Albany
ALBERT J. VALOCCHI, University of Illinois, Urbana-Champaign
WILLIAM J. WALSH, Pepper Hamilton LLP, Washington, DC
CLAIRE WELTY, Drexel University, Philadelphia, Pennsylvania

Staff

LAURA J. EHLERS, Study Director
ANIKE L. JOHNSON, Project Assistant

Preface

Under the auspices of the Water Science and Technology Board (WSTB), the National Research Council (NRC) established the Committee on Environmental Remediation at Naval Facilities in 1997 to study issues associated with the remediation of contaminated soil, sediment, and groundwater at Navy facilities. The committee was initially established to provide guidance on the following three main areas pertinent to characterization and remediation of Navy facilities:

1. **Risk-based methodology.** What are the strengths and weaknesses of risk-based methodologies for cleaning up contaminated sites, including (but not limited to) the Risk-Based Corrective Action Standard (RBCA) devised by the American Society of Testing and Materials (ASTM)?

2. **Innovative technologies.** What innovative technologies are appropriate to assist the cleanup efforts at Navy facilities?

3. **Long-term monitoring.** For Navy facilities that will not be able to meet regulatory standards for cleanup in the near future, what guidance can be given for establishing and maintaining long-term monitoring at such sites?

The project was supported by the U.S. Navy with the stipulation that the three study topics listed above would be funded incrementally.

The first report produced by the committee addressed risk-based methodologies (Task 1 above), providing a review of existing risk-based methodologies including ASTM's Risk-Based Corrective Action, a description of their strengths and weaknesses, and a set of recommendations on how the Navy should proceed. After publication of the first report in 1999, the NRC convened a workshop with some committee members and about 30 Navy remedial project managers and others to better define a scope of work for future studies. In doing so, a proposal was developed that deviated somewhat in content from the second and third tasks above.

The goals of the Committee on Environmental Remediation at Naval Facilities Phase 2 were to address the following items (among others) related to the latter stages of site cleanup, including remedy selection, remedial operation, long-term monitoring, and site closeout.

Systems engineering approach. The study will define a decision-making framework that is embodied within a "systems engineering approach" to site cleanup.

Innovative technologies. The study will review the state of development of technologies for cleanup of groundwater, sediment, and soils, discussing the top two or three technologies that should be considered for the three to four greatest problems encountered by the Navy.

Changing the remedy over time. The study will consider how innovative technologies can be introduced after the remedy has been selected and how remedies can be adjusted over time.

Defining milestones for site closure. At many Navy sites, the continued operation of remedies beyond a certain level may not yield a marked improvement in site conditions. The study will consider when, and using what criteria, technologies should be "turned off."

These issues were identified by Navy managers as important to the growing number of sites where remedial goals have not been met despite continued operation of selected remedies. Most of these sites were characterized as those with recalcitrant contamination, including dense nonaqueous phase liquids, metals, and other persistent contaminants.

To address these issues, a new committee was convened that included six members from the Phase 1 effort along with nine new members. Their areas of expertise spanned from environmental engineering and hazardous waste management to systems analysis, sediment contamination, and public participation. The new committee convened its first meeting in July 2000 and met five additional times over the next two years. The resulting report promotes using the concept of *adaptive management* to move forward at those sites where progress in reaching cleanup goals has halted. Although the concept of *adaptive site management* is particularly applicable to those sites that have reached the latter stages of site cleanup, it encompasses all stages of hazardous waste remediation, and it is consistent with current federal regulations (e.g., Superfund). I believe the committee's efforts provide useful guidance for some of the environmental restoration challenges of the Navy, which should also be relevant to a broader universe of sites and facilities. Adaptive site management is especially appropriate for remaining sites, which tend to be larger and more complex than those that have already been cleaned up.

The study benefited greatly from contributions from various individuals who made presentations at our meetings, including Stephen Eikenberry, Naval Facilities Engineering Service Center; Walt Kovalick, EPA Technology Innovation Office; Kevin Mould, EPA Federal Facilities Restoration and Reuse Office; Carol Bass and Ken Lovelace, EPA Superfund Office; Patty Lovera, Center for Health, Environment and Justice; Mike Maughon, Cliff Casey, and Steve Beverly, Southern Division NAVFAC; Frank Chapelle, U.S. Geological Survey; Rob Simcik, TetraTech NUS; Steve Rosansky, Battelle; Steve Tsangaris, CH2M Hill; Tom Sale, Colorado State University; Arun Gavaskar, Battelle; Deanna Spehn; Sabine Apitz and Victoria Kirtay, SSC San Diego; and Chuck Newell, Groundwater Services, Inc.

The committee was fortunate to have taken several field trips in conjunction with committee meetings. The following individuals are thanked for their participation in organizing and guiding these trips: Mike Maughon, Southern Division NAVFAC; Steve Rosansky, Battelle; Sam Ross, J. A. Jones; Ken Richter, Bart Chadwick, and Sabine Apitz, SSC San Diego; and Bill Collins, Southwest Division NAVFAC. The committee was ably assisted in these field trips and other administrative matters by Suzanne Benoit Albertsen, Naval Facilities Engineering Service Center.

The success of this report depended upon highly dedicated staff and the work of the committee members. I thank Laura Ehlers, the NRC study director for this project. Laura coordinated the committee meetings, gathered information, actively participated in the committee discussions, offered insightful comments and input, suggested alternative paths forward, and prepared copious minutes of the meetings. Laura worked with the committee members to maximize their contributions and written material, synthesized and edited the final report, and made the majority of revisions in response to reviewers comments. I appreciate the efforts of Anike Johnson who took care of the many mailings and made local meeting arrangements. I thank Gene Parkin who assisted me as vice-chair. Gene's positive spirit and intellect are much appreciated. I would like to thank the committee members for providing a stimulating environment for addressing the study issues. I have enjoyed immensely the opportunity to work with such a talented and articulate group of professionals. I especially appreciate their willingness to spend time researching, writing, and revising their contributions.

More formally, the report has been reviewed by individuals chosen for their diverse perspectives and technical expertise, in accordance with procedures approved by the NRC's Report Review Committee. The purpose of this independent review is to provide candid and critical comments that will assist the authors and the NRC in making the published report as sound as possible and to ensure that the report meets institutional standards for objectivity, evidence, and responsiveness to the study charge. The reviews and draft manuscripts remain confidential to protect the integrity of the deliberative process. We thank the following individuals for their participation in the review of this report: W. Frank Bohlen, University of Connecticut; Teresa S. Bowers, Gradient Corporation; Mario Ierardi, Air Force Base Conversion Agency; Aaron A. Jennings, Case Western Reserve University; Michael C. Kavanaugh, Malcolm Pirnie, Inc.; Kai N. Lee, Williams College; Garrick E. Louis, University of Virginia; Stavros S. Papadopulos, S. S. Papadopulos Associates, Inc.; Peter M. Strauss, P. M. Strauss & Associates; and C. Herb Ward, Rice University.

Although the reviewers listed above have provided many constructive comments and suggestions, they were not asked to endorse the conclusions or recommendations, nor did they see the final draft of the report before its release. The review of this report was overseen by Richard A. Conway, Union Carbide Corporation (retired). Appointed by the NRC, he was responsible for making certain that an independent examination of this report was carried out in accordance with institutional procedures and that all review comments were carefully considered. Responsibility for the final content of this report rests entirely with the authoring committee and the NRC.

Edward J. Bouwer,
Chair

Contents

xi

Summary

The number of hazardous waste sites across the United States has grown to approximately 217,000, with billions of cubic yards of soil, sediment, and groundwater plumes requiring remediation. Contamination at these sites ranges from relatively easy-to-clean petroleum hydrocarbon spills to complex multicomponent, multiphase, heterogeneous subsurface solute masses resulting from a variety of past industrial and commercial practices. Sites contaminated with more recalcitrant contaminants or with more complex hydrogeologic features have proved to be a significant challenge on every level—technological, financial, legal, and sociopolitical.

Like many federal agencies, the Navy is a responsible party with a large liability in hazardous waste sites. The Navy's Environmental Restoration Program encompasses a wide array of contaminants reflecting the military's multiple purposes over the past 100 years as well as a diversity of locations, including coastal environments and inland waterways. Because efforts to remediate hazardous waste sites began as much as 20 years ago, a large percentage of identified hazardous waste sites have reached the latter stages of cleanup (i.e., beyond remedy selection). As the Navy plans completion of the Environmental Restoration Program, several unresolved remediation issues have become evident. Most important, conventional remediation technologies such as pump-and-treat have been shown to be inadequate in meeting drinking-water-level cleanup standards for many of the complex sites typical of Navy facilities (NRC, 1994). For certain treatment technologies, it has often been observed that the removal rate of contaminant mass tends to decline over time to the point where further expenditure of resources appears to achieve little or no additional mass reduction. In many cases it is not clear how to change or terminate remedies that have proved ineffective or how to change cleanup goals.

To obtain advice in overcoming these obstacles, the Navy requested the National Research Council to study issues associated with the latter stages of remediation of contaminated soil, sediment, and groundwater at Navy facilities. In particular, the committee that was formed was asked to evaluate the unique technological and regulatory problems present at those sites for which chosen remedies have been in place but for which cleanup goals have not been met. The following specific tasks were given:

- define a decision-making framework that is embodied within a "systems engineering approach" to site cleanup,
- review innovative technologies for cleanup of groundwater, sediment, and soils, focusing on the top technologies that should be considered for the three or four greatest Navy problems,
- consider how remedies can be altered over time to introduce innovative technologies where the chosen remedy is no longer optimal because of changing site conditions, limited efficacy of technologies, or the discovery of new contamination and/or exposure pathways, and
- define logical endpoints and milestones for site closure.

In response, this report proposes a comprehensive and flexible approach, referred to as "adaptive site management," for dealing with difficult-to-remediate hazardous waste sites over the long term. Although adaptive site management is entirely consistent with the current cleanup paradigm used at federal facilities (as principally defined by Superfund), it has additional features that stress knowledge generation and transmittal and that complement more traditional cleanup objectives in order for progress to be made at sites where recalcitrant contamination prevents site closure and subsequent unrestricted land use.

Adaptive site management is responsive to the concern of large responsible parties that current technologies have proved to be ineffective in reaching cleanup goals for many types of contamination. Many studies and reports have documented that there are still no proven technologies for addressing hydrogeologically complex sites contaminated with dense nonaqueous phase liquids (DNAPLs) and metals, which are the contaminants of concern at many federal facilities. A variety of technical factors—such as geological and flow heterogeneity as well as slow mass transfer from solid phases and free phase contamination—limit remediation effectiveness and lead to the "asymptote" effect where further operation of the remediation system does not significantly reduce contaminant levels. At the present time, there is very limited regulatory or policy guidance on what to do when the asymptote is reached before cleanup goals have been met as long as the remedy remains protective of human health and the environment. The goals of adaptive site management are to facilitate decision making when the effectiveness of the remedy reaches an asymptote prior to reaching the cleanup goal and, if necessary, to facilitate implementation of long-term stewardship (long-term management in DoD terminology). This approach can accommodate different cleanup objectives, it provides guidance at key decision-making points, and it is a mechanism for dealing with the uncertainty inherent in many remedial strategies—both engineered technologies and institutional controls.

ADAPTIVE SITE MANAGEMENT DESCRIBED

The predominant paradigm for site restoration in the United States has until relatively recently involved a highly linear, unidirectional march from site investigation to remedial action and eventually to site closure. However, as sites have advanced through the restoration process and the need for site management over the long term has in many cases become apparent, there has been a growing recognition that a more iterative approach is needed. Thus, this report advocates a broad systems approach that promotes effective knowledge generation (monitoring and fundamental research) and use of that knowledge to provide a wider range of decision options and thereby improve overall site management. These characteristics are embodied in the concept of adaptive management—an approach to resource management in which policies are implemented with the express recognition that the response of the system is uncertain, but with the intent that this response will be monitored, interpreted, and used to adjust programs in an iterative manner, leading to ongoing improvements in knowledge and performance. The committee has coined the term "adaptive *site* management" (ASM) to refer to the application of the adaptive management concept to hazardous waste cleanup.

Within the environmental arena, adaptive management concepts are timely, given the observed limitations in remediation effectiveness and the increased use of remedies like containment and institutional controls that will leave residual contamination in place for long periods. To date, the principal use of adaptive management has been for applications to wildlife and ecosystem management, water resources planning, and global climate change assessment. However, the concept of adaptation is not foreign to hazardous waste cleanup, and there are certainly cases where project managers have modified remedial activities in response to poor system performance. Over the last decade, a number of formal approaches have been developed to introduce adaptation specifically into data collection and site characterization activities, although adaptive management has not yet been incorporated into the remedial design and implementation process as a whole.

ASM formalizes questions and decisions that the remedial project manager and remediation team should address and reach consensus on to readily adapt to changes in technology, remedy effectiveness, and other external influences that impact the management of contaminated sites. The main tenets of ASM are that it:

- is applicable at various stages of site restoration,
- is applicable to a wide variety of sites regardless of the contaminants being addressed or remedies envisioned,
- provides a mechanism for the optimization of existing remedies, changing ineffective remedies, and refining the site conceptual model,
- formalizes the routine examination of monitoring data and how to act upon the data,
- incorporates public participation,
- recognizes uncertainty and suggests approaches to dealing with it, especially when institutional controls are used,
- stimulates the search for new, innovative technologies to replace older or inefficient approaches,
- stresses the need for pilot programs to test both new technologies as well as modifications of existing technologies that might enhance their effectiveness, and
- recognizes the increasing role of long-term stewardship.

ASM encompasses the initial steps of site management, including the site conceptual model and risk assessment. Additional detail on these steps is provided in Chapter 2. This summary, however, focuses on the latter stages of ASM: remedy selection and implementation, monitoring remedy performance, adapting the remedy or management goals to accommodate changing conditions and improve cost-effectiveness, and completing the remedy and closing out the site. Figure S-1 shows the latter stages of ASM, which is characterized by management decision periods (MDP) designed to take advantage of the feedback loops embedded in ASM, such that uncertainties in site restoration can be addressed. These MDPs are also formal opportunities for the remedial project manager and other project managers, regulators, and interested stakeholders to evaluate incoming and existing data to determine if the remedial technology is meeting its objectives and, if not, to reach agreement on what additional management steps need to be taken. These decisions would take into account pilot-scale work, changes in land use or stakeholder needs, improvements in analytical resolution which might point to the presence of additional contaminants, and monitoring data and other intelligence that may lead the remedial project manager to refine and/or revise a management decision.

The purpose of the first decision period, MDP1, is to ensure that the remedy selected is practicable and implementable under site-specific conditions and that an appropriate, well-designed monitoring plan is developed. This can be important where there has been a long lag time (years) between remedy selection and implementation such that initial assumptions may no longer be valid. Subsequent to MDP1 and once the remedy is implemented, several actions can potentially occur as part of ASM. Along with operation of the remedy, there are ongoing monitoring activities. Several performance-related questions—lumped under MDP2—characterize this phase of cleanup.

Denoted alongside remedy implementation in Figure S-1 is evaluation and experimentation—an activity unique to ASM and one of the hallmarks of adaptive management in general. It refers to the conducting of experiments and other research activities in parallel with implementation of the chosen remedy. This activity occurs ideally at the level of an individual site, in which portions of the site are devoted to experimentation while others are undergoing the chosen remedy, although it may refer to collecting information about experiments going on elsewhere, the results of which are relevant to specific sites. The evaluation and experimentation track is an opportunity to test innovative, less certain, sometimes riskier remedies that were not well enough established to be chosen as the initial remedy in the Record of Decision (ROD).

Later management decision periods give remedial project managers an opportunity to use information gained during evaluation and experimentation and routine monitoring to optimize the existing remedy, change the remedy, or even change the remedial goal. Depending on the action chosen, MDP3 may lead back to the initial steps of site management, remedy selection, or remedy redesign. MDP3 is a critical juncture

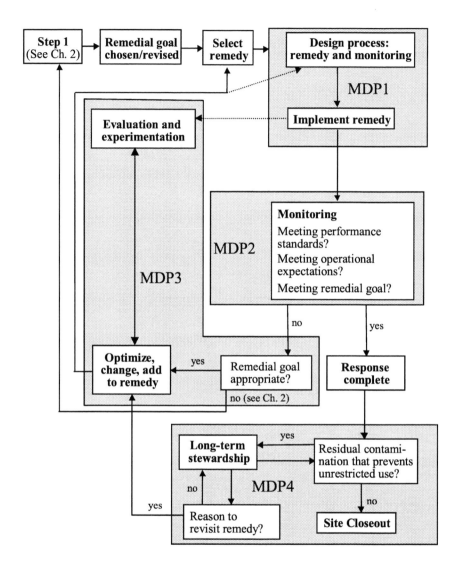

FIGURE S-1 Latter stages of adaptive site management: post-remedy selection. The shaded areas show the activities related to the management decision periods described in the text.

at which many current sites are stalled because of lack of information about alternatives and the absence of any regulatory incentive to change course.

The final major decision of adaptive site management is MDP4, during which sites with residual contamination levels above cleanup goals are periodically assessed. Like MDP3, this decision can lead to a change in remedy if it is found that alternative technologies exist that can help achieve cleanup goals. This also presents a departure from the current cleanup paradigm because the five-year review process that characterizes long-term stewardship does not support changing remedies unless the existing remedy is not protective of human health and the environment. When site managers, regulators, and the affected public have agreed that there are no unacceptable levels of contaminants left in place (i.e., the use is unrestricted), site closeout can proceed—the last step of ASM.

The following sections discuss different components of ASM in greater detail and provide key conclusions and recommendations. They correspond roughly to the organization of Chapters 3 through 6.

MONITORING AND DATA ANALYSIS DURING ADAPTIVE SITE MANAGEMENT

Management decision periods 2 through 4 require information in the form of quantitative data from a monitoring program and subsequent data analysis. For example, MDP2 involves the following key questions: (1) is the remedy meeting the performance standards (as set forth in the ROD or other binding documents), (2) are the operational expectations of the remedy being met (whether cost or other parameters that the remedial project manager and remediation team have set), and (3) is the remedial goal being met. Affirmative responses to these questions lead to "response complete" and eventually to MDP4, whereas negative responses lead to MDP3. Chapter 3 discusses in detail the parameters that should be measured to answer these questions, given the existing remedy and remedial goal, and several innovative monitoring techniques.

MDP3 allows for the remedy to be optimized, modified, or replaced entirely. Optimization of an existing remedy leads back to the "design process: remedy and monitoring" box, as denoted by the dashed line in Figure S-1. For those remedies that do not perform appropriately even after optimization, wholesale replacement may be required, necessitating a return to the "select remedy" box. Although a wide array of tools can help evaluate whether an additional remedial action or change is warranted once the point of diminishing returns has been reached, relatively simple graphical tools, described in Chapter 3, can be used. For example, in the case of groundwater contamination, contaminant concentration within the source area can be plotted over time; the need for a change may be evident when the slope of the line tangent to the performance curve approaches zero (the so-called asymptote) but the concentration remains above the site-specific remedial action goal. Such plots can also make it clear when continued operation of the existing remedy may incur substantial per unit costs with relatively little improvement in mass removal. These graphical techniques can also be utilized prior to initial remedy selection if enough information exists on the performance curves typical of certain treatment schemes. However, for most remedies, characteristic remedy curves and the predictive models that might generate such curves are not yet available. The following conclusions and recommendations from Chapter 3 concern the monitoring and data analysis aspects of ASM.

Plots of mass removal and/or concentration versus time or cost (or other metrics depending on the remedy) are objective and transparent tools for illustrating remedial effectiveness that should trigger when to either modify or optimize the existing remedy or to change the remedy. Such graphs should be used after remedy selection to address MDP2 and MDP3. Graphical representations should serve both to enhance stakeholder understanding of the options and to make better decisions about implementing or modifying remedies. At individual sites under investigation, the Navy, in consultation with all stakeholders, should select a unit cost for the continued operation of the remedial action, above which the existing remedy is no longer considered a tenable option.

The Navy should collect and analyze data to develop and validate predictive models of remedy performance. The remedy selection process could be made more quantitative and transparent with the provision of design guidance, charts, and models that summarize technology applications and predict their performance in different environmental settings.

Uncertainties in hydrogeological data, contaminant concentrations, and rates of remediation should be explicitly recognized in the development and application of performance plots. There are many sources of uncertainty in estimating the mass or risk reduction achieved by any remediation scheme. When sufficient site data are available, statistical methods can be used to estimate error or confidence bands on the performance plots. Site monitoring plans should be developed to ensure that the collected data serve to reduce uncertainty.

A concerted effort should be made to increase monitoring program effectiveness (and to reduce costs) by optimizing the selection of monitoring points, incorporating field analytics and innovative data collection technologies such as direct push, and adopting dynamic work plans and adaptive sampling and analysis techniques. Real-time *in situ* monitoring technologies should also be considered as those technologies mature. These techniques enhance the collection of information upon which ASM decision making is based. DoD should continue to support and foster research in chemical, physical, and biological techniques that would provide more rapid and adaptive approaches for monitoring remedy effectiveness.

EVALUATION AND EXPERIMENTATION

An essential feature of ASM is that it allows for a change in remedy—where the chosen approach is falling short of cleanup goals—that takes into account information about other potentially more effective remedies collected during evaluation and experimentation. Information from laboratory studies, on- or off-site pilot-scale activities, expert panel evaluations, literature reviews, or experience from other federal or private sector sites should be assessed on a regular basis to determine if a more effective remedy applicable to the site of concern exists. The evaluation and experimentation track of ASM specifically accommodates potential problems with remedy effectiveness by improving the understanding of the site (site conceptual model) and suggesting ways to enhance the performance of the existing remedy or guide the selection of an alternative remedy. Evaluation and experimentation can open up new opportunities to remediate and manage sites more effectively even when problems are not imminent. A more external benefit of evaluation and experimentation is that it can create an expanded database on the performance of remedial technologies. For a responsible party like the Navy that has a large number of hazardous waste sites, the external benefits of investing in learning (i.e., using what is learned in one place at other sites and in future decisions) can be substantial.

This parallel track is critical to overcoming the stalemate encountered at many sites where cleanup goals cannot be achieved. However, for this to succeed, potentially responsible parties in particular, and the federal government more generally, must make evaluation and experimentation an integral part of their overall remedial program. This feature of ASM differentiates it from the recent Navy guidance on remedy optimization, which does not explicitly specify the need for such activities.

There are numerous mechanisms for undertaking evaluation and experimentation at individual sites, and for obtaining relevant information and data externally (discussed in detail in Chapter 4). Some involve DoD agreements with U.S. Environmental Protection Agency (EPA) laboratories or offices, extramural grants with academic institutions or other non-governmental groups, or collaborative activities such as those conducted through the Remediation Technology Development Forum, a joint effort between EPA and private industry. Adoption of ASM would encourage the Navy to build stronger networks within the scientific and engineering communities in order to stay abreast of new technological developments that might prove applicable to future cleanup scenarios.

Although time will be required to test ideas and new technologies prior to a full-scale implementation, ASM should not be used as an argument for delaying important decisions while extensive analysis takes place. In fact, a hallmark of adaptive management is that more certain and sometimes simple actions are taken immediately while information is gathered about potentially more effective but less certain technologies. While evaluation and experimentation take place, the temporary inability to meet performance standards or other regulatory requirements should not be used as a basis for notices of deficiency or enforcement action. The following conclusions and recommendations address the role of evaluation and experimentation in ASM.

Evaluation and experimentation are integral to adaptive site management and should occur concurrently with remedy implementation. Improved understanding of a site through evaluation and experimentation can reduce the amount of uncertainty associated with the risk estimate, suggest ways to enhance the performance of the existing remedy, and help guide the selection of an alternative in case the remedy is ineffective in meeting cleanup goals. Employing evaluation and experimentation is most important for remedies likely to reach an asymptote prior to meeting the remedial goal, for sites with intractable contamination such as DNAPLs and metals, and where containment or institutional controls are used.

DoD should expand its programs that focus on developing and testing innovative remedial technologies and monitoring techniques. The lack of such research will result in DoD and others not having the new tools that can improve remedial programs and reduce long-term fiscal liabilities. Responsible federal agencies should collaborate closely with researchers in the public and private sectors to ensure that remedial project managers are trained and knowledgeable on innovative technologies that might be used to replace existing ineffective remedies.

Congress should make sure there are funds available to support the evaluation and experimentation track of adaptive site management. Although significant research efforts have been underway, unless the federal government provides new resources, only slow progress will be made toward finding cost-effective methods of reducing contaminant levels and meeting cleanup goals. Federal government support is needed to fill the gap left as a result of the lack of market incentives to develop innovative hazardous waste cleanup technologies.

Resource, timing, regulatory, and socioeconomic obstacles need to be overcome in order to fully adopt evaluation and experimentation as a component of ASM. Such obstacles include a lack of funds in federal cleanup programs beyond those needed to implement the chosen remedy; site manager perceptions that the results from research yield answers over time scales that are too slow to prove useful in optimizing existing remedies or in making informed decisions about when to replace a remedy; and the increasing use of containment and institutional controls, which has discouraged additional investment in the development of new remediation technologies.

INNOVATIVE TECHNOLOGIES

Chapter 5 reviews a variety of innovative technologies the Navy might consider during initial remedy selection, as replacements for existing remedies that have proved to be unsuccessful, or as additions to current remedies to better achieve cleanup goals or reduce cleanup time. Because the Navy identified its most pressing current problems as solvents and metals in soil and groundwater and sediment contamination, the focus is on these types of contamination and applicable remedial technologies, including *in situ* chemical oxidation, thermal treatment, permeable reactive barriers, enhanced bioremediation, technologies for treating contamination by inorganics, and several sediment management techniques.

Although all the technologies have their place, there is no clearly superior single remedy that can address even a small fraction of the Navy's contamination problems. In general, for the innovative technologies reviewed, there is a lack of refined evaluation procedures and peer-reviewed literature on their cost and performance—partly because their development is vendor-driven—making it impossible to fully evaluate their success or efficacy. Thus, further testing of innovative or new experimental technologies at selected sites is needed, both for site-specific application and if the results are likely to improve cleanup activities at other sites. In the evaluation of remedial options and technologies, the full life cycle of a technology and the management and disposition of all residuals that may be generated by the technology should be considered.

Optimization of existing remedies is also discussed in Chapter 5. Optimization can be as simple as ensuring that system components are still appropriate and are operating at design efficiency. Formal mathematical optimization can be used to evaluate well configuration and pumping rates in pump-and-treat or soil–vapor extraction systems for potential cost savings. In the course of taking such action, the degree of protectiveness of the remedial action at the site must not be reduced. More detailed instruction for site managers on how to optimize various remedial systems is required, because existing information in DoD guidance manuals is presented in very general terms and can be used only by persons who are already quite technically knowledgeable in the remediation field. In general, the reevaluation of the current remedy design for possible optimization should be a routine part of adaptive site management. The conclusions and recommendations below pertain to specific innovative technologies.

Site-specific analyses of the effectiveness of source removal technologies, including *in situ* chemical oxidation, thermal treatment, and enhanced bioremediation, are needed to better guide and justify remedy selection. Controlled field demonstrations are needed to evaluate the benefits (e.g., to groundwater quality) derived from partial mass removal from source zones and the compatibility of some technologies with natural attenuation. This should help in the determination of whether enough source mass can be removed to warrant the expense of implementing the technology.

Permeable reactive barriers can effectively treat a limited number of groundwater pollutants under well-defined hydrogeologic conditions. These pollutants include perchloroethylene, trichloroethylene, *cis*-dichloroethylene, and perhaps chromium (VI). The technology has been applied in the field for only seven years, so data on long-term performance are limited. Hydraulic capture remains a key issue in determining effectiveness, and the long-term integrity of these systems is unknown.

Because metal contaminants cannot be destroyed and their behavior and speciation are strongly coupled to site-specific conditions, remediation approaches for metal contaminants remain a challenge. Given that metals are frequently reported contaminants of concern at Navy sites, the Navy should devote resources to accelerate the development and field-scale testing of cost-effective technologies for mitigating risks from metal contaminants.

Presently, the only options that are routinely available for managing contaminated sediment include natural attenuation, capping either in situ or after dredged material removal, and dredging with disposal in confined disposal facilities or in upland landfills. Dredged material treatment options are under development and may be commercially available and viable in the future.

Treatment trains for the remediation of many contaminated sites are an important component of adaptive site management. Most sites are contaminated with multiple contaminants that may require different treatment processes. A common treatment train is source control in conjunction with monitored natural attenuation. This approach must be implemented with caution as certain source removal technologies can disrupt microbial metabolism via redox changes, removal of primary substrates, and creation of inhibitory conditions.

LONG-TERM STEWARDSHIP

Because many remedies today utilize containment and institutional controls rather than treatment of the contaminant source, residual contamination is expected to remain at these sites such that unrestricted use of soil, groundwater, and surface water will not be permitted. The activities needed to maintain such remedies collectively are called long-term stewardship, which is an integral part of ASM. Long-term stewardship starts when remediation, disposal, or stabilization is complete or, in the case of long-term remedial actions such as groundwater treatment, when the remedy is shown to be functioning properly. MDP4 during long-term stewardship provides the opportunity to ask the following questions: is there residual contamination that prevents unrestricted use, and is there a reason to revisit the remedy? The second of these questions represents a significant departure from the way many responsible parties usually conduct long-term stewardship. As shown in Figure S-1, this might lead to the replacement of containment or institutional controls with a more active remedial system. The motivation for asking this question is to be able to reach site closeout, which is not possible unless contamination is permanently reduced to levels that allow for unrestricted land use.

There are other reasons that site managers should reconsider remedies in place during long-term stewardship. Considerable cost savings may be possible if a new technology can alleviate the need for continual monitoring and/or maintenance. Also, there are substantial economic benefits to returning a site to unrestricted land uses. In the case of contaminants such as recalcitrant organic compounds, heavy metals, and radionuclides, land use controls may be required for hundreds or thousands of years. Over this timeframe, the cost and viability of land use controls is highly uncertain. Rarely is the complete future life-cycle cost of the original remedy compared to the life-cycle cost of implementing a new remedy. Clearly, an accurate assessment of the life-cycle costs is important to implementation of ASM.

The five-year review process of Superfund is the typical vehicle for assessing the protectiveness of remedies during long-term stewardship. However, as discussed in Chapter 6, the five-year review process currently does not support reconsideration of remedies during long-term stewardship if they are maintaining protectiveness of human health and the environment. Adoption of ASM would require expanding the scope of the five-year review process to include MDP4 and the basic elements of long-term stewardship—stewards, operations, public information, public participation, research, and information systems. This includes considering whether there are newly available technologies that could expeditiously lead to site closeout; if there were a more effective remedy available, the user would cycle back through the previous parts of ASM (see Figure S-1). Other improvements in the five-year review process are also suggested, particularly with regard to the lack of adequate public involvement in long-term stewardship, the performance and capability of the stewards, and the adequacy of funding for long-term stewardship.

During long-term stewardship, the remedy should be reconsidered as part of the five-year review, even if it is currently protective of human health and the environment. Because of changing conditions and the development of new technologies, there may be opportunities to achieve remedial goals for less money or in less time or achieve more aggressive remedial goals for the same money and time. Thus, it may be possible to replace land use controls with treatment remedies that will achieve unrestricted use and lead to site closeout. Only if unrestricted use levels are attained can the military and other agencies permanently remove sites from federal stewardship. The benefits of achieving site closeout include not only cost savings from reduced long-term operation and maintenance costs, but also increased taxes and minimization of potential future legal liability.

A government-wide policy for long-term stewardship (also known as long-term management) at federal sites is needed. Because all federal agencies with environmental restoration programs face this issue, ideally the Administration should convene an interagency task force for this purpose. This activity is needed to legitimize the basic elements of long-term stewardship and the expenditure of resources on these elements. As part of this effort, it will be important to develop a life-cycle cost estimating technique and appropriate discounting methods that reflect the timeframes for which long-term stewardship will be needed.

OVERALL CONCLUSIONS AND RECOMMENDATIONS

Adaptive management approaches are now being used by a number of public and private organizations to improve the quality of their operations and decisions. Like the domains of natural resource and business management where the principles of adaptive management have been applied, site cleanup planning, remediation, and stewardship involve significant uncertainty in system response. Despite these similarities, to the committee's knowledge adaptive management has never been formally used for hazardous waste cleanup. There is strong support for adaptive approaches already present in recent federal guidance on monitoring and remediation. For example, recently developed guidance for the Navy and Air Force recommends close scrutiny of existing remedies and monitoring data and informal optimization of remedies. The Navy guidance calls for an alternative strategy when a plot of cumulative mass removed versus time exhibits "an asymptotic condition" prior to attaining the cleanup goal. ASM goes further to suggest how to interpret the monitoring data, when to consider using new technologies, and how to reach site closure for all types of sites. The inclusion of evaluation and experimentation within ASM affords a way to manage uncertainty while moving forward with the cleanup process because conventional remedies can be implemented first while additional information is gained on innovative but more risky technologies.

The Navy and other federal agencies should adopt adaptive site management. The underlying statutes on hazardous waste management are consistent with adaptive site management, and existing regulatory guidance could be modified to be so. All the mechanisms for changing and modifying selected remedies—formal remedy amendments, RCRA permit modifications, contingency records of decision, five-year reviews, technical infeasibility waivers, and optimization studies, among others— can be encompassed by ASM. The Navy is currently drafting policy that will require periodic reviews of remedies, as prescribed by recent Navy guidance on optimization. Because ASM is broader in scope, it will be necessary for the federal agencies to develop guidance to further define the management decision periods that are inherent to ASM.

The responsible federal agency should solicit public involvement during each of the four management decision periods of ASM. Changes to the remedy, the remedial goals, and future land use should be issued only after consideration of public comments. Although many individual guidance documents mention public involvement, there is no coherent public involvement process described in existing guidance or practiced in the field *after* remedy selection. As part of the Restoration Advisory Board rule development process, DoD should work with regulators, public representatives, and other stakeholders to develop a menu of options for involving the public in the long-term oversight of cleanup programs at facilities where remedies or long-term stewardship activities are continuing.

Full-scale ASM that includes public participation during each decision period should be targeted to the more complex and high-risk sites where projected large costs are at stake. ASM is particularly appropriate for sites with multiple or recalcitrant contaminants and multiple stressors and heterogeneous hydrogeology because progress at such sites is likely to have stalled prior to reaching cleanup goals. Prior to widespread adoption, the Navy should consider pilot testing ASM at a limited number of high-risk, complex sites to allow Navy managers to better understand any transactional costs and delays that may accompany ASM implementation.

1
Introduction

The last 30 years have seen a rise in the nation's awareness of hazardous materials and how their discharge and ultimate disposal can affect public health and the environment. Approximately 217,000 contaminated sites that have as much as 31 million cubic yards of soil, 1.2 billion cubic yards of sediment, and 1.4 million acres of groundwater plumes require remediation to prevent adverse effects on public health and the environment from past military, industrial, agricultural, and commercial operations (EPA, 1997, 1998a). These sites range from those contaminated by relatively simple petroleum hydrocarbon spills to complex multicontaminant sites, of which there may be hundreds at federal facilities such as military bases. Table 1-1 lists the major classes of contaminants found at hazardous waste sites in the United States. The cost to clean up these 217,000 sites is estimated in EPA (1997) to be at least $187 billion, while the National Research Council's best guess of the present value cost is $280 billion using current cleanup policies, with a range from $140 billion if cleanup policies are less stringent to $630 billion if cleanup policy becomes more stringent (NRC, 1994)[1].

---

[1] Probst and Konisky (2001) provided unit cost estimates from actual Superfund expenditures, but did not produce a nationwide cleanup cost estimate that is comparable to either NRC (1994) or (EPA) 1997. Comparisons with other sources of information indicate the NRC values provide a reasonable estimate of the order of magnitude of cleanup costs likely over the next 30 years. OMB estimated the cost to cleanup property owned by the federal government of $234–389 billion over the next 75 years (Federal Facilities Policy Group, 1995). None of these cost estimates include munitions, chemical weapons, or other nonhazardous waste problems. No studies were found that specifically addressed the impact of the greater use of containment and institutional controls.

TABLE 1-1 Types of Contaminants Found at Hazardous Waste Sites

Contaminant Category	Example Constituents
Nonhalogenated volatile organic compounds (VOCs)	BTEX (benzene, toluene, ethylbenzene, xylene) Acetone Methyl ethyl ketone Methyl tert butyl ether (MTBE)
Halogenated VOCs	Tetrachloroethylene Trichloroethylene Cis-1,2-dichloroethylene Vinyl chloride;1,1,1-trichloroethane 1,1-Dichloroethane
Nonhalogenated semivolatile organic compounds (SVOCs)	Phthalates such as n-bis(2-ethylhexyl)phthalate 2-Nitroaniline Benzoic acid Polynuclear aromatic hydrocarbons (PAHs) such as naphthalene, anthracene, benzo(a)pyrene Non-halogenated pesticides/herbicides such as parathion
Halogenated SVOCs	Polychlorinated biphenyls (PCBs) Dioxins/furans Halogenated pesticides/herbicides such as 4,4'-DDD and 4,4'-DDT
Fuels	Gasoline range hydrocarbons Diesel range hydrocarbons Residual range hydrocarbons
Inorganics	Heavy metals such as lead, zinc, mercury, copper, cadmium, beryllium Nonmetallic elements such as arsenic Asbestos Inorganic cyanides Perchlorate
Radionuclides	Radium-224, -226 Cesium-134, -137
Explosives/propellants	Trinitrobenzenes (TNB) 2,4-Dinitrotoluene (2,4-DNT) 2,4,6-Trinitrotoluene (TNT) Nitrocellulose Hexhydro-1,3,5-trinitro-1,3,5-triazine (RDX) Octahydro-1,3,5,7-tetranitro-1,3,5,7-tetraocine (HMX)
Unexploded ordnance	NA

SOURCE: Adapted from FRTR (1998).

Growing public awareness of hazardous waste issues was triggered by key incidents in the 1970s at locations such as Love Canal, New York, and Times Beach, Missouri. In response to public concerns, two important environmental statutes were written into law: the Resource Conservation and Recovery Act (RCRA) of 1976 (42 USC 6901) and the Comprehensive Environmental Response, Compensation, and Liability Act (CERCLA) of 1980 (42 USC 9601). CERCLA (also known as Superfund) and RCRA mandate the identification of hazardous waste sites, their assessment for contamination and risk to humans and ecological receptors, and the process by which they should be remediated. The Superfund Amendments and Reauthorization Act (SARA) of 1986 brought all military facilities under the authority of the Superfund program. These laws and corresponding state statutes have instigated a massive effort to clean up thousands of hazardous waste sites across the country. In general, most of the sites that have been successfully cleaned up to "background" levels were relatively simple, with well-defined contamination or releases of predominantly degradable petroleum hydrocarbons to a subsurface area characterized by relatively homogeneous hydrogeology (NRC, 1994). Sites contaminated with more recalcitrant contaminants or with more complex hydrogeologic features have proved to be a significant challenge on every level—technological, financial, legal, and sociopolitical.

Figure 1-1 shows the steps in the CERCLA process—the cleanup paradigm used at the most complex hazardous waste sites, particularly those located on federal facilities (see NRC, 1999a, for a detailed description of CERCLA). The first half of the CERCLA process involves site characterization and risk assessment; the second half includes a variety of risk management activities, including selection and implementation of a remedy. The Department of Defense (DoD) and other federal (e.g., RCRA) and some state cleanup programs have developed their own terminology for individual steps in the cleanup process, although all include investigation, remedy evaluation and selection, the site-specific remedy design and construction, and ongoing remedy operation. This report predominately uses DoD and CERCLA terminology, and the reader is referred to each program for other nomenclature. As shown in Figure 1-2, the latter stages of site cleanup at military facilities are characterized by milestones such as remedy in place, response complete, and site closeout.

Given the time that has passed between the signing of RCRA and CERCLA and the present, a large percentage of identified hazardous waste sites have reached the latter stages of cleanup—that is, after selec-

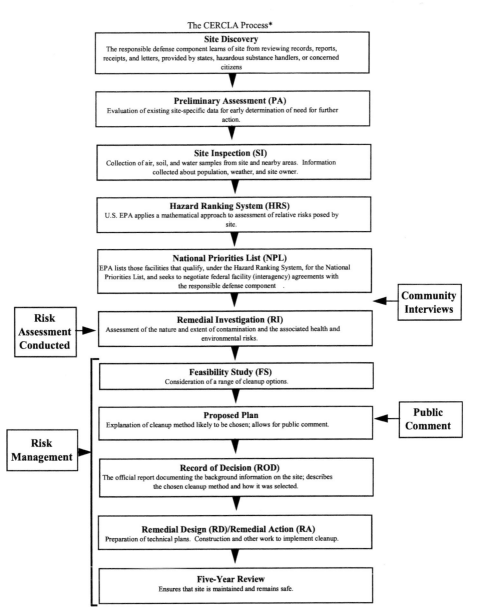

FIGURE 1-1 The steps of the CERCLA process. Each box describes the actions taken during the sequential phases. SOURCE: Adapted from EPA (1992).

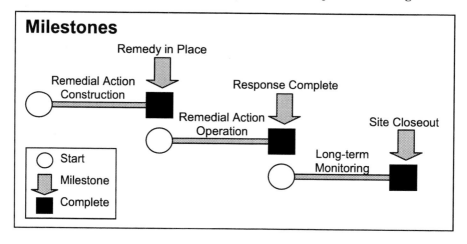

FIGURE 1-2 Milestones of the Defense Environmental Restoration Program. SOURCE: Office of the Deputy Under Secretary of Defense (1998).

tion of the remedy (as codified in a Record of Decision). For example, as of September 2000, 82 percent of the 1,509 sites on the National Priorities List (NPL), which lists many of the nation's most contaminated areas, had moved beyond remedy selection to remedy design, remedy construction, and construction completion (www.epa.gov/superfund/).[2] Only 3 percent of sites had not yet begun the remedial investigation process. Fourteen percent of the sites had studies underway or a remedy selected.

Several National Research Council (NRC) reports have addressed the cleanup of hazardous waste sites, primarily with an emphasis on contaminated soil and groundwater but more recently on contaminated sediment (NRC, 1993, 1994, 1997, 1999a,b, 2000, 2001). These reports have largely focused on risk assessment and treatment strategies applicable to the earlier phases of CERCLA or similar cleanup paradigms. At the request of the U.S. Navy—a responsible party with a large liability in

---

[2] Individual sites are evaluated based on the degree of hazard presented using the Hazard Ranking System. If a site receives a high enough score, usually the entire facility is placed on the NPL, even though there may be many other sites at the facility that do not pose as great a hazard. It should be noted that EPA defines site as the "entire facility, installation, unit", whereas DOD defines "site" as a discrete area of contamination on an installation or facility.

hazardous waste sites—this report specifically addresses the latter stages of site cleanup. In particular, it evaluates the unique technological and regulatory problems present at those sites for which chosen remedies have been in place but for which cleanup goals have not (for a number of reasons) been met. A comprehensive and flexible approach, known as adaptive site management, is proposed as a mechanism for dealing with such difficult sites over the long term.

The hazardous waste challenges facing the Navy are similar in nature to those facing many potentially responsible parties (PRPs) in the United States, with some important distinctions as mentioned below. As of September 30, 2001, the Navy identified 3,656 contaminated sites at active facilities and 1,020 sites at BRAC (Base Realignment and Closure) facilities (Navy, 2002). The majority of Navy sites are in the latter stages of cleanup; in fact, the number of sites that have reached "response complete" is 2,797—about 60 percent of all sites identified. Table 1-2 lists the number of Navy contaminated sites that are presently at each stage of the cleanup process. The Navy estimates that the remaining cumulative cost to complete remediation in today's dollars is $4.77 billion (Navy, 2002). [Since its inception in 1986, $3.81 billion has been spent in the Navy's Environmental Restoration Program (Navy, 2002.)]

CHARACTERISTICS OF NAVY FACILITIES

The goals of the Navy's Environmental Restoration Program are many, including to (Navy, 2002):

- Comply fully with federal, state, and local requirements;
- Act immediately to eliminate human exposure to contamination that poses an imminent threat, including removing or containing the contamination;
- Across the nation, first clean up sites posing the greatest risk to human health and the environment;
- Develop partnerships with the U.S. Environmental Protection Agency (EPA), state, and local regulatory agencies;
- Involve the local community through Restoration Advisory Boards. Encourage participation with timely information and opportunities for public comment, and take all comments into account when making decisions;

TABLE 1-2 Number of Navy Contaminated Groundwater, Soil, and Sediment Sites by Phase of Cleanup as of late 2001

Phase[a]	Groundwater[b]	Soil[c]	Sediment[c]	Soil/ Sediment[d]
Preliminary Assessment/ Site Inspection	93	439	22	447
Remedial Investigation/ Feasibility Study	467	834	226	908
Remedial Design	211	113	33	118
Remedial Action– Construction	234	110	24	114
Remedial Action– Operation	317	115	44	129
Response Complete	737	988	163	1,058
Long-Term Monitoring	259	63	28	68
Total[e]	1,894			2,842

SOURCE: NORM database, which is an internal database of contamination problems at Navy installations.
[a]See Figures 1-1 and 1-2 for descriptions of the phases.
[b]Column entries do not equal total because a site may be in multiple phases and thus counted more than once.
[c]A site may have both contaminated soil and sediment and would be counted on both lists. Thus, the soil/sediment column is not a total of the previous two columns.
[d]Unlike in groundwater column, sites with overlapping or multiple phases are classified under the earliest phase. Thus, the entries do equal the total.
[e]The total from this table (4,736) is larger than the number of Navy sites quoted in the main text (4,676) because some sites have both contaminated groundwater and soil/sediment and are counted twice in Table 1-2.

- Expedite the cleanup process and demonstrate a commitment to action; and
- Consider current, planned, and future land use when developing cleanup strategies.

Two major factors differentiate the Navy's Environmental Restoration Program from typical contaminated sites. The first is the wide array of contaminants reflecting the military's multiple purposes over the last 100 years. Some private industrial sites that have operated for decades may have released a vast array of contaminants, especially if the industrial owner and products have changed over time. However, the sheer number and diversity of Navy facilities and activities have led to a greater array of complex contaminants than any one industrial owner typically faces. Second, Navy facilities encompass a diversity of locations, including a high prevalence in coastal environments. Although other branches of the military and the Department of Energy have a comparable diversity and number of contaminant types and facility locations, the higher prevalence of Navy facilities along major coastal or inland waterways brings sediment contamination to the forefront as an important issue for the Navy. A hypothetical, yet "typical," Navy facility is shown in Figure 1-3 to illustrate the diversity of challenges present at these facilities.

Navy facilities are contaminated by an array of compounds that reflect a variety of activities. Most Navy facilities provide services, materials, and equipment to support aircraft, submarines, and ships. Large-scale transportation and industrial activities associated with this mission have resulted in contamination by marine and aviation fuel, solvents, and heavy metals. Chlorinated solvents have been widely used for equipment cleaning and degreasing. At some facilities, large-scale industrial activities such as designing and manufacturing weapons systems have introduced explosives, fuels, chlorinated solvents, and metals. Painting activities release heavy metals and solvents, and the discharge or spill of bilge water results in oil contamination. Site types associated with these activities include industrial landfills, waste disposal pits, above-ground and underground storage tanks, and spill sites. Groundwater and soil contaminant sources also include those associated with urban centers, such as municipal solid waste landfills, wastewater treatment plants, hospitals, laundries, golf courses, and underground storage tanks for automobile and truck fuels. Other potential sources of contamination include personnel training activities, such as "fire pits" where fire-fighting techniques have been practiced.

Of those listed in Table 1-1, certain contaminants either because of their sheer volume or their recalcitrance are more prevalent at hazardous waste sites both across the United States and at Navy facilities. Table 1-3 lists the top ten organic and inorganic compounds on the 1999 CERCLA Priority Hazardous Substances for the nation's most contaminated sites, which is based on frequency of occurrence, toxicity, and potential for human exposure. Arsenic and lead are the inorganic compounds of greatest concern, while vinyl chloride, benzene, and polychlorinated biphenyls are the most problematic organic compounds.

Determining whether these contaminants are present at Navy sites is hampered by the lack of a central, comprehensive compilation of data on Navy hazardous waste sites (although contaminant mass and concentration data for individual sites are collected at each facility). NRC (1999a)

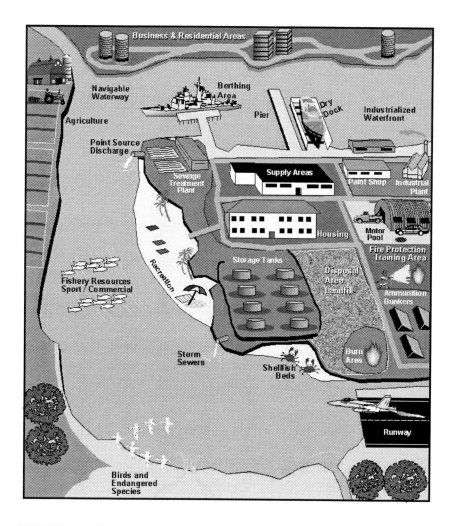

FIGURE 1-3 Typical Navy facility. SOURCE: Courtesy of the U.S. Navy, Marine Environmental Support Office.

TABLE 1-3 Top Ten Inorganic and Organic Contaminants on the 2001 CERCLA List of Priority Hazardous Substances (rank number is out of 275)

Inorganic Constituent	Rank	Organic Constituent	Rank
Arsenic	1	Vinyl chloride	4
Lead	2	Polychlorinated biphenyls	5
Mercury	3	Benzene	6
Cadmium	7	Benzo[a]pyrene	8
Chromium,	18	Polycyclic Aromatic Hydrocarbons (PAHs)	9
hexavalent		Benzo[b]fluoranthene	10
Beryllium	38	Chloroform	11
Cobalt	49	1,1,1-Tricholoro-2,2-bis(p-chlorophenyl)	12
Nickel	53	ethane (DDT)	
Zinc	73	Aroclor 1254 (PCB)	13
Chromium	76	Aroclor 1260 (PCB)	14

SOURCE: ATSDR (2002). Note that TCE is ranked 15, dibenzo(A,H)anthracene is 16, dieldrin is 17, chlordane is 19, and hexachlorobutadiene is 20.

indicates that organic contaminants are the most common contaminants found at Navy facilities. Petroleum, oil, and lubricants and hydrophobic organic contaminants exist at over half of all facilities, and pesticides are found at almost a quarter. Metals are found at over 42 percent of Navy facilities. It was not possible to determine from readily available data whether the reported contaminants exist as mixtures at a given site.

Subsequent to the NRC (1999a) evaluation, the Navy was asked to provide more detailed information about the contaminants found at its facilities and at individual sites. Surprisingly (given the 1999 evaluation), the NORM database (a database internal to the Navy which contains information on contamination at Navy installations) revealed that metals are the most frequently encountered contaminant type for all site types (groundwater, soil, and sediment). Lead, zinc, and barium are among the top five frequently encountered constituents at both groundwater and soil sites, with lead being the most common contaminant at all site types. Nonhalogenated volatile organic compounds (VOCs) are the second most frequent contaminant type at groundwater sites on Navy facilities, with benzene and toluene being the most frequently encountered constituent in this category. Benzo[a]pyrene and pyrene are the most frequently encountered organic compounds at soil and sediment sites, respectively. It is not clear how accurately these data reflect conditions at Navy facilities. If metals are the most prevalent contaminants at Navy sites, then remedial actions will need to be designed to better address this fact.

Although there are apparent similarities between the NORM database data and the data in Table 1-3, the data from the NORM database provided to the committee indicated prevalence only, whereas Table 1-3 also considers factors such as toxicity and mobility. If it has not already been accomplished, the Navy should consolidate its contaminant information into a single database, determine relative risk for all of its contaminated sites,[3] and then identify appropriate response strategies for those contaminants and sites posing the highest risks.

The Navy identified four scenarios as characteristic of their contaminated sites: petroleum hydrocarbons in soils and groundwater, chlorinated solvents in groundwater, metals in soils and groundwater, and persistent contaminants in sediment. These scenarios are generally consistent with the most frequently occurring contaminants at Navy facilities (see Table 1-1 in NRC, 1999a). The chemical properties and the fate-and-transport mechanisms for these four contamination scenarios are discussed below to provide background on their ease of remediation and the innovative technologies that are discussed later. Unexploded ordinance, radioactivity, and other less prevalent compounds are not considered further in this report.

Figure 1-4 shows the universe of pathways of human exposure to hazardous waste. At any given site, some pathways will predominate over others and control both the risk assessment and the remedial goal chosen. Ecological receptors are the primary driver for risk assessment at many hazardous waste sites, particularly where contaminated sediment is involved.

Petroleum Hydrocarbons in Soil and Groundwater

Although generally not considered high-risk or difficult to remediate, sites contaminated with petroleum hydrocarbons remain a concern because of their sheer number. Petroleum hydrocarbons include components of gasoline [benzene, toluene, ethylbenzene, and xylene (BTEX) and oxygenates such as MTBE] as well as other fuels. When free-phase hydrocarbons are released to soil, they are retained on soil pores until sufficient hydrocarbon has spilled to saturate the soil. Once the soil is saturated, such nonaqueous phase liquids (NAPLs) typically accumulate

---

[3] This could be accomplished, for example, using the qualitative Relative Risk Site Evaluation Framework developed for the Navy by Anderson and Bowes (1997).

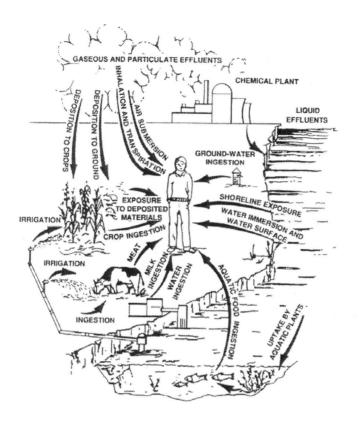

FIGURE 1-4 Pathways of human exposure to hazardous waste. SOURCE: Reprinted, with permission, from the National Research Council (1991). © (1991) National Academies Press.

in a layer on the water table. The more soluble constituents will be transported with the groundwater. Volatilization also may be a significant transport mechanism for the lighter hydrocarbons near the soil surface. A cartoon of these processes is shown in Figure 1-5. Inhalation of vapors from the vadose zone (as in confined areas like a basement) or direct ingestion of soil are frequently considered exposure pathways for petroleum hydrocarbon contamination.

Lighter petroleum hydrocarbons are relatively mobile and are more readily biodegradable than are other types of organic contaminants because they can serve as the primary substrate for many bacteria widely distributed in nature. The biodegradation rate and the metabolic products produced are controlled primarily by the types of hydrocarbons present and the availability of electron acceptors and nutrients required by the microorganisms. Generally, biodegradation is more rapid if oxygen is present to serve as the electron acceptor. Under anaerobic conditions, microorganisms use alternative electron acceptors, including nitrate, iron, sulfate, and carbon dioxide. Heavier petroleum hydrocarbons, including waste oils and crude oils, contain polyaromatic hydrocarbons (PAHs) that have relatively lower degradability. The lower degradation rate of PAHs is partly a consequence of their structural complexity and partly a consequence of their limited solubility in water and strong tendency to sorb to solids, which limits their bioavailability to microbes compared to the more soluble compounds.

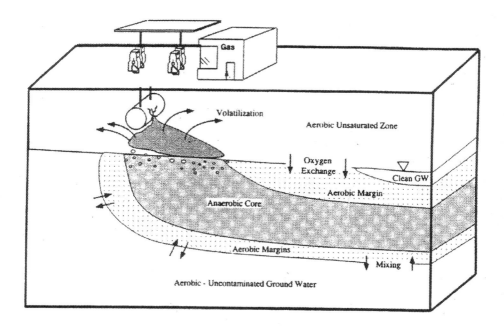

FIGURE 1-5 A schematic of petroleum hydrocarbon contamination of soil and groundwater. SOURCE: Reprinted, with permission, from Norris and Mathews (1994). © (1994) Lewis Publishers.

Chlorinated Solvents in Groundwater

Compared to petroleum hydrocarbons, chlorinated solvents are bio-degradable under a more limited set of environmental conditions. Because chlorinated solvents have a relatively high oxidation state, they are not easily susceptible to oxidation reactions and are biodegraded more easily through reduction reactions under anaerobic conditions where the compounds act as an electron acceptor (reductive dechlorination). Achieving complete dechlorination is critical because many of the chlorinated intermediate products, such as vinyl chloride, are as toxic as or more toxic than the parent compound. Transformation of chlorinated solvents requires the presence of electron-donor substrates and a consortium of microorganisms (Bouwer, 1992; NRC, 2000). If contaminants such as BTEX are present to act as electron donors, microbial reduction of chlorinated solvents is possible as long as the groundwater remains anaerobic. However, most groundwater systems tend to be organic carbon-poor, making it difficult to achieve complete transformation of solvents (Chapelle, 1993). In addition to reductive dechlorination, trichloroethylene (TCE) and other chlorinated VOCs are susceptible to cometabolic oxidation by aerobic microorganisms that have oxygenases with broad substrate specificity. Methanotrophs—microorganisms that primarily oxidize methane for energy and growth using methane monooxygenases—are one group of aerobic bacteria that have been shown to transform TCE through cometabolic oxidation (Little et al., 1988; Tsien et al., 1989).

Chlorinated solvents are also more difficult to remediate than petroleum hydrocarbons because free-phase chlorinated solvents (dense nonaqueous phase liquids or DNAPLs) are denser than water and can migrate deep into the saturated zone, which tends to lessen the effectiveness of conventional cleanup technologies, especially in fractured rock environments. Chlorinated solvents are sufficiently soluble and weakly sorbed to solid phases such that long dissolved plumes often form from DNAPL pools, further complicating remediation. The many phases characteristic of DNAPL contamination—from vapor and soil-bound to free-phase and dissolved in groundwater—are shown in Figure 1-6. The leaching-to-groundwater pathway primarily drives risk assessment for sites contaminated with chlorinated solvents.

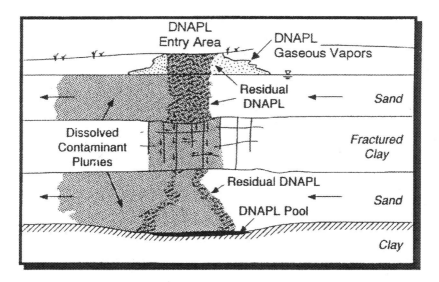

FIGURE 1-6 Fate of DNAPLs in the subsurface following a spill event. SOURCE: Reprinted, with permission, from Cohen and Mercer (1993). © (1993) C. K. Smoley and Sons.

Inorganics in Soils and Groundwater

Cleanup of metals is challenging because, unlike most organic contaminants, metals cannot be destroyed by chemical or biological reactions. In addition, the speciation of metals (as determined by the geochemistry of the water system) significantly affects their mobility and toxicity. Metals that form anions or oxyanions in solution, such as chromium, are often mobile in oxic environments (depending on the other constituents present) but can form relatively insoluble mineral precipitates in reducing environments. These metals also commonly form coprecipitates with iron and sulfide under reducing conditions. Metals that form cationic dissolved species, including cadmium, copper, lead, mercury, and zinc, are mobile in acidic environments. These metals form relatively insoluble carbonate, hydroxide, or sulfide minerals at moderate to high pH. Sorption onto mineral surfaces present in aquifers or bottom sediments also affects the mobility of metals. Because common hydroxide and silicate mineral surfaces carry a negative charge at near-neutral pH conditions, they will strongly sorb many cationic metals and thus reduce their mobility. In contrast, if the system is acidic, cationic metal ions tend not to sorb and tend to be very mobile. Arsenic presents a particularly complex situation. Arsenate, which dominates in aerobic environments, generally binds tenaciously to solids within soils and sediments, particularly hydrous oxides of ferric iron. Arsenite also forms strong complexes on iron (hydr)oxides and iron-sulfide minerals but it has a narrow adsorption envelop centered around pH 7, and it does not partition extensively on aluminum-hydroxide or aluminosilicate minerals (e.g., kaolinite). Thus, in non-sulfidic systems where ferric (hydr)oxides are absent or undergoing degradation, or where the pH deviates appreciably from neutrality, one can expect arsenic to partition to the solution phase. When more than one metal contaminant is present at a site, conditions that lower the mobility of one metal may enhance the mobility of another. Finally, the presence of organic compounds also affects the mobility of some metals through the formation of organic complexes. These organic complexes, such as those formed with arsenic and mercury, tend to be more toxic than the inorganic forms. Equilibrium modeling of elemental speciation is now a commonplace practice. There is growing realization, however, that in many environments speciation is under kinetic rather than thermodynamic control. Analytical methods capable of documenting speciation are therefore especially important.

Persistent Contaminants in Sediment

Because of the active hydrologic, geomorphic, and biogeochemical conditions found in sediment environments, only certain highly persistent classes of contaminants are considered problematic when associated with sediments, including metals and hydrophobic organics that have low solubility and a strong tendency for sorption. Numerous metals fall into this category such as lead, arsenic, and tri-butyl tin. The organic contaminants include PAHs, polychlorinated biphenyls (PCBs), and pesticides. PAHs are neutral, nonpolar organic molecules that contain two or more benzene rings and may also contain alkyl substituents or nitrogen, oxygen, or sulfur substitution for an aromatic ring carbon. PCBs are synthetic compounds composed of the biphenyl structure with 1–10 chlorine atoms, resulting in 209 different congeners. Although PCBs have been banned in the United States, they were once used widely as capacitor dielectrics, transformer coolants, heat transfer fluids, plasticizers, and fire retardants.

Unfortunately, the same characteristics that lead to their accumulation in sediment—immobility and resistance to chemical and microbial transformation—greatly limit degradation of these contaminants *in situ*. Microbial degradation of PAHs is minimal because of the anaerobic nature of most sediments. In addition, aged or weathered sediments often contain a resistant fraction of PAH that is not bioavailable for microbial degradation (NRC, 2001). PCBs have very limited solubility in water, are nonvolatile, and have very slow microbial degradation rates, making them stable under ambient conditions. Pesticide fate-and-transport properties are extremely complex and variable, and depend on the type of pesticide, on how the pesticide entered the environment, and on environmental conditions at the site.

CHALLENGES ASSOCIATED WITH REMEDIATION OF NAVY FACILITIES

Most Navy installations are located in coastal areas, where contaminated groundwater, soil, and sediment are close to environmentally sensitive habitats and surrounding communities. Of the 67 Navy facilities that are on the NPL, 43 percent are located in the coastal areas of California, Florida, Virginia, and Washington. These Navy facilities include Atlantic, Pacific, and Gulf Coast settings, resulting in considerable complexity in the suite of climatic, geomorphic, hydrogeologic, and ecosystem characteristics that affect characterization and remediation. Beyond those facilities on the NPL, Navy contaminated sites are located in Hawaii, Alaska, Guam, and Puerto Rico, resulting in the Navy's having a high diversity of locations to address.

Given that many of its facilities are located in coastal areas and are near bodies of navigable water, the Navy has a large liability in contaminated sediments. As many as 110 facilities have identified sediment contamination, including the Pearl Harbor Naval Complex, Hawaii; the Long Beach Naval Complex, the Alameda Naval Air Station, and the San Diego Naval Complex, all in California; and facilities along the Chesapeake Bay (Apitz, 2001). Many coastal, harbor, and estuary hazardous waste sites are still in the remedial investigation and feasibility study stages of remediation. This is partly because the cost and size of the sediment problem is very large and was consequently deferred throughout most of the 1970s and 1980s, and partly because the affected receptors are ecological rather than human.

The remediation of contaminated sediment poses unique challenges compared to the remediation of soil and groundwater. First, most soil and groundwater remediation projects are based on human health risk assessments, while most sediment studies begin with ecological risk assessment—a field that is less well developed (in terms of scientific methods and procedures), is less familiar to Navy project managers, and inherently complex due to food chain interactions (as illustrated in Figure 1-7). Second, hydrodynamics (tides, wave action, and currents), sedimentation and erosion, and human activities such as dredging and channelization can affect contaminant distribution either through direct transport of contaminated sediment or dissolved constituents or through mobilizing contaminants previously bound to sediment. A third complication

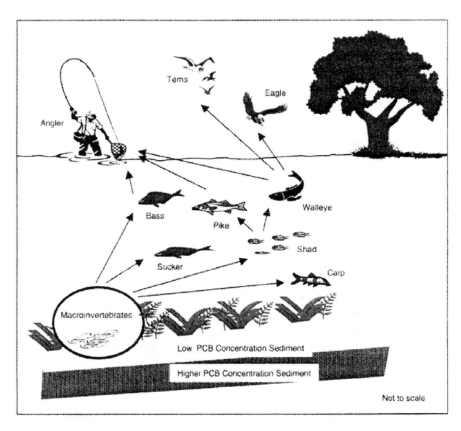

FIGURE 1-7 Sediment contaminants like PCBs transfer between multiple levels of a food chain, and bioaccumulate in certain species, making ecological risk assessment and remediation of sediments a challenge. SOURCE: Reprinted, with permission, from the National Research Council (2001). © (2001) National Academies Press.

involves the area of sediment affected by the contamination and the duration of that impact. Heavy metals and synthetic organic compounds, which are common at Navy sites, tend to accumulate in sediments and may persist at detectable concentrations for years or decades. Contamination that is sufficient to impair biological processes across various trophic levels exerts a particularly widespread effect that may be felt in areas and receptors far distant from the facility that was the original source of the contamination. Finally, differentiating between the relative contributions of various sources to sediment contamination is a challenge because sediments are integrators of multiple sources. If multiple sources are actively contributing to contamination, it can be difficult to determine whether a proposed remedial action at, for example, a Navy facility will lead to an improvement in ecological condition (Stahl and Swindoll, 1999; Swindoll et al., 2000). Unfortunately, there is limited guidance on assessing so-called multiple stressors or conducting comparative ecological risk assessment, although Suter (1999) suggests that frameworks in existence today can be applied to assessing risks from multiple activities.

One of the primary complications inherent at hazardous waste sites located in coastal areas are the numerous exposure pathways and ecological receptors that must be taken into consideration. For example, there is the potential for the discharge of groundwater contaminants to wetlands and surface water, which greatly complicates site characterization, risk assessment, and remedy selection beyond that normally encountered at inland groundwater sites (Winter et al., 1998; EPA, 2000a). Sharp gradients in sediment organic carbon concentrations and mineralogy, microbial activity, and porewater redox characteristics found in locations where surface water and groundwater meet can result in extreme changes in abiotic and biotic transformation of organic contaminants (Lendvay et al., 1998; Lorah and Olsen, 1999a,b) and inorganic contaminants (Benner et al., 1995; Harvey and Fuller, 1998) over small spatial scales.

Ecological risk assessment at hazardous waste sites in coastal environments must encompass an enormous diversity of potential receptors. Important commercial stocks of finfish and shellfish are highly dependent upon the estuarine environment to provide spawning and nursery grounds (Tait and De Santo, 1975). Many bird species utilize coastal areas for habitat, feeding, resting, or nesting, including *Gaviiformes* (divers), *Podicipediformes* (grebes), *Pelecaniformes* (pelicans, cormorants, anhingas), *Ciconiiformes* (herons, ibises, and spoonbills), and *Charadriiformes* (snipes, plovers, oyster catchers, gulls, terns, and skimmers)

(Howard and Moore, 1991). Navy facilities also are likely to support important habitats for resting, feeding, and reproduction of various mammalian species, both aquatic (otters, seals, and sea lions) and terrestrial (mice, voles, shrews, and muskrat) in the near-shore and wetland areas.

Facilities located in Florida, Hawaii, and Alaska present special and challenging cases. Assessing impacts on coral-dominated ecosystems in Hawaii, Florida, and Puerto Rico may require methods not routinely applied at other contaminated sites. Likewise, in Alaska there are habitats (tundra) and potential receptors (brown bear) that are not found in the lower 48 states. The life-history and other important basic biological data necessary for risk assessment may be missing for some bird species that are found only in the Hawaiian islands, necessitating the collection of these data on rare or highly localized species at particular facilities. In Alaska, the large number of migratory birds residing there during the summer months while raising their offspring must be considered. As with estuarine and marine aquatic species using the estuary for a rearing area, there may be a significant proportion of juvenile migratory birds inhabiting areas subject to contaminant releases from coastal facilities.

Although urbanization is dominant outside Navy facilities in many coastal areas, there may be large terrestrial areas within a facility that have not been altered since the facility was occupied originally and may be functioning as habitat refuges. Because of their importance from a national security perspective, many Navy facilities restrict human movement onto the facility and offset the perimeter from residences or commercial entities. These areas may be important for feeding, nesting, and resting for terrestrial species. The same may be true of some of the aquatic environments on the site. More important, some larger facilities may contain remnants of ecosystems that have vanished elsewhere over the last 50–80 years and, as such, may represent unique biological reservoirs worthy of protection. Even if there are no such remnants present, these habitat islands may still provide refuge necessary to the continued existence of some regionally important rare plants and animals.

The high prevalence of wetlands at Navy facilities is an additional area of concern because of the importance of freshwater and estuarine wetlands in providing various services, including diverse food webs and nutrient cycling functions, transport and degradation of contaminants, and provision of breeding grounds for important commercial species (NRC, 1995). Estuarine and marine wetlands along the Atlantic, Gulf, and Pacific Coasts have been greatly reduced in area since European settlement (Dahl, 1990), such that their associated habitats and biological communities will be in need of careful consideration during hazardous waste management at Navy facilities.

TRENDS IN REMEDY SELECTION

As the nation's remediation efforts under CERCLA and RCRA have matured over the last two decades from investigation to implementation, several trends in remedy selection are evident. According to an EPA evaluation of 757 construction completion sites, the most commonly utilized cleanup approaches are (1) excavating hazardous soil and solid waste (352 sites), (2) capping (348 sites), and (3) pumping and treating contaminated groundwater (284 sites) (EPA, 2001a). These statistics reflect the fact that more than one technology can be used at an individual site. Most remedies have been chosen to treat so-called principal threats; thus, from 1982 to 1999 treatment of groundwater was selected at a majority of sites rather than containment or offsite disposal (EPA, 1996, 2001a). In fact, treatment at all or a portion of the source areas at Superfund sites increased from 14 percent to 30 percent in the 1982–1986 period to a peak of 74 percent in 1993 and then decreased to around 45 percent in 1999 (Figure 1-8a,b). Concomitantly, containment of the source area followed an opposite trend. The use of institutional controls, monitoring, and other remedies (beyond containment and treatment) has increased steadily over the 1982–1999 period, such that institutional controls are now part of the remedy at 368 of 757 construction completion sites (EPA, 2001a).

Monitored natural attenuation (MNA) alone or in conjunction with other remedial actions increased from 0 percent in 1982 to between 13 percent and 25 percent in the 1997 to 1999 period (Figure 1-8c). A common reason cited for selecting MNA at contaminated soil and groundwater sites is "low and decreasing concentrations of contaminants at the site."

Innovative technologies, defined by EPA as those technologies or applications of technologies that have had limited full-scale application, have been selected in only 19 percent of the cases in which treatment is involved (EPA, 2001b). The rate of selection of innovative treatment technologies has decreased consistent with the overall trend in the selection of treatment technologies. Partly for this reason, there is effectively little traditional economic incentive for the small business entrepreneurial research sector to develop innovative cleanup technology (NRC, 1997). As a result, research on innovative treatment technologies is sponsored almost exclusively by federal agencies and, in some circumstances, by individual companies and industry groups that have joined with federal agencies in seeking more cost-effective solutions to common problems.

These long-term trends in remedy selection should not be surprising. In the early 1980s, knowledge of the technical capabilities of permanent remedies for contaminated sites was limited. After CERCLA was amended in 1986 giving a preference for permanent remedies and attainment of drinking water standards in groundwater, the number of treatment remedies dramatically increased. The primary treatment technology for contaminated soils, solid waste, and some contaminated sediments was high-temperature incineration, which is the most expensive method of treatment (EPA, 2000b). As a result, the unit cost of hazardous waste cleanups and the estimates of the long-term remediation costs escalated dramatically.

By the early 1990s, new knowledge about the limitations of technology became available. NRC (1994) reported that it is not feasible to reduce groundwater concentrations to drinking water standards or health-based cleanup goals with existing technology in a reasonable time frame (decades) at a large number of contaminated sites. Similarly, the DoD Inspector General concluded that 78 pump-and-treat systems operated as of 1996 remediate contamination slowly, cost $40 million annually, and will not attain cleanup goals within a reasonable period of time[4] (DoD, 1998). The report noted that as of 1998, these pump-and-treat systems "continued to operate without any form of review to determine their efficiency and effectiveness." The cumulative long-term cost of these systems was estimated to be as much as $2.3 billion in the year 2020 assuming 97 systems. As discussed in Chapter 5, optimizing the operation of these pump-and-treat systems will be critically important to the success of remediation at many facilities, including Navy sites.

In response to the rising costs of contaminated-site cleanups and the growing recognition of the limitations of technology, federal and state regulatory agencies issued a number of explicit policies that led to the acceptance of more containment, as reflected in the trends discussed above. EPA's 1990 Superfund remedy rules state that even though permanent remedies are preferred, EPA expects to use treatment to address the principal threats posed by a site, wherever "practicable," and engi-

---

[4] The determination of a reasonable period of time varies both within and between federal and state agencies. It has been previously noted to be 30 years (EPA, 1988) and 70 years (EPA, 1989) but no exact determination exists.

A

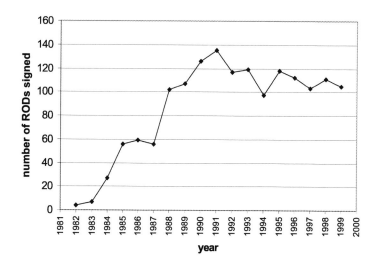

B

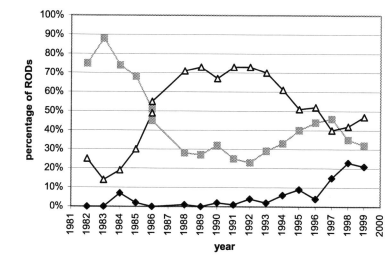

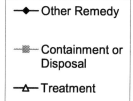

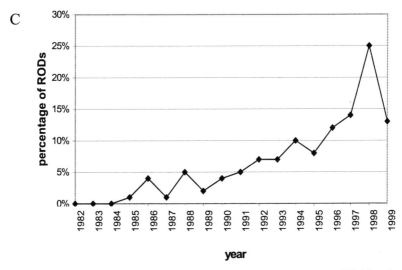

FIGURE 1-8 Trends in remedy selection, 1982 to 1999. (A) Number of RODs signed. (B) Percentage of different types of RODs. (C) Percentage of RODs with monitored natural attenuation as the remedy. SOURCE: EPA (2001b).

neering controls, such as containment, for sites that pose a relatively low long-term threat (EPA, 1991). Indeed, EPA's polychlorinated biphenyl (PCB) disposal rule allows soil contaminated with low levels of PCBs and other wastes to remain at a site as long as human health and the environment are protected from an unreasonable risk (EPA, 1998b).

The increased use of containment and monitored natural attenuation is likely to continue and will have several important implications. Because these remedies result in contamination remaining onsite, continual monitoring including the five-year review process will be required. Indeed, a recent report (NRC, 2000) highlights the complexity of assessing the performance of natural attenuation and emphasizes the need for long-term monitoring. In addition, groundwater remedial actions and monitoring activities at CERCLA, RCRA corrective action, and non-state sites cannot legally be terminated unless the chemicals remaining at the site are no longer a significant threat to human health or the environment.

At the vast majority of containment sites where the groundwater cleanup goal remains a drinking-water maximum contaminant level (MCL), cleanup times will extend from a few years to thousands of years, with the actual treatment time highly uncertain (NRC, 1994). Reducing the time required for remedy operation and monitoring is the motivation for developing innovative technologies, many of which are expected to focus on source removal.

STATEMENT OF TASK AND REPORT ROADMAP

Although 2,797 out of 4,676 Navy sites have achieved "response complete" (Navy, 2002), these sites consist mainly of petroleum hydro-carbon contamination and other problems that are relatively easy to address. These numbers may even include sites that were found, on closer inspection, to not be contaminated. Thus, the bulk of the difficult sites remain to be completely remediated. This is reflected in the Navy's cleanup budget, which is disproportionately allocated to the most contaminated sites. According to data provided by the Navy in 1998, the 59 percent of sites that ranked as high risk[5] comprised 81 percent of total cleanup costs, while low risk sites comprised 25 percent of the ranked sites but only 8 percent of the cost.

As the Navy plans for completion of the Environmental Restoration Program, several issues have become evident. First, the average time for completion of a cleanup remedy at Superfund sites (once a site has been placed on the NPL) is 10.6 years (GAO, 1998). Despite the fact that environmental remediation at Navy facilities has proceeded for a short period of time compared to the decades of military operations that are the source of present-day contamination, there is pressure to reduce the time needed to restore these sites, particularly when property is slated for transfer of ownership under BRAC. Second, conventional remediation technologies, such as pump-and-treat for groundwater cleanup, have been shown to be inadequate in meeting drinking-water-level cleanup standards for many of the complex sites typical of Navy facilities (NRC, 1994). Finally, decision making during the latter part of cleanup is unstructured, partly because the number of complex sites reaching this stage has been relatively low. For example, it is not clear how to change or terminate remedies that have proved to be ineffective or how to change cleanup goals.

---

[5] According to the Relative Risk Site Evaluation model.

To obtain advice in overcoming these obstacles, the Navy requested the NRC to study issues associated with the remediation of contaminated soil, sediment, and groundwater at Navy facilities and provide guidance on risk-based methodologies, innovative technologies, and long-term monitoring. The NRC committee's first report (NRC, 1999a) reviewed existing risk-based methodologies, described their strengths and weaknesses, and recommended a risk-based decision-making approach for the Navy. As a follow-up activity, the committee was asked to provide guidance on the latter stages of site remediation, including remedy selection, remedy operation, long-term monitoring, and site closeout. In particular, the committee was asked to define a decision-making framework that is embodied within a "systems engineering approach" to site cleanup. It was asked to review innovative technologies for cleanup of groundwater, sediment, and soils, focusing on the top technologies that should be considered for the three or four greatest problems encountered by the Navy. It was also asked to consider how remedies could be altered over time to introduce innovative technologies. This would be applicable in cases where the chosen remedy is no longer optimal because of changing site conditions, the limited efficacy of technologies, or the discovery of new contamination and/or exposure pathways. Finally, the committee was asked to define logical endpoints and milestones for site closure, including determinations of technical impracticability.

This report proposes a decision-making approach for site cleanup—adaptive site management (ASM)—that considers the entire lifecycle of a remedial project. The components of ASM include site characterization, risk assessment, selection and implementation of the remedy, monitoring performance of the remedy, adapting the remedy or management goals to accommodate new knowledge and changing conditions in order to improve performance and cost efficiency, long-term stewardship, and site closeout. ASM facilitates making decisions about when remedies can be changed due to ineffectiveness, when to incorporate a new technology, when remedies can be discontinued, and when site cleanup goals should be revised. Chapter 2 introduces the multiobjective nature of cleanup and the ineffectiveness of current remedies, which are problems that can be accommodated by ASM. The specific components of ASM are then described. Chapter 3 discusses the environmental monitoring needed to support adaptive site management and the interpretation of monitoring data. Adaptive site management is characterized by an evaluation and experimentation track that occurs in parallel with remedy operation. This activity is discussed in Chapter 4. Innovative technologies of relevance to Navy sites are the focus of Chapter 5. Finally, Chapter 6 considers the nontechnical barriers to the use of adaptive site management, such as regulatory constraints, and the roles of public participation and long-term stewardship in adaptive site management. Because there are similarities between the Navy cleanup program and those of other potentially responsible parties, the conclusions and recommendations are applicable to a broad universe of sites, including those at any federal facility. Thus, the report is intended not only for Navy remedial project managers but also for higher level managers and decision makers within the Navy Environmental Restoration Program and their counterparts in other federal agencies and private organizations that have a sizable cleanup liability.

REFERENCES

Anderson, J. L., and M. D. Bowes (Center for Naval Analysis). 1997. Appendix A: Relative Risk, from Department of the Navy Environmental Cleanup Program: Cost Control.

Apitz, S. E. 2001. Navigating in uncertain waters: the science, technology and politics of managing contaminated marine sediments. Presentation to National Research Council Committee on Environmental Remediation at Navy Facilities. February 28, 2001. San Diego, CA.

Agency for Toxic Substances and Disease Registry (ATSDR). 2002. 2001 CERCLA list of priority hazardous substances. Available at http://www.atsdr.cdc.gov/clist.html.

Benner, S. G., J. N. Moore, and E. Smart. 1995. Geochemical processes in a transition zone between surface water and acidic, metal-rich groundwater. Environ. Sci. Technol. 29:1789–1795.

Bouwer, E. J. 1992. Bioremediation of organic contaminants in the subsurface. Pp. 287–318 In: Environmental Microbiology. R. Mitchell (ed.). New York: John Wiley.

Chapelle, F. H. 1993. Ground-water microbiology and geochemistry. New York: John Wiley & Sons.

Cohen, R. M., and J. W. Mercer. 1993. DNAPL site evaluation. Boca Raton, FL: C. K. Smoley.

Dahl, T. E. 1990. Wetlands losses in the United States: 1780s to 1980s. Washington, DC: U.S. Fish and Wildlife Service.

Department of Defense. 1998. Evaluation of DoD waste site groundwater pump-and-treat operations. Report No. 98-090, Project No. 6CB-0057. Washington, DC: Office of the Inspector General.

Environmental Protection Agency (EPA). 1988. Guidance for conducting remedial investigations and feasibility studies under CERCLA (interim final). EPA 540/G-89/004. Washington, DC: EPA OSWER.

EPA. 1989. Risk assessment guidance for Superfund (RAGS). Volume 1– Human health evaluation manual. EPA/540/1-89/002. Washington, DC: EPA.

EPA. 1991. See 40 C.F.R. § 300.430 (a)(1)(iii), cited in: A guide to principal threat and low level threat wastes (OSWER Directive 9380-.3-06FS, 1991). Available at: http://www.epa.gov/superfund/resources/gwdocs/threat.pdf.

EPA. 1992. Understanding Superfund risk assessment. USEPA 9285.7-06FS. Washington, DC: EPA.

EPA. 1996. Memorandum from Elliott P. Laws, Assistant Administrator of the Office of Solid Waste and Emergency Response, to Superfund, RCRA, UST, and CEPP National Policy Managers Federal Facilities Leadership Council and Brownfields Coordinators, Re: Initiatives to promote innovative technology in waste management programs. April 29, 1996.

EPA. 1997. Cleaning up the nation's waste sites: markets and technology trends (1996 edition). EPA 542-R96-005. Washington, DC: EPA Technology Innovation Office.

EPA. 1998a. The incidence and severity of sediment contamination in surface waters of the United States. Volume 1: National sediment quality survey. EPA 823-R-97-006. Washington, DC: EPA Office of Water.

EPA. 1998b. Disposal of polychlorinated biphenyls (PCBs) (final rule). Federal Register 63:35384–35386.

EPA. 2000a. Proceedings of the ground-water/surface-water interactions workshop. EPA/542/R-00/007. Washington, DC: EPA.

EPA. 2000b. Land disposal restrictions; advance notice of proposed rulemaking. Federal Register 65:37932–37938.

EPA. 2001a. Focus on construction completion, 757th completion. EPA 540-F-00-023. Washington, DC: EPA.

EPA. 2001b. Treatment technologies for site cleanup: annual status report (10th edition). EPA 542-R-01-004. Washington, DC: EPA.

Federal Facilities Policy Group. 1995. Improving federal facilities cleanup. Washington, DC: OMB Council on Environmental Quality.

Federal Remediation Technologies Roundtable (FRTR). 1998. Guide to documenting and managing cost and performance information for remediation projects. EPA 542-B-98-007. Washington, DC: EPA.

General Accounting Office (GAO). 1998. Superfund: times to complete site listing and cleanup. GAO/T-RCED-98-74. Washington, DC: GAO.

Harvey, J. W., and C. C. Fuller. 1998. Effect of enhanced manganese oxidation in the hyporheic zone on basin-scale geochemical mass balance. Water Resources Research 32(8):2441–2451.

Howard, R., and A. Moore. 1991. A complete checklist of the birds of the world (2nd edition). San Diego, CA: Academic Press.

Lendvay, J. M., S. M. Dean, and P. Adriaens. 1998. Temporal and spatial trends in biogeochemical conditions at a groundwater/surface water interface: implications for natural bioattenuation. Environ. Sci. Technol. 32:3472–3478.

Little, C. D., A. V. Palumbo, S. E. Herbes, M. E. Lidstrom, R. L. Tyndall, and P. J. Gilmer. 1988. Trichloroethylene biodegradation by a methane-oxidizing bacterium. Appl. Environ. Microbiol. 54(4):951–956.

Lorah, M. M., and L. D. Olsen. 1999a. Degradation of 1,1,2,2-tetrachloroethane in a freshwater tidal wetland: field and laboratory evidence. Environ. Sci. Technol. 33(2):227–234.

Lorah, M. M., and L. D. Olsen. 1999b. Natural attenuation of chlorinated volatile organic compounds in a freshwater tidal wetland: field evidence of anaerobic biodegradation. Water Resources Research 35(12):3811–3827.

National Research Council (NRC). 1991. Environmental exposure: report of a symposium. Washington, DC: National Academy Press.

NRC. 1993. In situ bioremediation: when does it work? Washington, DC: National Academy Press.

NRC. 1994. Alternatives for ground water cleanup. Washington, DC: National Academy Press.

NRC. 1995. Wetlands: characteristics and boundaries. Washington, DC: National Academy Press.

NRC. 1997. Innovations in ground water and soil cleanup: from concept to commercialization. Washington, DC: National Academy Press.

NRC. 1999a. Environmental cleanup at Navy facilities: risk-based methods. Washington, DC: National Academy Press.

NRC. 1999b. Groundwater and soil cleanup: improving management of persistent contaminants. Washington, DC: National Academy Press.

NRC. 2000. Natural attenuation for groundwater remediation. Washington, DC: National Academy Press.

NRC. 2001. A risk management strategy for PCB contaminated sediments. Washington, DC: National Academy Press.

Navy. 2002. Environmental Restoration, Report for Fiscal Years 2002–2006.

Norris, R. D., and J. D. Matthews. 1994. Handbook of bioremediation. Ann Arbor, MI: Lewis Publishers.

Office of the Deputy Under-Secretary of Defense (Environmental Security). 1998. Management guidance for the Defense Environmental Restoration Program. Washington, DC: Department of Defense.

Probst, K. N., and D. M. Konisky. 2001. Superfund's future: what will it cost? Washington, DC: Resources for the Future.

Stahl, R. G., Jr., and C. M. Swindoll. 1999. Invited perspective: the role of natural remediation in ecological risk assessment. Human and Ecological Risk Assessment 5(2):219–223.

Suter, G. W. 1999. A framework for assessment of ecological risks from multiple activities. Human and Ecological Risk Assessment 5(2):397–413.

Swindoll, C. M., R. G. Stahl, Jr., and S. Ells (eds.). 2000. Natural remediation of environmental contaminants: its role in ecological risk assessment and risk management. Pensacola, FL: SETAC Press.

Tait, R. V., and R. S. De Santo. 1975. Elements of marine ecology. New York: Springer-Verlag.

Tsien, H., G. A. Brusseau, R. S. Hanson, and L. P. Wackett. 1989. Biodegradation of trichloroethylene by *Methylosinus trichosporium* OB3b. Appl. Environ. Microbiol. 55(12):3155–3161.

Winter, T. C., J. W. Harvey, O. L. Franke, and W. M. Alley. 1998. Ground water and surface water: a single resource. USGS Circular 1139. Denver, CO: U.S. Geological Survey.

2

Adaptive Site Management

MOTIVATION FOR ADAPTIVE SITE MANAGEMENT

The Navy's request for guidance on its environmental cleanup program was prompted in part by the ineffectiveness of current remedies to meet cleanup goals at their major remaining hazardous waste sites, particularly those high-risk sites contaminated with recalcitrant chlorinated solvents and metals, often in complex hydrogeologic or sediment settings. Remediating and reaching closure for these types of sites has proved to be elusive in the context of current technologies. In addition to the ineffectiveness of many remedies, the Navy is also struggling with how to balance and meet different remediation goals, such as risk reduction, attainment of drinking water standards, and complete removal of the source of contamination. This chapter first explores these two basic problems (the multiobjective nature of cleanup and the ineffectiveness of current remedies to meet cleanup goals) and then introduces adaptive site management—an approach that can address these problems while encompassing all stages of cleanup.

Multiobjective Nature of Site Cleanup

Contaminated sites can pose multiple hazards to human and ecological health, natural resources, and the economic and social welfare of surrounding communities. In a similar vein, the objectives for site cleanup and restoration are multidimensional and often evolve over time. The eight key objectives are:

1. To protect the health and safety of those on the site and in surrounding communities,

2. To ensure the ecological viability and health of native plants and animals, and migratory species,

3. To protect and restore natural land and water resources,

4. To promote positive economic value and development in the area of the site,

5. To comply with all applicable laws and regulations governing the site and the cleanup process,

6. To promote positive participation and communication with the local community and other affected stakeholders,

7. To advance the understanding of site contamination and cleanup processes (technical, managerial and social), and

8. To accomplish each of these objectives in an affordable, cost-effective, and efficient manner.

These objectives are usually pursued with different emphasis and urgency in different phases of a site cleanup effort. Following site discovery, the first priority is to eliminate immediate threats to human health and safety. Open contamination is enclosed, leaks are plugged, and, if necessary, local residents are switched to alternative sources of drinking water and other protective measures are implemented. Similarly, acute risks to wildlife and aquatic species are controlled to eliminate fish kills, animal poisonings, and other effects that could threaten the viability of ecosystem populations on or near the site. Virtually all Department of Defense (DoD) and other federal sites in the United States have passed beyond this initial phase of site discovery and "emergency response." Following the control of immediate site hazards, cleanup and management can emphasize different remediation goals and objectives. A broad range of operational objectives have evolved over the last 20 years, from complete soil, aquifer, or sediment restoration to use of a technology-based approach to goals based on minimizing long-term risk to humans and the environment ("risk-based" objectives).

The objectives listed above are closely related to the set of nine criteria established in the National Contingency Plan (NCP) for the evaluation of a proposed remediation plan.[1] Thus, for example, the first NCP criterion of overall protection of human health and the environment is

---

[1] The NCP criteria (EPA, 1990) include (1) overall protection of human health and the environment; (2) compliance with the chemical-specific standards that are considered the statutorily required "applicable or relevant and appropriate requirements" (ARARs); (3) long-term effectiveness and permanence; (4) reduction of toxicity, mobility, or volume through the use of treatment; (5) short-term effectiveness; (6) implementability; (7) cost; (8) state acceptance, and; (9) community acceptance.

divided into two separate objectives—one for human health (objective 1 above) and one for ecosystem protection (objective 2 above)—because the steps needed to pursue these objectives are not always fully coincident.

The NCP criterion for compliance with the chemical-specific standards that are considered the statutorily required "applicable or relevant and appropriate requirements" (ARARs) is equivalent to this report's fifth objective for regulatory compliance with Superfund and applicable state requirements. For contaminants in groundwater, a typical ARAR would be the maximum contaminant level (MCL) for that compound, if one exists. Some states may adopt a "complete restoration" and thus more stringent goal as the site-specific cleanup objective, with which the Navy would need to comply. Superfund and some state regulatory programs allow nonresidential land use assumptions to be considered in the selection of cleanup levels and remedies, so long as selected remedies are protective of human health and the environment.

The three NCP criteria of long-term effectiveness and permanence, short-term effectiveness, and implementability are not specifically noted in the list on page 2 because these features of a remediation plan are all essential to ensure that the other objectives are met. Similarly, reduction of toxicity, mobility, or volume through the use of treatment is the operational objective of a site cleanup needed to accomplish the broader objectives, and is addressed in a subsequent discussion.

The seventh NCP criterion regarding cost is equivalent to this report's eighth and final objective, because it constrains the extent to which all other objectives can be met. Cost minimization is a key objective in any public or private endeavor, although the weight placed on cost depends on the relevant statute and site-specific factors (EPA, 1996a, 1997a). Given the long-term requirements of site cleanup and stewardship at many sites, estimating costs over the full life cycle of a project is difficult. Approaches that appear cost-effective because of lower capital and initial operating costs may in the long term be more costly, especially if unanticipated problems arise in remediation performance and/or site conditions. Better anticipation of such problems, both initially and through ongoing data collection and evaluation, and ensuring that flexibility is maintained for improving or changing remediation technologies when needed, are key elements of the adaptive site management approach proposed later in this chapter. As discussed in the recent *Guidance for Optimizing Remedial Action Operation (RAO)* report for the Navy (NAVFAC, 2001), careful assessment of operation and maintenance costs for site cleanup plans can reveal many opportunities for cost reduction (see Box 2-1).

The eighth and ninth criteria of the NCP (state acceptance and community acceptance) are related to this report's fifth objective for regulatory compliance and sixth objective for positive participation and communication with the local community and other affected stakeholders. Positive participation includes community involvement in the development of remediation proposals rather than implying (as "community acceptance" does) that the community be involved only after remedial plans are proposed. As recognized by EPA (1999a, 2000a), effective public participation and input into the planning of a site cleanup are both a means—to ensure that the remediation plan can be implemented without costly delays and conflict—and an end—because public participation is a core value of a democratic society. Promoting participation and communication with the local community and other affected stakeholders applies to all aspects of a military site's operation, but it is especially important in dealing with health and safety risks to the public, where trust is easy to lose, but very difficult to regain (Slovic, 1993).

BOX 2-1
Important Elements of Long-Term Site Remediation Operation and Maintenance Costs, and Opportunities for Reducing These Costs
(from NAVFAC, 2001, Table 6-1)

Labor—Labor costs can be minimized through the use of remote and automated data-acquisition systems; the use of base personnel for routine operation and maintenance; and the contracting of the operation and maintenance for similar systems in bulk packages, achieving economies of scale and reducing administrative burden.

Analytical Costs—Long-term, frequent, and spatially extensive analysis of many chemical and biological parameters is expensive, but can be reduced by focusing on data needed to track remediation effectiveness; by using onsite analyses for measurements taken frequently; by seeking bulk analysis discounts for coordinated sampling events; and by reducing regulatory sampling frequencies if compliance is demonstrated on a consistent basis.

Power/Utilities—Energy and utility efficiency can be improved by the proper sizing of equipment; the use of periodic, pulse modes for *in situ* operations; and creative, onsite use of treated water for cooling water, landscaping, fire response supply, etc. (thereby reducing the need for purchasing such supplies).

Repairs—System repair costs can be controlled by using standardized system designs with common replacement parts and by maintaining careful records to ensure full use of vendor warranties.

Our list of objectives includes three that are not explicitly mentioned in the NCP remedy criteria: to protect and restore natural land and water resources (objective 3 above), to promote positive economic value and development in the area of the site (objective 4 above), and to advance site-specific and more general scientific knowledge (objective 7 above). The inclusion of objectives 3 and 4 is based on the fundamental importance of these issues in environmental and economic policy and on the committee's professional experience as to what is important to states and communities. These objectives are especially likely to arise as a key component of long-term site stewardship and management efforts, once the more immediate threats to health and safety are addressed. The restoration of land and water resources and the return of economic value may or may not be linked for a given site. For some uses, ensuring that there are no (or minimal) risks to health and safety may be sufficient to allow surrounding economic development (including use of lands for recreational or species preservation purposes, if this is the locally targeted objective) to proceed, even though some land and water (especially groundwater) contamination remains. In other locations, planned uses may dictate a more complete cleanup. When site cleanup is critical to an economic or community development plan for a region, strong community and political pressure will be brought to bear both to identify cleanup criteria that can be met in a timely manner and to proceed with the needed effort to reach this objective.

Our seventh objective of advancing knowledge during a site cleanup effort—both knowledge of the site itself and broader insights applicable to other sites—is usually secondary, and as a practical matter may be hard to justify to site managers and the public alike. However, because the science of cleaning up hazardous waste sites is often insufficient to attain even risk-based remediation goals, advancing scientific knowledge must be a component of site remediation. That is, such learning is essential if the other cleanup objectives are to be met in an effective manner. Although scientific study cannot be the principal driver for site cleanup (taking precedence over essential health, safety, and economic objectives), failure to take advantage of opportunities to use data and experiences acquired as part of a cleanup program to enlighten and guide subsequent efforts is in itself wasteful and dooms many of these later efforts to repeat mistakes that could otherwise have been avoided. Indeed, for responsible parties with large numbers of hazardous waste sites, the benefits that accrue from scientific study can be captured by using what is learned in one place at other sites and in future decisions. More focused and explicit building, cataloging, and transmission of knowledge during remedial investigation, remedy implementation, and monitoring is a key feature of the proposed adaptive site management process.

Given the close correspondence and dependence of the objectives put forth here with those identified in the NCP for remedial selection, the Navy and other agencies can view this charge for ongoing management as fully consistent with the existing directives and goals.

Risk Reduction Versus Mass Removal Objectives

The eight broad objectives discussed above can help guide the overall context and goals of a site remediation plan. However, it can be difficult to translate these into specific programs and activities for site cleanup (especially all at once). To assist in this effort, two more specific site cleanup metrics, *contaminant mass removal* and *risk reduction*, are often used to define the specific operational objectives of a remediation program. Like the eight broader objectives identified above, these two metrics are consistent with previous NCP guidelines (i.e., for the reduction of toxicity, mobility, or volume) and with well-established procedures already used by the military and other federal agencies to track and evaluate cleanup. These specific operational objectives can promote the broader goals of site cleanup to different degrees. For example, contaminant mass reduction may (in some cases) be especially important for achieving objectives 3 and 4 (natural resource protection and economic development), with risk reduction being central to the first two objectives (protecting human and ecological health and safety).

Evaluating the potential for risk reduction—central to a risk-based approach to site cleanup—has been used with increasing frequency in recent years. This approach defines the objectives for site cleanup as solely, or at least principally, to minimize human health and/or ecological risk (NRC, 1999a). Although socioeconomic impacts and risks to community welfare are sometimes considered in a broader framing of risk issues, they are rarely included in a formal risk assessment.[2] Depending on the current or potential hazard to human and/or ecological health and safety, a risk-based approach may lead to full-scale remedial activities (e.g., complete removal of the contaminant source); to more limited on-site engineering and control activities (e.g., containment measures); or

---

[2] See NRC (1996) pp. 45–47 for a discussion of socioeconomic and community-welfare risks, including effects on property values, increased community emergency preparedness costs and insurance premiums, community stigma and disruption, and concerns for environmental justice and equity.

even to no onsite remediation (e.g., use restrictions and other institutional controls). Thus, risk-based approaches as defined do not place inherent value in soil and groundwater resources, unless human or ecological health is directly threatened by contamination of those resources. As a consequence, risk-based approaches are more likely than strategies aimed at natural resource restoration to result in remedies that leave contamination in place.

Box 2-2 describes risk-based approaches, which have gained growing acceptance as a basis for site management decisions despite controversy surrounding the risk assessment process. As noted in the box, a broader definition of risk, combined with effective community input and communication, can help to ensure that risk assessments are appropriately structured and implemented to include the key values and concerns of the affected parties at a site.

Although risk reduction at a site is always sought, source removal (also called mass removal) at the site can be an objective in and of itself. This is because a site that contains significant quantities of remnant contamination may require continued limitations on human use or ecological function, leading to a loss of natural resource value and economic benefit. Surrounding property values and local or regional development may be impaired by the presence and stigma of remaining contamination and perceived risks, even if the actual risks of exposure have been minimized (e.g., Edelstein, 1988; Zeiss and Atwater, 1991; Gregory et al., 1995). Furthermore, breaches of the containment or loss of institutional controls could lead to actual exposures and risks for future generations.

It is sometimes the case that technologies that achieve some (or even a high degree of) mass removal may have little effect on exposure concentrations. Aggressive mass removal can even lead to increased pollutant release and mobilization in the surrounding environment, at least for the short term. This concern is especially important when considering large-scale excavation of contaminated soils or active dredging of sediments.

Although risk reduction and mass removal are not the only targets for hazardous waste cleanup, they are common operational objectives; thus, their relationship is explored in greater detail. There are distinct tradeoffs between different treatment strategies in terms of meeting mass removal and risk reduction goals of cleanup over time. Figure 2-1 schematically shows both the contaminant mass removal (A) and the exposure or potential risk (B) using alternative cleanup and management methods over time. Two major types of remediation strategies are illustrated in Figure 2-1. The first type, designated as an "M" strategy, seeks

BOX 2-2
Risk Reduction as a Basis for Site Management

A focus on risk reduction for site management implies that the presence of contaminants onsite or in a specific medium does not necessarily constitute unacceptable risk when exposure occurs at a level below which potential harm can occur. Thus, actions are focused on alleviating or reducing risks and not necessarily on full source mass removal (although it may be that removal is one of the risk management actions taken). On an historic level, risk assessment was developed to provide some quantitative measure of potential harm and was considered essential to developing a practical cleanup program (GAO, 2000).

There is sometimes opposition from environmental groups and affected residents over the use of a risk-based approach to cleanup versus one that stresses complete cleanup to natural or background levels. Critics argue that risk assessment is a tool that can be easily manipulated (Woodhouse, 1995; Andrews, 1997; Sexton, 1999; O'Brien, 2000). In its worst application, some say it is used to justify a specific, preferred (usually less costly) action, such as leaving in place large amounts of contaminants in soils or sediments. Risk-based approaches may appear to be biased or arbitrary to some, because the same data may be viewed, interpreted, and applied differently by different scientists, and the nonscientist often may not have a sufficient background to choose between "dueling" experts. Similarly, there are few black and white decisions in the "manage the risk approach," whereas in the mass removal approach, the decision could be relatively straightforward—"remove a certain percentage of the mass."

There are important benefits that accrue with a risk-based approach, so long as it is broadly defined to include the full range of important human health, ecological, and socioeconomic impacts. First, a risk-based approach gives the decision maker the ability to prioritize areas for action so that the most important or high-risk areas are addressed first. Being able to prioritize also equates to a more efficient and timely use of funds. More important, if the approach were simply based on mass removal, some actions may be taken that lead to no concrete improvements in human health, the environment, or community welfare. The risk-based approach uses data to help guide the risk management action. Once such actions are implemented, scientists have an established mathematical basis and a database on which to build a monitoring plan to ensure that these actions have the intended result.

A complicating factor in the risk-based approach is public perception. Often it can be difficult for scientists and decision makers to explain readily (or clearly) why leaving contaminants in place does not pose an unacceptable risk to human health or the environment. This difficulty feeds public skepticism, and the risk-based decision may face great difficulty in being accepted (see, for example, NRC, 1989; Kasperson et al., 1992; Renn, 1999). It can be much easier to convey the objective of mass removal than to convey the scientific reasons for leaving contaminants in place. Despite these limitations, the risk-based approach to environmental decision making has developed considerably over the last ten years, for situations in which both humans and ecological receptors are affected (Pittinger et al., 1998; NRC, 1999a; Sexton, 1999; Stahl et al., 1999, 2001).

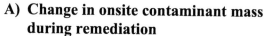

A) Change in onsite contaminant mass during remediation

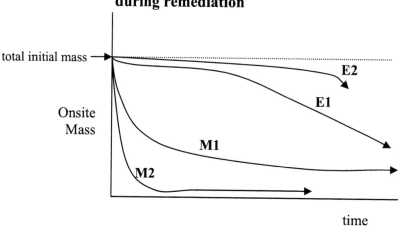

B) Change in exposure and potential risk during remediation

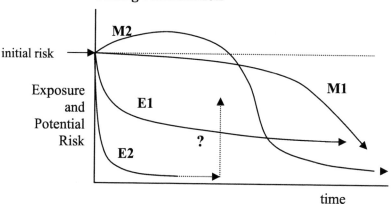

FIGURE 2-1 Two representations of the same cleanup situation. (A) shows the amount of contaminant mass remaining at a site as cleanup progresses, while (B) shows the amount of exposure or potential risk for the same cleanup plan. M strategies emphasize contaminant mass removal, while E strategies focus on exposure and risk reduction. The question mark refers to the possibility that onsite containment or an institutional control fails, leading to a sudden increase in exposure and potential risk.

first to remove contaminant mass. M1 represents an in situ mass removal strategy, which is typified by an asymptotically diminishing capture, since the last portions of remaining mass are often difficult to access and remove and may thus remain in place (Figure 2-1A). As shown in Figure 2-1B, M1 also displays a limited reduction in exposure concentrations and potential risk until a significant portion of the onsite mass is removed. The M1 curve represents the known behavior of certain *in situ* technologies, such as chemical oxidation or active bioremediation. Vapor extraction and conventional pump-and-treat might also yield results of this type of curve in the case where much (even if not all) of the pollutant mass is found in, or is readily transferred to, the captured fluid. Strategy M2 represents a more aggressive mass removal strategy, such as soil excavation or sediment dredging, which achieves results over a shorter time period (Figure 2-1A) but which might lead to short-term increases in exposure and risk during the period of implementation (Figure 2-1B).

The second type of remediation strategy, indicated with an "E," places first priority on reducing exposure to contamination. E1 represents an approach like plume containment, reactive barrier walls, or natural attenuation where the contaminant source zone is not targeted and the focus is on exposure reduction at some compliance point. E2 represents a pathway intervention strategy that would be implemented through institutional controls or onsite containment leaving the bulk of the contamination in place. The dotted upward arrow for E2 in Figure 2-1B signifies the possibility that the remaining onsite contamination could become exposed and impose a potential risk in the future, if the containment is breached or the institutional control is lost.

Figure 2-2 combines the progress in mass removal and risk reduction into a single, multiobjective graph. With this representation, the origin represents the starting point of site cleanup, when there is significant contaminant mass present at the site and affected populations are subject to significant exposure and potential risk. Progress in achieving cleanup and restoration is shown by moving along one of the paths over time toward the upper right-hand corner (i.e., total risk reduction and complete mass removal). Although the ultimate goal of cleanup and restoration is to move to this point in as rapid and efficient a manner as possible, this is not always feasible. Indeed, in many cases the costs are prohibitive, and the objectives of complete mass removal and/or exposure and risk elimination may simply be unachievable with current technology and policy options.

As discussed in Chapter 3, visualizations like Figures 2-1 and 2-2

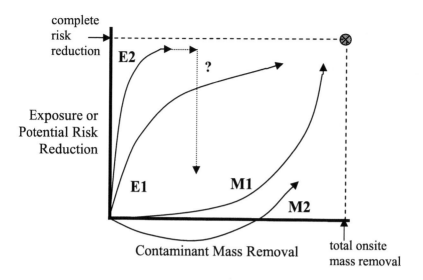

FIGURE 2-2   A multiobjective representation of alternative remediation and site management strategies in terms of contaminant mass removal and exposure or risk reduction. This graph combines Figures 2-1 A and B. The question mark refers to the possibility that containment or an institutional control fails, leading to a sudden change in exposure and potential risk.

can be used as a means for assessing and tracking the effectiveness of facility management options as a program for site restoration and stewardship evolves over time. By collecting the data and information necessary to record progress to date and by predicting (using mathematical models) the possible future outcomes for the objectives displayed in these figures, a more coherent and responsive effort can be planned and executed for the adaptive site management program recommended in this study.

Effectiveness of Remedies

Adaptive site management is needed not only to handle multiple and sometimes conflicting objectives, but also to provide flexibility when cleanup remedies are not entirely effective. Particularly at sites characterized by complex hydrogeology and contaminated with recalcitrant hydrophobic compounds and metals, the potential need for changing a remedy over time can be high. The following section describes conditions typically seen at complex hazardous waste sites in which the chosen remedy has been unsuccessful in meeting cleanup goals.

Technical Factors Influencing Remedy Effectiveness

Many methods for remediation of contaminated soil, sediments, and groundwater are characterized by an initial phase of relatively high effectiveness, followed by a prolonged period of much lower effectiveness. An obvious reason why mass removal rates of a remedial system may decrease over time is that contaminant mass is depleted in the vicinity of the extraction or treatment points. However, observing this behavior does not mean that complete mass removal *across the entire site* has been achieved (unless the initial contaminant mass is known with a high degree of certainty, which is highly unlikely). The more likely causes of decreased mass removal rates and hence decreased remedy effectiveness over time have been well documented for pump-and-treat systems (Mackay and Cherry, 1989; NRC, 1994) and soil vapor extraction (Travis and Macinnis, 1992). These technical factors are summarized below and include geological heterogeneity, flow heterogeneity, slow desorption, slow dissolution from nonaqueous phase liquids (NAPLs), erosion–deposition processes in contaminated sediments, and thresholds for microbial degradation. Although some of the factors described below apply across the range of contaminated media (e.g., slow desorption), others (e.g., geological and flow heterogeneity) apply only to *in situ* remediation of soil and groundwater.

Geological heterogeneity. The subsurface environment is heterogeneous, and soil permeability can vary orders of magnitude over short spatial scales. Substantial quantities of contaminants can thus be trapped in lower-permeability strata that are bypassed during conventional pump-and-treat or soil vapor extraction. Transfer out of these strata into the faster-moving fluid is controlled by molecular diffusion, which is a very slow process that can take years. Heterogeneities can also exert a strong influence on NAPL migration following spills. For example, the controlled field study reported by Kueper et al. (1993) showed that downward perchloroethylene (PCE) migration was hindered by small, finer-grain lenses on the centimeter scale. These lenses enhanced lateral spreading and caused a highly variable distribution of PCE pools and residuals that were trapped in relatively coarser-grained horizontal lenses. Both geological and flow heterogeneity (discussed below) can limit the effectiveness of many strategies for active remediation of soil and groundwater, such as pump-and-treat, soil vapor extraction, *in situ* chemical oxidation, and air sparging.

Flow heterogeneity. In addition to geological heterogeneity, there are several other important factors that can cause spatial variability of groundwater flow. Even in relatively homogeneous porous media, the spatial arrangement of extraction or injection wells can result in flow that bypasses certain regions of the aquifer (Javandel and Tsang, 1986; Christ et al., 1999). Water flow through material having high dense nonaqueous phase liquid (DNAPL) residual saturation will be hindered, and in fact the water flow patterns will change as the DNAPL saturation decreases due to dissolution (Nambi and Powers, 2000). Flow heterogeneity can have a significant impact upon the performance of recently developed technologies for aggressively treating DNAPL source zones. Some of these techniques require injection of fluid containing reacting chemicals (e.g., potassium permanganate, surfactants) that must mix with ambient groundwater or trapped DNAPL. Because of flow heterogeneity, this mixing may be incomplete, and in some instances the injected fluid can push contaminated groundwater outside of the treatment zone. For *in situ* air sparging, injection of air bubbles through a series of regularly spaced wells results in removal of volatile organic compounds that are dissolved in the groundwater. However, several studies have demonstrated that the bubbles may follow preferential airflow channels, thus bypassing a significant fraction of the contaminated zone (Brooks et al., 1999; Elder and Benson, 1999).

Slow desorption. Many hazardous compounds tend to sorb onto aquifer solids and sediments. The degree of sorption depends upon the properties of both the contaminant and the solid phase; however, certain classes of compounds like metals, polychlorinated biphenyls (PCBs), and polyaromatic hydrocarbons (PAHs) tend to sorb much more strongly than others (e.g., light petroleum compounds) (NRC, 2003). Although some of the sorbed mass is readily desorbed, a significant fraction of some organic compounds that have been in long-term contact with aquifer materials containing diagenetically aged carbon will undergo very slow desorption as a result of hindered molecular diffusion through microporous solids (Weber et al., 1991; Pignatello and Xing, 1996; Luthy et al., 1997). This mechanism may be partially responsible for the long tailing observed in contaminant breakthrough curves during active remediation of groundwater and soil, and it may also affect the performance of *ex situ* separation operations like soil washing. Slow desorption may be a benefit for certain kinds of contaminated sediment management strategies (e.g., capping) that aim to isolate contaminants *in situ*. Many studies suggest that the slowly desorbing fraction of an organic contaminant pool has reduced bioavailability and thus may present less of a potential risk (NRC, 2003a).

Slow dissolution from nonaqueous phase liquids. Many organic contaminants were released into the subsurface as a separate organic liquid phase. Liquids such as light petroleum products are less dense than water and tend to form pools above the water table, whereas organic solvents and coal tar that are more dense than water can sink below the water table and be trapped as small ganglia and lenses. These pools, ganglia, and lenses serve as reservoirs of contaminants that dissolve very slowly into the flowing groundwater. It is well known that NAPL-contaminated sites are among the most difficult to remediate. NRC (1994), using simplified dissolution rate equations, computed that it could take more than 100 years for moderately sized NAPL spills to completely dissolve.

Erosion–deposition processes in contaminated sediments. Natural attenuation processes in sediments may lead to decreases in water column contaminant concentrations over time. These processes include deposition of clean sediment, which tends to stabilize and separate contaminated sediment from the overlying water, as well as contaminant degradation and transformation processes. However, other processes such as erosion or resuspension of the sediment may serve to reintroduce contamination to the water column from the sediment. Erosion of sediment during high-flow conditions may lead to replenishment of surficial sediment concentrations, such that significant water column concentrations are maintained over long periods.

Threshold for microbial degradation. At low concentrations, some contaminants may no longer be able to act as a carbon or energy source or as an electron acceptor to support microbial communities. This is generally because the enzymatic machinery responsible for contaminant transformation requires a certain contaminant concentration in order to be activated. Thus, microbial degradation may slow or stop, leading to an asymptotic limit on the decrease in concentration. Sokol et al. (1998), for example, reported that below 35–45 ppm, microbial transformation of PCBs in Hudson River sediments effectively ceased. This value is generally above both screening levels and remedial goals frequently set for these compounds (EPA, 1997b; Buchman, 1999).

There are numerous case studies that document how the effectiveness of remediation decreases over time, one of which is highlighted in Box 2-3. Also, see Appendix B and over 270 additional case studies described at the Federal Remediation Technologies Roundtable (FRTR) website (http://www.frtr.gov) and in FRTR (1995, 1997, 1998a, 2000, 2001). The case presented here exhibits the so-called "asymptote effect" where the contaminant concentration or mass decreases over time and

---

BOX 2-3
Lawrence Livermore National Laboratory Site 300 GSA Operable Unit

Lawrence Livermore Site 300 is a Department of Energy experimental test facility with VOC contamination in soil and groundwater. Two different types of operations were conducted at the eastern and central portion of the site, resulting in maximum recorded groundwater TCE concentrations of 74 µg/L in the eastern portion and 240,000 µg/L in the central portion. Pump-and-treat using three groundwater extraction wells was the remedy selected for the eastern portion of the7 General Services Area (GSA). In the central portion, pump-and-treat with 19 extraction wells was used in addition to soil vapor extraction (SVE) with seven extraction wells. Operations at the eastern location commenced in 1991, and those at the central location in 1993.

Figure 2-3 demonstrates the performance of the pump-and-treat system for the eastern portion. Above is the total mass removed by the extraction wells, and below is the TCE concentration in the extracted groundwater. It can be seen that the rate of mass removal decreases toward the end of the reporting period, and the TCE concentrations show a rapid initial decrease followed by a long "tailing" period of slow decline. The data are similar for the combined pump-and-treat and SVE operations for the central region. In the six years of operation, the maximum TCE concentrations measured in groundwater monitoring wells have been reduced from 74 to 13 µg/L at the eastern site and from 240,000 to 33 µg/L at the central site. Despite these impressive reductions, concentrations are still above the cleanup goal of 5 µg/L at several monitoring locations.

Continued

BOX 2-3 Continued

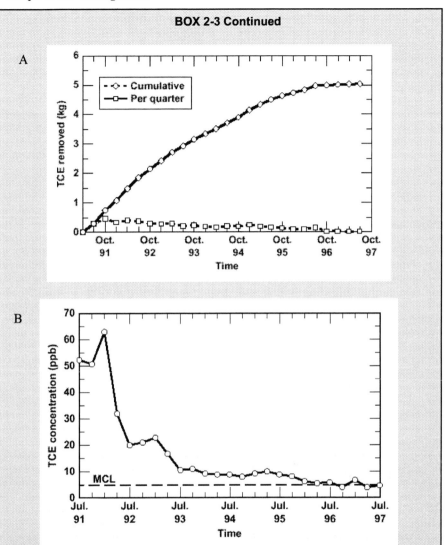

FIGURE 2-3 Effectiveness of pump-and-treat at Lawrence Livermore Site 300. (A) Mass of TCE removed from groundwater at the Eastern GSA. (B) TCE concentration in groundwater treatment system influent in the Eastern GSA. Note that the y-axis in panel B reflects the TCE concentration in the groundwater treatment (GAC) influent—that is, it is the concentration in the extracted groundwater, which is diluted due to mixing of groundwater with high and low TCE concentrations. Technically, the cleanup standard of 5 μg/L should not be drawn on this graph because it applies only to concentrations in the groundwater, that is measured in monitoring wells.

SOURCE: FRTR (1998a).

levels off at a finite value. For a treatment technology, such an asymptote may be indicative of system ineffectiveness *if the asymptotic value is greater than the cleanup goal.* (The many technical reasons for this were described above.) However, for a containment technology, such an asymptote could be indicative of effectiveness. Some technologies (e.g., permeable reactive barriers or some forms of institutional controls) do not exhibit the asymptote effect, because contaminant concentration or exposure may decrease "instantaneously" to acceptably low values.

The behavior shown for the Livermore site in Box 2-3 is typical of that of many additional case studies (see the FRTR website noted above and Appendix B) and consistent with the experience gained in attempting to clean up sites using many different approaches. The general trend is shown conceptually in Figure 2-4, which demonstrates how the effect-

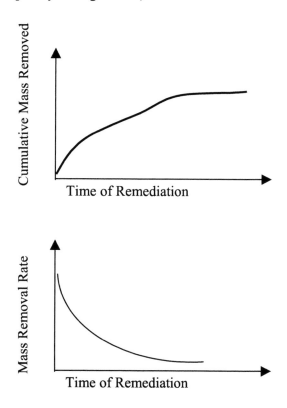

FIGURE 2-4 Schematic graphs showing typical changes in cumulative mass removed and the mass removal rate over time. In cases where costs increase linearly with time, the x-axes may also represent cost.

effectiveness of typical cleanup technologies in removing contaminant mass decreases over time. On the left is shown the cumulative mass removal over time, and on the right is shown the mass removal rate. Not all of the case studies reviewed by the committee reached an asymptotic value as shown in Figure 2-4. However, many did, and others are likely to do so in the future. Most of the pump-and-treat systems that did reach such a limit did so within five years of commencing operations (FRTR, 1998b,c).

Regulatory Options When Remedies Are Ineffective

Congress was not aware of the level of complexity involved in attaining health-based cleanup goals when the Comprehensive Environmental Response, Compensation, and Liability Act (CERLA) was enacted in 1980 and amended in 1986. Thus, there is limited guidance in the law regarding what to do when remedies are not able to meet cleanup goals. For Superfund sites, Congress requires that each remedy attain the requirements provided in environmental law, for example, health-based drinking water standards. (There are exceptions to this requirement as discussed in Chapter 6—e.g., if compliance can be shown to be technically impracticable or if the remedy will lead to greater risk.) In many cases, treatment technologies reach an asymptote prior to reaching this goal, although protectiveness can often be maintained by cutting off exposure pathways via containment or institutional controls. Except for guidance on technical impracticability, there is minimal if any guidance on how to deal with situations in which cleanup goals are not being met after prolonged operation of the remedy (which necessarily prevents site closeout).

Private sector cleanups may shed some light on how sites are dealt with when remedies are no longer effective. The U. S. Environmental Protection Agency (EPA) requires private sector potentially responsible parties (PRPs) to implement the remedial action "until the Performance Standards [remedial action goals] are achieved and for so long thereafter as is otherwise required."[3] EPA may modify the work when it "is necessary to achieve and maintain the Performance Standards or to carry out and maintain the effectiveness of the remedy set forth in the ROD [Record of Decision]"; a "modification may only be required ... to the extent

---

[3] EPA model language for remedial design and remedial implementation consent decree at Paragraph 13, available at: http://es.epa.gov/oeca/osre/docs/rdra-cd.pdf. The vast majority of consent decrees use this model language.

that it is consistent with the scope of the remedy selected in the ROD." Clearly, then, changes are allowed if a remedy is proving to be ineffective in meeting cleanup goals. However, the PRP is allowed to seek a remedy change only if EPA determines that the remedial action is not protective of human health and the environment. This creates a problem because many remedies that are ineffective in reaching cleanup goals might be very effective at protecting human health and the environment through plume containment. The situation is different for Resource Conservation and Recovery Act (RCRA) corrective action sites. Here, EPA policy is to consider implementing additional, more effective remedial technology if it becomes available (EPA, 1993). The committee, however, could find no examples of this policy having yet been implemented.

Base Realignment and Closure policy also limits the implementation of additional remedial action to situations where the selected remedy is no longer protective of human health and the environment because the remedy (including institutional controls) did not perform as expected, or because there has been a discovery of additional contamination attributable to the Department of Defense (DoD) (DoD, 1997). This same guidance does not explicitly state whether failure to attain a health-based groundwater or soil remedial action goal constitutes a failure to "perform as expected."

Thus, except for technical impracticability, there is no widely accepted policy for addressing situations where cleanup goals are not being met after extended operation of the remedy. The lack of explicit policies for addressing the large number of hazardous waste sites reaching this "point of diminishing return" is most likely a reflection of the fact that such situations have not become common until recently.

***

The need to respond to a set of multiple, sometimes conflicting, objectives; the ineffectiveness of current technologies in reaching cleanup goals for contamination at complex, high-risk sites; and the limited guidance on what to do when an asymptote is reached prior to meeting cleanup goals as long as the remedy remains protective of human health and the environment are pressing problems at many federal facilities. The remainder of this chapter describes an approach to cleanup that can accommodate these issues, that provides guidance at key decision-making periods, and that deals with the uncertainty inherent in many remedial strategies—both engineered technologies and institutional controls.

BACKGROUND ON ADAPTIVE MANAGEMENT

The predominant paradigm for site restoration in the United States has until relatively recently involved a highly linear, unidirectional march from site investigation to remedial action and eventually to site closure, which reflects our natural and understandable desire to "deal with the contamination and put it behind us." The paradigm is implicit in a recent DoD description of key milestones for site restoration programs, as shown in Figure 2-5. This figure depicts three targeted and sequential milestones of "remedy in place," "response complete," and "site close-out" that are consistent with this linear approach to site cleanup. Nowhere does the schematic allow for sites to cycle back through previous stages, although it indicates that some sites may need to be reevaluated prior to closeout. As sites have advanced through the restoration process, there has been a growing recognition that more iterative procedures are needed, with ongoing site stewardship and reevaluation of monitoring and remediation efforts at many sites. Because of the complexity of the subsurface environment, often incomplete identification of contaminant sources, and the long timeframes required for remediation, site cleanup must not be viewed as a one-time event or an action that ends once a remedy is implemented.

The need for iterative, adaptive approaches to site restoration and stewardship is supported by much of the recent literature on risk assessment and risk management. Most recently published frameworks and approaches to understanding and addressing risks to human health or the environment incorporate a high level of public participation and deliberation in which iterative steps are proposed for problem formulation, process design, option identification, information gathering, synthesis, decision, implementation, and evaluation (NRC, 1996). Such an iterative risk management framework was developed by the Presidential–Congressional Commission on Risk Assessment and Risk Management (1997a) (see Box 2-4). It is applicable to federal and other facilities and was recommended for use on PCB-contaminated sediments (NRC, 2001a). Hallmarks of such frameworks are that ongoing learning and feedback are used to address and incorporate scientific knowledge, new technological capabilities, and changing socioeconomic conditions into action plans over time, thereby informing new analysis, institutional learning, and public participation.

The overall environmental planning and management system for federal site restoration involves a hierarchy of decisions, many of which are stimulated by the changing conditions (economic, technological, pub-

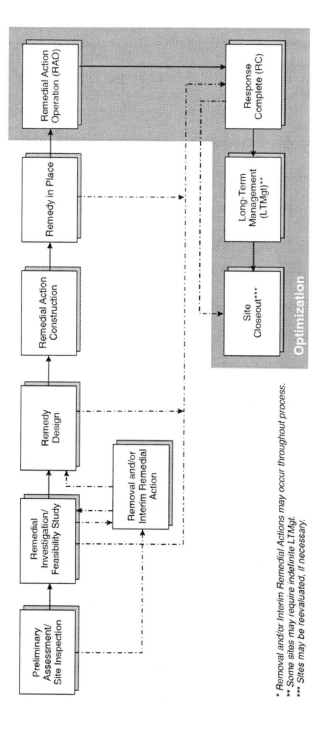

* Removal and/or Interim Remedial Actions may occur throughout process.
** Some sites may require indefinite LTMgt.
*** Sites may be reevaluated, if necessary.

FIGURE 2-5 Navy environmental restoration process for CERCLA sites. SOURCE: NAVFAC (2001).

BOX 2-4
President's Risk Commission Framework

The framework developed by the Presidential–Congressional Commission on Risk Assessment and Risk Management in its 1997 report *Framework for Environmental Health Risk Management* has many of the features of adaptive site management (ASM). In particular, as shown in the figure below it recognizes the need to engage stakeholders early and continuously throughout the risk management process. The participation of all stakeholders, including the public, regulatory groups, and responsible parties, is especially important in identifying the problem and placing it in context. The framework suggests assessing risks and evaluating options via an equally open process, and, if necessary, revisiting these steps as new information becomes available. Note that this goes beyond the traditional CERCLA view of public comment on remedial investigation and feasibility study reports by encouraging active participation by all stakeholders in these efforts. One advantage of the open, participatory process is an understanding of the economic, societal and cultural risks posed by the environmental problem, which is broader than simply understanding the risk to human health and the environment. Also important is the emphasis of the framework on stakeholder participation in decision making and the equal weight given to implementation and evaluation of whatever actions are undertaken. The framework recognizes proper implementation as necessary to achieve the desired goals and evaluation as necessary to validate that achievement.

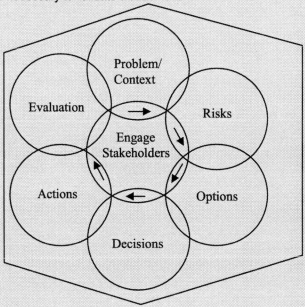

lic needs) encountered over the years. Choices among management alternatives may be constrained by the limitations of technical and institutional capabilities; however, these constraints are not necessarily fixed. Rather, investments in research and experimentation may allow development or improvement of technology to overcome current limitations, and thus provide DoD with more alternatives for risk management than existed initially. A broader systems approach that promotes effective knowledge generation (monitoring and fundamental research) and use of that knowledge (adaptation) can provide a wider range of decision options and thereby improve site management over the long term.

The characteristics of a broader systems approach described above are embodied in the concept of adaptive management. Adaptive management is an innovative approach to resource management in which policies are implemented with the express recognition that the response of the system is uncertain, but with the intent that this response will be monitored, interpreted, and used to adjust programs in an iterative manner, leading to ongoing improvements in knowledge and performance (Holling, 1978; Walters, 1986, 1997; Walters and Holling, 1990; Lee, 1993).

As noted by Lee (1999), "Adaptive management is learning while doing; it does not postpone action until 'enough' is known but acknowledges that time and resources are too short to defer … action." As such, adaptive management provides a structured approach for addressing uncertainty, making decisions in the face of it, and seeking to improve these decisions in an iterative manner by actively acquiring the knowledge necessary to reduce uncertainty. Part of this process can involve formal hypothesis testing, which is an important theme in the adaptive management literature. That is, hypotheses are formulated about the future events of interest, and then experiments or other activities (sometimes including statistical analysis) are conducted that will either confirm or reject the hypotheses. Adaptive management is also enhanced both by formal analysis/optimization methods (e.g., Williams, 2001) and by public participation (Shindler and Cheek, 1999).

The committee has coined the term "adaptive *site* management" (ASM) to refer to the application of the adaptive management concept to hazardous waste cleanup. Within the environmental arena, adaptive management concepts are relatively new but are particularly timely, especially given the observed limitations in remediation effectiveness and the increased use of remedies that will leave residual contamination in place for long periods. ASM is a flexible and iterative approach designed to allow decision makers to evaluate new information as it is received and adjust cleanup procedures or other management options over time.

ASM formalizes questions and decisions that the remedial project manager (RPM) and remediation team should address to readily adapt to changes in technology, remedy effectiveness, and other external influences that impact the management of contaminated sites. It follows closely the approach outlined in the President's Risk Commission Framework, and it builds on proposed Navy guidance for maximizing remedial effectiveness and cost efficiency (NAVFAC, 2001). It is also consistent with recent Department of Energy (DOE) interpretation of EPA guidance—*Using Remedy Monitoring Plans to Ensure Remedy Effectiveness and Appropriate Modifications* (DOE, 1998). At its heart is a formal decision process that stresses the collection and evaluation of information on remedial performance, provides options in the face of uncertainty, and embeds linked feedback loops so that action is taken quickly to change or optimize remedies that are unlikely to attain site-specific cleanup goals within a reasonable period of time. The main tenets of ASM are that it:

- is applicable at various stages of site restoration,
- is applicable to a wide variety of sites regardless of the contaminants being addressed or remedies envisioned,
- provides a mechanism for the optimization of existing remedies, changing ineffective remedies, and refining the site conceptual model,
- formalizes the routine examination of monitoring data and how to act upon the data,
- incorporates public participation,
- recognizes uncertainty and suggests approaches to dealing with it, especially when institutional controls are used,
- stimulates the search for new, innovative technologies to replace older or inefficient approaches,
- stresses the need for pilot programs to test both new technologies as well as modifications of existing technologies that might enhance their effectiveness, and
- recognizes the increasing role of long-term stewardship (which is synonymous with long-term management used in DoD terminology and in Figure 2-5).

Adaptive management provides a way of moving forward with site cleanup and stewardship programs in the face of uncertainty. As noted in

Chapter 3, uncertainty is inevitably present in the type, amount, and location of contaminants; in the response of contaminants to changes in physical, chemical, or biological conditions brought about by the remediation technology; in the response of on- and offsite species to these changes; and in the evolving socioeconomic conditions of the surrounding community that can affect the community's preferences. Chapter 4 advocates that more certain steps be taken in the short term based on available data, while data collection and evaluation of more uncertain elements of the overall plan proceed in parallel, in order to make progress toward cleanup.

Adaptive Management Approaches in Other Fields

There is growing experience with adaptive management approaches in a number of other public and private domains. The principal use of adaptive management has been for applications to wildlife and ecosystem management (see, for example, provisions for the U.S. Forest Service's Land and Resources Management Plans,[4] recent decisions by the Alaska Department of Fish and Game,[5] and the EPA/Environment Canada-sponsored Lake Superior Lakewide Management Plan.[6] NRC (1999b) reviewed the growing use of adaptive management for ecosystem resources in the Grand Canyon. In particular, it noted the need for (1) a long-term monitoring program in the Grand Canyon and (2) a strategy for scientific evaluation of policy alternatives, both in terms of ecological outcomes and the values of stakeholder groups. The authors recognized that effective adaptive management in the Grand Canyon will require tradeoffs among objectives favored by different groups as well as mechanisms for equitable weighting of these objectives. Similarly, NRC (2002) supported adaptive management to enhance scientific inquiry and policy formulation about the Missouri River ecosystem. The key tenets of the approach were outlined as (1) programs to maintain and restore ecosystem resilience, (2) recognizing and adapting to uncertainty, (3) interdisciplinary collaboration, (4) models to support collaboration and decisions, (5) meaningful representation of a wide array of interest groups, and (6) ecosystem monitoring to evaluate the impacts of management actions.

Adaptive management is also an important element of water re-

---

[4] http://www.fs.fed.us/forum/nepa/decisionm/p2.html
[5] http://www.state.ak.us/adfg/geninfo/press/2001/3-2-01.htm
[6] http://www.epa.gov/glnpo/lakesuperior/lamp2000/

sources planning (e.g., Mays and Tung, 1992) and strategies for dealing with global climate change (McCarthy et al., 2001). Thus, for example, NRC (2001b) recommends an adaptive implementation of strategies for determining and implementing Total Maximum Daily Loads under the Clean Water Act, including immediate actions, an array of possible long-term actions, success monitoring, and experimentation for model refinement.

Adaptive management has a similarly strong foundation in business strategies and planning approaches that utilize "feedback loops" to help respond to new information (Ayres, 1969). Iterative programs with ongoing performance evaluation and improvement are also a key component in the recent design of corporate environmental management systems (Crognale, 1999). These environmental management systems focus on planning for continuous improvement in environmental compliance and include the identification of performance metrics, comparison to goals, consideration of costs and benefits, and institutional and personnel steps needed to ensure that the process is self-sustaining. A review of a recent advance in methods used for adaptive management of private and public investment decisions is presented in Box 2-5.

BOX 2-5
New Approaches for Adaptive Management for Business, Investment, and Resource Management Decisions: Real Options

A number of businesses and business researchers have advocated the use of flexible, adaptive approaches for management of the modern firm (Nonaka and Takeuchi, 1995; Collins and Porras, 1997; deGeus, 1997; Hax and Wilde II, 1999). One of the most powerful innovations in investment theory in support of adaptive approaches has been the development of real options theory (Merton, 1973; Dixit and Pindyck, 1994; Trigeorgis, 1996). Real options are physical facilities, investments, or programs that allow for adjustments in response to changing conditions. Real option theory values the flexibility these provide, and invokes a corresponding penalty for irreversible decisions. Particular examples of real options relevant to energy and pollution control include the provision of dual combustion units in electric power plants to allow real-time fuel switching to take advantage of changing fuel prices (Kulatilaka, 1993), and the ability to adjust facility construction plans in response to changing expectations for environmental regulation (Kaslow and Pindyck, 1994). Implementation of the real options method requires explicit identification of future decision points and alternative courses of action that can be taken based upon ongoing performance assessment and outcome evaluation.

Like the business management models discussed above, adaptive management requires explicit identification of future decision points and alternative actions that can be taken depending upon ongoing performance assessment and outcome evaluation, all of which are embodied in the ASM approach. Nonetheless, there are important issues related to applying adaptive management to hazardous waste cleanup that make it different from business applications. Site management is constrained by particular legal requirements, though consideration of changes in regulations that could better enable flexibility is also important. Furthermore, as noted before and expounded upon below, site management is a multiobjective problem, often involving incommensurate measures of health, safety, ecological quality, cost, natural resources value, and the social and economic well-being and satisfaction of the surrounding community. Finally, federal site management cannot afford obvious failures in remediation. Every site is unique and important, and its aggregate management is expected to be successful. However, not every individual project explored for the site need necessarily be successful. This is both expected and accepted when the full range of possible outcomes for each option is considered, and effective contingency plans and alternatives are available for program adjustment. Absent the valuable data collected during evaluation and experimentation outlined in the ASM approach, it is unlikely that new and innovative approaches to cleanup will be developed and implemented.

The concept of adaptation is not foreign to CERCLA and RCRA activities. There are certainly cases where project managers have modified remedial activities in response to poor system performance. Over the last decade, a number of formal approaches have been developed to introduce adaptation specifically into data collection and site characterization activities. Examples of these include Expedited Site Characterization (ASTM, 1999), Adaptive Sampling and Analysis Programs (DOE, 2001), the Observational Approach (Baecher and Ladd, 1997), and, most recently, EPA's TRIAD program (Crumbling et al., 2001). Expedited Site Characterization has emphasized the application of appropriate onsite technical expertise, rapid data collection activities, and dynamic work plans to subsurface characterization problems. Adaptive Sampling and Analysis Program work has primarily focused on the application of in-field decision making and field analytics to contaminated soil issues. The Observational Approach has its roots in geotechnical engineering. As discussed in Box 2-6, it explicitly acknowledges that environmental decision making often involves irresolvable uncertainty. This uncertainty is addressed by contingency planning and flexible designs that can be

---

BOX 2-6
The Observational Approach: A Type of Adaptive Management

The observational approach provides a mechanism for dealing with inherent uncertainty that attempts to design around the most likely system outcome (Baecher and Ladd, 1997). Uncertainty is handled by identifying potential departures from this likely outcome that would be of significant concern, and developing monitoring approaches and contingency plans so that significant deviations from these expectations can be expeditiously recognized and responses taken. Manufacturers and other practitioners concerned with quality control use a similar approach based on "control charts" (Deming, 1993; Montgomery, 1996). Pierce and Larsen (1993) discuss the use of control charts and the observational approach in the evaluation of soil quality in agriculture.

As a simple example of the observational approach, consider contaminated soil excavation work that is ongoing, but can be modified based on data collected during the course of the excavation. A traditional approach to uncertainty in this case would be to estimate a conservative excavation footprint, e.g., one that provides 95 percent confidence that the excavation will capture the full extent of contamination (EPA, 1989, 1996b). The excavation proceeds based on that footprint unless visible evidence of contamination outside the footprint is found during the excavation process. An observational approach would instead determine an initial ("best guess") excavation footprint based on available data, and then modify the targeted area based on new data collected as the work proceeds. The amount invested in additional data collection would depend on the consequences of errors of different types. If excavation and disposal costs are high, a significant investment in data collection might be justified to reduce uncertainty (and the risk of over-excavation) as work proceeds. Conversely, if excavation and disposal costs are relatively low, but the environmental implications of under-excavation are high, uncertainty could be addressed by deliberate over-excavation. ASM in general, and adaptive data collection programs in particular, are consistent with the observational approach to project management.

---

modified based on changing field conditions as they are encountered. Finally, the EPA's TRIAD program mixes systematic planning with dynamic work plans and onsite analytical tools to streamline environmental data collection programs. Many of these sampling techniques are discussed in greater detail in Chapter 3. ASM builds on these experiences to provide a framework that weaves adaptability into the remedial design and implementation process as a whole.

By explicitly recognizing the role of ongoing research and information collection, remedial performance assessment, and evaluation of alternative remedies, ASM provides greater flexibility than the current, more linear approach to site management, with the expectation of an overall improvement in the performance of the management system.

ADAPTIVE SITE MANAGEMENT DESCRIBED

ASM involves two main steps. Step 1 focuses on understanding and conceptualizing a site problem and identifying risks to human health and the environment. Step 2 focuses on the selection and implementation of a remedy, monitoring performance of the remedy, adapting the remedy or management goals to accommodate changing conditions and improve cost-effectiveness, and, finally, completing the remedy and closing out the site. Although many Navy sites may have progressed past the point of selecting a remedy, the entire process (Steps 1 and 2) is shown because it is important that it be visualized by and articulated to all remediation team members so that key tasks are understood and not overlooked. This is absolutely essential at the onset of the Problem/Context phase and should continue regularly throughout implementation of the remedy. Note that the text gives greater weight to those steps within ASM that differentiate the approach from current practice. This is not meant to impart greater importance to any particular activity, but rather to provide detail where there is otherwise less information.

Recent Air Force (2001) and Navy (NAVFAC, 2000, 2001) guidance documents contain a number of the precepts imbedded in the proposed ASM approach. For example, NAVFAC (2001) requires evaluation of system performance (how well a remedy is meeting design criteria), system suitability (how likely it is the remedy will attain cleanup goals), and whether there are life-cycle limitations (i.e., whether the remedy will reach the point of diminishing returns). The guidance calls for an alternative strategy when a plot of cumulative mass removed versus time exhibits "an asymptotic condition" prior to attainment of the cleanup goal. The alternative strategy may include (1) modifying an existing system to improve cost-efficiency and cost-effectiveness, (2) implementing different remediation (including the sequencing of several remedial technologies to achieve cleanup goals) when the remedial action cannot be modified to achieve cleanup goals, or (3) changing the cleanup goals. Various places in the following discussion highlight where ASM builds upon or reinforces NAVFAC (2001) and where it differs.

There is also new and explicit guidance from the DoD Under Secretary of Defense that supports ASM. In particular, updated Defense Environmental Restoration Program guidance states that the evaluation does not end once a response action is implemented (DoD, 2001). Continued activities should include optimizing the overall performance and effectiveness of the remedy, assessing if the ROD objectives have been achieved and if the treatment system is still needed, and determining if a different remediation goal is needed or if an alternative technology or approach is more appropriate. The guidance also suggests that technology development efforts should be supported, with a goal of increasing the overall effectiveness of response activities, including the validation and certification of emerging technologies.

It should be noted that "optimization" is used in the remediation literature in different contexts. In the broadest sense, optimization means implementation of any change to a remedial system to make it work more efficiently toward the cleanup goal, where operating costs are reduced as a result of making a change, or where the desired asymptotic cleanup condition is reached more quickly. The change could involve making a single technology more efficient, adding components to a treatment train (as discussed in Chapter 5), or switching cleanup remedies. Extensive literature, including the military documents cited above, is available to provide guidance and criteria for such "optimization." Within this report and within ASM, however, "optimization" is used specifically to refer to making a single technology more efficient.

Step 1: Pre-Remedy Selection

Step 1 of the ASM process (Figure 2-6) is specific to those tasks that occur before a remedy is selected. The importance of this pre-remedy selection step is also discussed in existing Navy guidance (e.g., NAVFAC, 2001), internal Navy memos (e.g., Pirie, 1999), and previous NRC reports (e.g., NRC, 1999a). In some cases, sites may have progressed to Step 2—remedy selection and implementation—without adequately completing Step 1, which presents another series of problems that ASM can help to address. The individual tasks of Step 1 include developing the problem/context and the site conceptual model and conducting risk assessments. It should be noted that there is nothing about the Step 1 process outlined in Figure 2-6 that is not already encompassed by the CERCLA cleanup paradigm.

Problem/Context Formulation

During the problem/context formulation phase, decision makers,

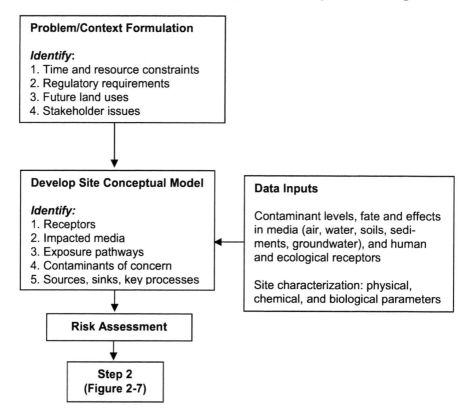

FIGURE 2-6 Step 1 of adaptive site management: pre-remedy selection.

stakeholders, and other interested parties identify and define important issues to help formulate the problems at an individual site. Even for sites that have progressed beyond the problem/context phase, it may be necessary to return to this point if new public issues, risks, or problems arise. That is, if site conditions and other important influences on management actions should change, the remediation team will need to return to the problem/context phase. As shown in Figure 2-7, ASM allows users to revisit Step 1 as new information is obtained, as site conditions change, or where it is otherwise necessary to review previous tasks.

One of the first issues that should be defined during the problem/context phase is the amount of time available for undertaking the requisite studies and information gathering, as well as any resource constraints (personnel or financial) that might hinder this activity. It is also important to understand any regulatory requirements that may influence the kinds of data and information collected and data analysis needed. There may be more than one set of regulations operating at a given site, and some requirements may be mandatory while others may be left to the discretion of site managers.

Another important issue is to determine and clearly understand what the future land use will be for the site, or whether there might be multiple land uses envisioned. The latter may occur where large sections of a site may be relatively unimpacted by operations and require relatively little restoration before they can be utilized for commercial or recreational purposes. The Navy has issued memoranda (Navy, 2001) to ensure that future land use activities remain compatible with the land use restrictions imposed on the property during the remediation and restoration process. Future land use has a direct influence on the type of data needed from environmental or engineering studies, on the long-term stewardship undertaken, as well as on the level of cleanup that is required. In addition, there is a need to work with local stakeholders and the public, particularly since these groups will likely have a vested interest in the disposition of the land once the remedial action has been completed and/or the site is closed out.

Both active military bases and those that are closing are of substantial interest to the local public and, in some cases, to the larger public in a particular geographic region. Thus, it is imperative that site managers identify individuals or groups that should be consulted on future land use issues and on the remedies being considered to achieve cleanup consistent with the planned land use. Stakeholders will need to understand their role in the decision-making process and have a mechanism with which to articulate their expectations to the RPM.

Site Conceptual Model Development

The site conceptual model (SCM) is an "illustration" of the site that details the location, concentration, and pathways by which contaminants are thought to be moving through the environment and how/why humans and ecological receptors are being exposed to those contaminants (NRC, 1997). That is, the SCM provides an initial assessment of the hydrogeologic environment, contaminant sources and sinks, and key processes (such as environmental fate) that are potentially operating at the site, and illustrates how these could influence the need for or types of remediation.

With regard to contaminant sources, the SCM should identify both known sources such as landfills and underground storage tanks as well as impacted media such as groundwater, surface water, soils, sediments, and air. The SCM documents exposure pathways or routes through which humans, non-humans, and habitats come in contact with contaminant(s) and links those pathways with a list of contaminants of concern found at the site that might pose potential risks. Processes that may alter contaminant form such as hydrolysis, photolysis, and microbial degradation should be assessed (Peyton et al., 2000). Finally, the SCM should identify those receptors—both human and non-human—that are exposed to the contaminants, including organisms in affected aquatic and terrestrial habitats. Further guidance on developing a site conceptual model can be found in NRC (1994), USACE (1996), EPA (1997c), and NAVFAC (2001).

From a more formal perspective, the site conceptual model and its companion risk assessment can be thought of as a set of linked hypotheses, with the remediation process that follows constituting a test of the validity of these hypotheses. If the experience acquired during remediation brings some of these hypotheses into question, the site conceptual model may need to be revised. For example, the SCM will need to be changed to reflect a discovery of new contaminants during implementation of the remedy.

Data Inputs

Although not a separate activity per se from problem/context formulation and site conceptual model development, data collection is nonetheless a key element of ASM. Quantitative data are needed to support remedy selection, to determine if the remedy is effective, and to reveal how the remedy should be implemented or modified to achieve optimal performance (NAVFAC, 2000, 2001). During this activity, the remediation team collects and analyzes data on contaminant fate and effects in media. This includes determining what contaminants are present (at what levels and in which media, considering air, surface water, soils, sediments, and groundwater) as well as which receptors (human, non-human, habitats) are likely to be exposed to and adversely impacted by these contaminants. More detailed information can be found in NRC (1999a) and EPA (1997c).

Risk Assessment

Risk assessment is an important step in the site cleanup processes proscribed under CERCLA and RCRA. Briefly, the process results in a mathematical estimate of the potential risk faced by humans or ecological receptors exposed to the contaminants of concern. A similar framework applies to both human health and ecological risk assessment. The risk assessment can be purely quantitative, purely qualitative, or some combination of the two. Risk assessments integrate information on the physical conditions at the site, the nature and extent of contamination, the toxicological and chemical/physical characteristics of the contaminants, the current and/or future land use conditions, and the dose–response relationship between projected exposure levels and potential toxic effects. The results of the risk assessment, coupled with the expected future land use, are key in determining what level of cleanup will be necessary for a particular site. Because risk assessment is not the focus of this report, the reader is referred to the seminal report describing this process (NRC, 1983) and to other reports (EPA, 1992; Calabrese and Baldwin, 1993; Maughan, 1993; Suter, 1993; EPA, 1997c; Committee on Environment and Natural Resources, 1999; NRC, 1999a).

Step 2: Remedy Selection and Subsequent Activities

As shown in Figure 2-7, the second part of ASM involves selecting, implementing, monitoring, adapting, and completing the remedy. Step 2 of ASM also links readily with the elements of Step 1 because it is expected that there will be situations where it will be necessary to return to Step 1 to refine the site conceptual model, collect additional information, refine the risk assessment, or change the remedial goal.

A key element of ASM is the formalization of "management decision periods" (MDP) at which decisions are made based on pilot-scale work, on changes in land use or stakeholder needs, and on monitoring data and other intelligence that may lead the RPM to refine and/or revise a management decision. Such management decisions provide an opportunity for periodic check-ups to determine if the remedial technology is meeting its objectives and, if not, whether adjustments are needed. In this respect, ASM differs from recent cleanup guidance for Navy facilities (NAVFAC, 2001), which does not formalize these decisions nor stress the need for stakeholder involvement. These management decision periods are similar to "scientific management decision points" detailed in

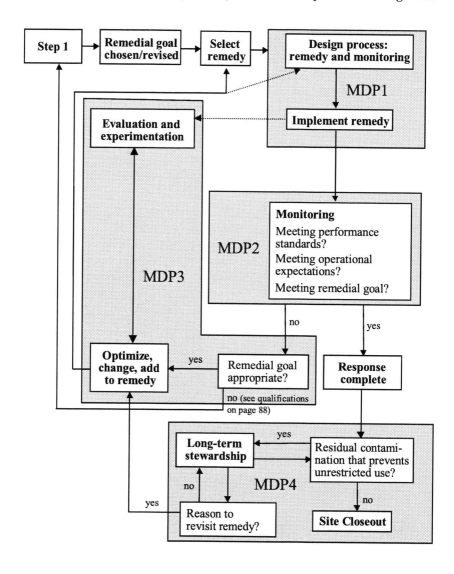

FIGURE 2-7 Step 2 of adaptive site management process: post-remedy selection. The shaded areas show the activities related to the management decision periods described in the text.

EPA guidance for Superfund (EPA, 1997c).

The management decision periods are designed to take advantage of the feedback loops embedded in ASM, such that uncertainties in site restoration can be addressed. They also represent formal opportunities for the RPM and other project mangers, regulators, and interested stakeholders to evaluate incoming and existing data and to reach agreement on what additional management steps, if any, need to be taken. Having formal management decision periods does not preclude routine discussion among groups or individuals involved with the remediation process; in fact, it should encourage greater dialogue among groups so that important issues are managed as they arise. For simplicity, the management decision periods are focused on the four main categories of actions associated with Step 2: implementation, monitoring, adaptation, and long-term stewardship.

Management Decision Period #1: Remedy Selection, Design, and Implementation

The purpose of MDP1 is to ensure that the remedy selected is still practicable and implementable under site-specific conditions and that an appropriate, well-designed monitoring plan is developed. At the onset of Step 2, a remedial goal is chosen that takes into consideration the risk assessment results, expected or desired future land use, public and stakeholder concerns, and technological capabilities. The goal can be established to protect human health, ecological receptors, or both, and will be site-specific. In some cases there may be multiple goals set for sites where there will be multiple future land uses, as may be the case at larger federal facilities. ASM affords the flexibility to return to Step 1 in order to revise the remedial goal in cases where the goal has been determined to be unattainable (see discussion below).

Subsequently, an initial remedy is selected and either published as a ROD or finalized under another regulatory context (perhaps as a Consent Agreement, Consent Order, or federal facilities cleanup agreement). Details on specific remedies of importance at complex sites including the use of treatment trains can be found in Chapter 5, along with recommendations on when specific remedies are applicable and when they are not.

Designing the remedy requires data on contaminant concentrations and movement and other exposure-related parameters that are generally collected before remedy selection. However, because there can be a long lag time (years) between remedy selection and implementation, there may be a need for additional information in these same areas or in areas specific to engineering or construction needs so that final implementation is based on valid assumptions. Thus, the remedy design should take into account new information on the types of contaminants detected in affected media and on specific engineering or construction requirements that was not available when the initial remedy was selected. For example, placing a sheet-pile wall in a shallow aquifer, coupled with a pumping regime to gain control over the contaminant plume, could be viewed as the most likely remedy for groundwater remediation. The ROD could be written to reflect this view and performance standards put in place when the ROD is issued. A finite amount of data would have been collected to support this remedy decision, but there may have been uncertainties with the data set. Later, as remedy design begins, it is discovered that the geological conditions or hydrologic scheme are such that the wall either will not work as envisioned or will require extensive modification for proper performance. That is, the hypothesis that the wall would be the appropriate remedy for the conditions present is disproved by subsequent data collection. The RPM and remediation team now have a specific engineering hurdle to overcome prior to implementing a remedy. The value of the MDP1 analysis prior to full-scale implementation is that it helps to ensure that the remedy will indeed meet the requirements set forth in the ROD.

Another early decision point not specifically noted in Figure 2-7 but available within the CERCLA process is called value engineering. Value engineering is not a regulatory process per se, but entails the application of proven engineering analysis to the implementation and functioning of a remedy. In this process the full details of what the remedy is expected to do, the conditions at the site, and all other relevant data are presented to a group of experts who may have had little or no prior contact with the site. These experts "peer review" the proposed remedy and provide an analysis of the potential for the remedy to work and, more importantly, they identify the situations they believe could cause the remedy to fail. The review can help to determine the level of uncertainty in the remedy's performance and the steps that should be taken to reduce that uncertainty. For any independent review to be successful, site managers and regulators must agree beforehand to abide by the results of the expert panel. In this way both groups gain the benefit of additional advice absent the potential bias that may occur were the review to be conducted by experts from one group or the other. [It should be noted that EPA created a National Remedy Review Board to provide technical and policy review of remedial decisions (EPA, 1996c, 2001a), although the independence of this review process is questionable because the team is composed of the EPA supervisory personnel involved in the Agency's cleanup program.]

Another important task prior to implementation is to design an appropriate monitoring program (see Chapter 3 for details), especially in situations where the remedy is likely to require significant operation and maintenance costs over an extended period of time. Long-term monitoring, which is likely when pump-and-treat is the preferred remedy for groundwater, will be an important source of quantitative data highly relevant to determining whether the remedy is effective or not. The monitoring plan should be designed in light of the site conceptual model so that the data not only will illuminate the remedy's performance, but will also be valuable in refining the site conceptual model at later stages of site restoration.

Developing a monitoring program will require that the site managers and regulators agree on the kinds of parameters to be monitored and how those parameters fit with the performance standards established in a ROD or other regulatory document. An even more important element of MDP1 is reaching agreement on how the monitoring data will be used in decision making, particularly because the potential future decisions involve modifying or changing the remedy based on the results of the monitoring program.

Subsequent to MDP1 and once the remedy is implemented, several actions can potentially occur as part of ASM. In addition to operation of the remedy, there are ongoing monitoring activities, as discussed below under MDP2. A third activity—evaluation and experimentation—is denoted in Figure 2-7 by a dashed line. This activity is unique to ASM and is one of the hallmarks of adaptive management in general. It refers to conducting experiments and other research activities in parallel with implementation of the chosen remedy. This activity may occur at the level of an individual site, in which portions of the site are devoted to experimentation while others are undergoing the chosen remedy, or it may refer to collecting information about experiments going on elsewhere, the results of which are relevant to specific sites. Evaluation and experimentation may consist of pilot-scale studies done at one of the national demonstration sites. The evaluation and experimentation track is an opportunity to test innovative, less certain, sometimes riskier remedies that were not well enough established to be chosen as the initial remedy in the ROD. The data and information gathered on this parallel track can then be used later to optimize or change the remedy at MDP3 if performance standards or the remedial goal are not being met. This explicit evaluation and experimentation track of ASM is discussed in greater detail below under MDP3.

Management Decision Period #2: Monitoring

MDP2 consists of a series of key questions regarding the output of the monitoring program—quantitative information on the effectiveness of the remedy. Affirmative responses to these questions lead to "response complete" and eventually to MDP4, whereas negative responses lead to MDP3. MDP2 was developed to ensure that at regular intervals, the monitoring data are evaluated and judged against the operational parameters agreed to or imposed by a regulatory construct. Throughout the monitoring period, and specifically at these intervals, the RPM should ask three sequential questions: (1) is the remedy meeting the performance standards (as set forth in the ROD or other binding document), (2) are the operational expectations of the remedy being met (whether cost or some other parameter that the RPM and remediation team have set), and (3) is the remedial goal being met.

The first question listed above is perhaps the simplest—is the remedy meeting performance standards? The data to make this judgment can derive both from the monitoring program and from other relevant studies that might have been completed (e.g., geotechnical studies to further define site conditions). If the remedy is not meeting performance standards, the RPM should initiate a thorough review of the remedy, the site conceptual model, and other relevant data to ascertain why. An important consideration is to determine whether the remedy has been in place long enough to have had time to equilibrate and operate properly. If not, then the RPM may wish to continue with the remedy and collect performance data for an additional period to ensure the remedy is not changed prematurely. The decision to continue the remedy and monitor it should have a finite time limit so that an ineffective remedy is not operated indefinitely.

If the system is meeting the performance standards, the second question is whether it is meeting operational expectations. These expectations may take the form of cost, up time, or some other metric, depending on the remedy and the needs of the RPM with respect to long-term stewardship and fiscal responsibility. For this comparison, the latest scientific information on technology performance could be combined with updated site characterization data to determine operational expectations. Generally these will be metrics selected by the RPM and others to ensure that the system is functioning *as they want it to*, above and beyond what might be required by the ROD or other binding regulatory provision. Indeed, a system may be meeting regulation-specific performance standards, yet may not be particularly cost-effective, may require excessive maintenance or down time, or may require constant reoptimization.

If the remedy is found to be meeting performance standards and is operating as efficiently as expected, then the third question is whether remedial goals are being met. For example, goals may be risk-based, such as reduction of contaminant concentration to a level below that which poses a risk to human health or the environment. Remedial goals may be based on containment, as in preventing a groundwater plume from moving offsite, or they may be based on mass removal, as for many groundwater extraction and sediment and soil excavation remedies. If the remedy is not meeting the remedial goal, then site managers are faced with two important considerations discussed in MDP3: how the remedy can be modified, optimized, or changed to meet the performance standards, and whether the remedial goal is inappropriate and in need of change. If sufficient data are available to indicate the remedy has achieved its stated remedial goal, then another decision period (MDP4) is reached.

It is important to allow sufficient time to pass before deciding whether a system is meeting remedial goals. The necessary amount of time will be highly dependent on the chosen remedy and even more so on the hydrogeological conditions at the site. For example, in the case of bioremediation, enough time must pass for the microorganisms to acclimate to seasonal changes (in temperature, for example). Some aggressive source removal strategies like dredging and *in situ* chemical oxidation produce changes within weeks. However, in most cases additional monitoring is needed to determine the permanence of the result and the potential for rebound. This can be problematic with *in situ* oxidation, which usually consists of only one injection and perhaps an additional injection if rebound occurs within months. If rebound occurs later or is not monitored for at all, false conclusions about the performance of the technology may arise. Months are needed to properly evaluate the performance of soil/vapor extraction, and years are required for pump and treat, air sparging, monitored natural attenuation, or permeable reactive barriers. Time frames for performance determinations may also be imposed by the regulatory authority.

Monitoring data used in this fashion allow initial hypotheses about remedy performance to be tested and either confirmed or rejected. As discussed previously, these data should also be used to evaluate whether the hypothesis of environmental conditions represented by the site conceptual model remains valid. Updating the model will reduce uncertainty in the current remedy's effectiveness, and it may lead to the development of new management options. As a matter of course, changes in the site conceptual model should stimulate site managers to review ASM Step 1 to ensure that other potential issues (e.g., changes in risk assumptions) are not overlooked.

Management Decision Period #3: Adaptation

MDP3 illustrates most clearly the adaptive nature of ASM. Prior to this point, the site managers are focused on obtaining monitoring data and asking specific questions about remedy performance. Now, in MDP3, they must turn their attention to analyzing those data and any other relevant information to determine what future management steps are appropriate.

Reviewing, Modifying or Changing a Remedial Goal. After determining that the remedy is not meeting remedial goals, the RPM and remediation team, in coordination with regulatory agencies and interested stakeholders, should address whether the remedial goal is still appropriate. If it is not, the site managers would return to Step 1 of ASM. There are a number of issues that must be taken into consideration to ensure that this decision is made deliberately. First, it is possible that the goal was inappropriate in the first place. For example, a remedial goal may be inappropriate when a health-based cleanup goal, such as an MCL for an organic contaminant, is applied to groundwater that is nonpotable because of naturally occurring aquifer constituents. Second, a remedial goal could become inappropriate over time because changes occurring subsequent to the signing of the ROD—for example, a land use change or the discovery of new contamination—have altered the site conceptual model. In the case of a land use change, a goal designed for commercial/industrial land use would be inappropriate if in fact the future land use is residential. A third issue is whether a sufficient amount of time has been allowed for the remedy to function before reaching a determination of "goal inappropriate." As discussed above, the time period will vary depending on the site, regulatory constraints, stakeholder concerns, the chosen remedies, and other influences. Site managers should recognize that changing a remedial goal is not a simple task and will require that they engage the relevant stakeholders to ensure that the new remedial goal is acceptable and compatible with the expected future land use.

Assuming that site conditions have not changed and sufficient time has passed since remedy implementation, the inability to attain remedial goals is largely controlled by the complexity of the site and the types of technologies that can be applied. At sites with either highly heterogeneous stratigraphy or fractured media aquifers and highly refractory contaminants such as DNAPLs, cleanup to background or health-based standards may be technically infeasible. Thus, one option at MDP3 is to pursue a technical impracticability (TI) determination, which for the most part documents the inability to achieve a particular remedial goal regardless of the technology applied (see EPA, 1993, and Chapter 6 for details on technical impracticability waivers). TI waivers, which are granted for groundwater contamination only, result in the selection of a new least-cost remedial goal (such as a new feasible concentration level). The remedy is then modified to achieve that new goal (e.g., to achieve containment rather than removal). This option is not the equivalent of site closeout because activities to protect human and ecological receptors such as land use restrictions, groundwater monitoring, and five-year reviews must continue.

If it is determined that the remedial goal should be modified, prompting a return to Step 1, a number of additional changes will result that should be discussed openly so there are no misconceptions. Where no existing technology can meet the initial remedial goal, the result is likely to be a less stringent goal, which may result in greater contaminant concentrations being left on site. This situation may necessitate a change in the future land use, perhaps from residential use to commercial/industrial use, which may in turn require modifications to deeds or restrictions on current or future property access agreements. Because changing the remedial goal automatically leads site managers back to remedy selection (see Figure 2-7), the remedy and monitoring program will have to be revisited.

Optimizing, Changing, or Adding a Remedy. If it is determined that the remedial goal is appropriate, but the remedy is not achieving the goal after sufficient time has passed, then the remedy should be optimized, added to, or changed entirely. Details on optimizing, modifying, and adding remedies are provided in Chapter 5 and NAVFAC (2001). At a minimum, optimization techniques, particularly for groundwater systems, should be undertaken whenever a remedy is not performing appropriately. Optimization of an existing remedy leads the site manager back to the "design process: remedy and monitoring" box. For those remedies that do not perform appropriately even after optimization, the RPM should consult with the remediation team, recognized experts, and others to help them ascertain if the remedy can be further modified to achieve proper performance. If after further evaluation and consultation it is determined that the remedy cannot be modified adequately, the remedy may be unsalvageable and require wholesale replacement. Such a determination will, as illustrated in Figure 2-7, require a return to the "select remedy" box in Step 2.

Although a wide array of tools can be used to evaluate whether an additional remedial action or change is warranted once the point of diminishing returns has been reached, a relatively simple, graphical test can be used. In the case of groundwater contamination, contaminant concentration within the source area can be plotted over time; the need for a change may be evident when the slope of the line tangent to the performance curve approaches zero (the so-called asymptote) but the concentration remains above the site-specific remedial action goal. Such plots can also make it clear when continued operation of the existing remedy may incur substantial per-unit costs with relatively little improvement in mass removal. In order for these plots to be useful, the remediation team and the regulatory agencies must agree on a unit cost for the continued operation of the remedial action at the site under investigation, above which the existing remedy is no longer considered a tenable option. Information on the types of data to plot and their analysis are found in Chapter 3 and NAVFAC (2001).

Evaluation and Experimentation. Site managers should strive to incorporate new information collected during the evaluation and experimentation track of ASM into decisions about optimizing, adding, or changing remedies. As discussed previously, laboratory studies, pilot-scale activities conducted on- or offsite, expert panel evaluations, literature reviews, or newly acquired experience from other federal or private-sector sites should be assessed on a regular basis to determine if a more effective remedy applicable to the site of concern exists. For example, a selected remedy of containment might be replaced with an innovative treatment technology that would allow unrestricted use[7] of the site and the associated economic and health benefits if the incremental costs were reasonable. This parallel track in which site managers adapt remedies to

---

[7] The term "unrestricted use" means EPA's definition of "unlimited use and unrestricted exposure," that is, the selected remedy will place no restrictions on the potential use of land or other natural resources. In general, if the selected remedy relies on restrictions of land and/or groundwater use by humans and/or ecological populations in order to be protective, then the use has been limited and a five-year review should be conducted (EPA, 2001b).

information from internal and external sources is critical to overcoming the stalemate encountered at many sites where cleanup goals cannot be achieved. However, in order for this to succeed, potentially responsible parties, the Navy in particular and the federal government more generally, would have to make evaluation and experimentation an integral part of their overall remedial program. This adaptive feature of ASM differentiates it from the recent Navy guidance (NAVFAC, 2001), which does not specify an explicit need for ongoing evaluation and experimentation. It is also an extension of the report's seventh objective, which stresses the role of knowledge generation and transmittal.

As discussed in Chapter 4, there are numerous mechanisms for undertaking evaluation and experimentation at an individual site and for obtaining relevant information and data externally. Some could involve current DoD agreements with EPA laboratories or offices, extramural grants with academic institutions or other nongovernmental groups, or collaborative activities such as those conducted through the Remediation Technology Development Forum (RTDF), a joint effort between EPA and private industry. Adoption of ASM would encourage the Navy to build stronger networks to the scientific and engineering communities in order to stay abreast of new technological developments that might prove applicable to existing or future cleanup scenarios.

The committee recognizes that time will be required to test ideas and new technologies prior to a full-scale implementation. It is not the committee's intent that ASM be used as an argument for delaying important decisions while extensive analysis takes place (so-called "paralysis by analysis"). In fact, a definable characteristic of adaptive management is that more certain and sometimes simple actions are taken immediately while information is being gathered about potentially more effective but less certain technologies. That information should then be used to periodically revise the original action. In order for the concept to succeed, both tracks must operate simultaneously. However, in recognition of the many existing sites for which remedies have been ongoing for some time and for which no evaluation and experimentation were done, Figure 2-7 shows an upward arrow from "optimize, change, and add to remedy" to "evaluation and experimentation." This suggests that it can be useful to conduct the latter activities even after a chosen remedy has stalled in meeting cleanup goals. While evaluation and experimentation take place, the temporary inability to meet performance standards or other regulatory requirements should not be used as a basis for notices of deficiency or enforcement action. Ideally, ASM should foster frequent interactions between site managers, regulatory agencies, and other stakeholders that will improve overall communications, build trust and credibility, improve flexibility, and ultimately lead to greater efficiency in the restoration of federal sites.

Management Decision Period #4: Long-Term Stewardship

The purpose of MDP4 is to provide a clear road map and describe actions necessary to achieve a site closeout designation. Crucial elements of MDP4 include planning for long-term stewardship and monitoring (if required) and agreeing on time intervals at which the site status can be formally reviewed by the RPM, remediation team, regulators, and interested stakeholders. The CERCLA process provides an explicit mechanism for doing this through the five-year review process, and five years should be viewed as the maximum time period between reviews. For non-CERCLA sites, the RPM and remediation team may wish to establish their own timetable for formal reviews and ensure that the regulatory agencies and interested stakeholders are involved.

Once a remedy has been in place for sufficient time and monitoring data provide measurable evidence that the remedial goal has been met, a site is designated (under military terminology) as "response complete." Depending on the nature of the site, a variety of actions may still be required. If the remedial goal was based solely on mass removal and the required mass was removed, then the RPM is no longer required to take additional action if long-term monitoring is not required and if residual contamination (if any) poses no risk to human health or the environment. That is, any remaining contamination must be present at levels below those that allow for unrestricted use. In many cases where contamination is left in place, additional action is needed to ensure protection of human health or the environment. Examples include where a series of remedies are in place and only one of several has reached completion, where the remedy has changed from an active one (pump-and-treat) to a more passive remedy (monitored natural attenuation), or where a passive remedy was selected initially as the remedy of choice and a long-term monitoring plan was put in place.

The monitoring and oversight actions necessary to ensure protectiveness at such sites are lumped under the umbrella activity of long-term stewardship. It is the primary required activity once it has been determined that residual contamination has been left in place at levels above those required for unrestricted use, or when the remedy is one that requires monitoring and maintenance (pump-and-treat/containment of groundwater plume; institutional controls). MDP4 presents an opportunity to make forward progress with long-term stewardship and eventually reach site closure. For sites where there is still some ongoing remediation, such as passive technologies or monitored natural attenuation, contaminant concentrations may steadily decrease, albeit slowly. In these cases, the site manager should periodically ask whether there is still residual contamination present in amounts that preclude unrestricted use and thus pose a human or ecological health risk. If not, it may be possible to proceed to site closure, as described below.

MDP4 introduces an opportunity to modify actions at those sites where residual contamination persists above unrestricted use levels. Periodically during long-term stewardship, the site manager should ask whether there is a reason to revisit the chosen remedy. If the answer is yes, then ASM affords the flexibility to optimize, modify, or replace a remedy with something that may be more effective in removing the residual contamination. As shown in Figure 2-7, this may eventually lead (via remedy selection and implementation) to site closeout if the new treatment is successful. There are several reasons that site managers should reconsider remedies in place during long-term stewardship. State law may require complete restoration; therefore, attaining these goals may be mandatory. Also, considerable cost savings may be possible if a new technology can alleviate the need for continual monitoring and/or maintenance. In addition, there are substantial economic benefits to returning a site to unrestricted land uses. This path is most likely to succeed if site managers stay abreast of recent developments in new treatment technologies, as discussed previously under the evaluation and experimentation track. As discussed in Chapter 6, the five-year review process currently does not support reconsideration of remedies during long-term stewardship if the remedies are maintaining protectiveness of human health and the environment.

It is important to clarify the meaning of the term "site closeout" in Figure 2-7. The term can have many connotations within the hazardous waste cleanup world, and may imply sites that have been cleaned up to, for example, industrial land uses. Indeed, at some sites that are, from EPA's standpoint, considered "closed," "deleted from the National Priorities List (NPL)[8]," and perhaps redeveloped, a variety of remedial actions

---

[8] It should be noted that deletion of sites from the NPL is not necessarily coincident with site closeout. The NCP (40 CFR 300.425(e)) states that a site may be deleted when no further response is appropriate. If monitoring to determine the need for future response action is ongoing, deletion is premature. However, EPA (2000b) states sites with ongoing operation and maintenance obligations may be deleted, which can occur before the five-year review.

(some very expensive) may continue until or unless the site meets unrestricted use (EPA, 2000b). Such actions include continued operation and maintenance of the remedy (e.g., the cover inspected and repaired, pumps replaced) and monitoring. Maintenance of institutional controls requires inspections and verification that land use has not changed. However, throughout this report the term "site closeout" is used to mean that residual contamination has been removed to levels below that which allows for unrestricted use, which is consistent with the DoD usage of the term. More specifically, "site closeout" refers to the "point at which the DoD will no longer engage in active management or monitoring at an environmental restoration site, and no additional environmental restoration funds will be expended unless the need for additional remedial action is demonstrated." According to interagency guidance, "for practical purposes site closeout occurs when cleanup goals have been achieved that allow unrestricted use of the property (i.e., no further long-term monitoring, including institutional controls, is required)."

Key activities during site closeout (that are sometimes overlooked) include the decommissioning of monitoring wells, treatment systems, and pipelines, and the termination of institutional use controls. Sufficient evidence should have been collected to ensure that conditions will not be reversed in the future. For example, quantitative evidence of the absence of a reversal in site conditions—such as a rebound in groundwater plume concentrations when the pump-and-treat system is shut down—should be gathered. If this requires additional monitoring, there should be agreement on the timing of this monitoring so it is well defined and finite. Only when the above conditions are addressed to the satisfaction of the RPM, remediation team, regulatory agencies, and interested stakeholders does the site move to the status of closed.

MAJOR CONCLUSIONS AND RECOMMENDATIONS

Adaptive management approaches are now being used by a number of public and private organizations to improve the quality of their operations and decisions. Adaptive management recognizes that uncertainty is inherently present when predicting the effects of new policies and programs, and that including directed testing, evaluation, and learning as part of these programs can build the knowledge base for ongoing improvements in decisions. Like the domains of natural resource and business management where the principles of adaptive management have been applied, site cleanup planning and stewardship involve significant uncertainty in system response. Given this, and the strong support for adaptive approaches already present in recent federal guidance on monitoring and remediation, we propose adopting ASM for site cleanup and long-term stewardship.

ASM is an iterative, flexible approach to improving federal site cleanup. It builds on recently developed guidance for the Navy (NAVFAC, 2001), but provides a much broader and more well-defined series of tasks to ensure that remediation is cost-effective. Whereas the recent Navy guidance also recommends close scrutiny of existing remedies and monitoring data, ASM goes further to suggest how to interpret the monitoring data, when to consider using new technologies, and how to reach site closure for all types of sites. The differences between current cleanup practice and ASM with regard to monitoring and data analysis, evaluation and experimentation, and long-term stewardship are elaborated in Chapters 3, 4, and 6, respectively.

A critical aspect of ASM is its call for evaluation and experimentation, and the coupling of that information with the adaptation of remedial programs so that ineffective or inefficient remedies are replaced quickly. This approach presents a way to manage uncertainty while moving forward with the cleanup process because conventional remedies can be implemented first while additional information is being gained on innovative but more risky technologies.

ASM formalizes discrete management decision periods to provide an explicit mechanism for communication (between the RPM and remediation team, the regulatory agencies, and interested stakeholders), to allow for critical evaluation of information, and to guide the determination of new management actions. MDP1 ensures that a selected remedy is indeed the right one to implement, while MDP2 details how to assess remedy effectiveness based on monitoring data. MDP3 draws upon monitoring data as well as information from evaluation and experimentation and stakeholder input to optimize, modify, or replace remedies or revise remedial goals. MDP4 provides the road map for long-term stewardship and site closure for sites where residual contamination remains in place. In addition, MDP4 suggests how to make forward progress at sites where remedies, such as pump-and-treat systems and monitored natural attenuation, require substantial financial resources for monitoring, operation, and maintenance. Feedback loops are present throughout in order to revisit different points when new information warrants such an examination. A final important feature of ASM is its applicability to sites at any stage of cleanup.

There is little more than anecdotal evidence about the difficulty or ease which with remedies can be changed, although it is likely that there are disincentives for RPMs to optimize or change in-place remedies in some cases. ASM institutionalizes the concept of being open-minded about chosen remedies, and its success will depend on the creation of incentives to promote this mindset. The recent trend within the Navy of optimizing existing remedies and changing ineffective remedies, but without the benefit of evaluation and experimentation (see example in Box 2-7), suggests that the Navy and other federal agencies would have a need for and an interest in ASM.

Given its many discrete decision points and the evaluation and ex-perimentation track, it is possible that ASM will be time-consuming and (over the short-term) more expensive than the current practice. Thus, full-scale ASM that includes public participation during all decision pe-riods should be targeted to the more complex (e.g., multiple contami-nants and stressors, heterogeneous hydrogeology) and high-risk sites where projected large costs are at stake. An example would be where DNAPL contamination threatens a sole source aquifer. Indeed, these are the sites where cleanup goals are not being achieved and where innova-tive technologies are needed to provide new avenues for treatment. A substantial number of DoD sites, including Navy sites, fall into this high-risk/high-cost category. If targeted in this manner, the ASM approach is expected to lead to an optimum solution from both a cost and perform-ance perspective, as well as to a solution acceptable to stakeholders. The benefits of ASM are expected to be less at smaller, low-risk, low-cost remediation sites (e.g., a BTEX spill at an UST site) where there is greater certainty about the ability of remedies to reach cleanup goals.

Because of the enormous variability in site conditions across Navy facilities, it is not appropriate for this report to suggest a distinct cost ba-sis for assessing whether or not to use ASM—for example, transactional costs should only represent a certain percentage of the total costs—although this may be possible in the future following more in-depth analysis by the Navy. Nonetheless, it is anticipated that up-front cost increases associated with implementing ASM will be balanced by the benefits of evaluation and experimentation, which include optimization of remedies and more expeditious achievement of cleanup goals. In many cases, the costs associated with ASM may be exceeded by the long-term savings that result from switching to a more efficient and ef-fective technology or by overall life-cycle savings. These issues, along with pertinent examples, are discussed in greater detail in Chapters 4 and 6, respectively.

Finally, current understanding within the military of what and for how long information needs to be collected, catalogued, and maintained is too inconsistent at present to fully support the ASM concept (M. Ierardi, Air Force Base Conversion Agency, personal communication, 2002). Information is currently found in many different information systems for records management, financial management, contract management, real-estate management, progress reporting, and technical data systems, most of which are not integrated. Greater understanding of the value of the information associated with the cleanup program will be needed to support ASM. Specific recommendations regarding data collection on remedy performance at federal facilities are discussed in Chapters 3, 4, and 5.

The Navy and other federal agencies should adopt adaptive site management. To our knowledge, ASM has never been formally used for hazardous waste cleanup. ASM will enable site managers to use new data and innovative technologies when they become available, both during active implementation of remedies and during long-term stewardship. The Navy is currently drafting policy that will require periodic reviews of remedies, as prescribed by the recent NAVFAC (2001) guidance on optimization (R. Kratke, NFESC, personal communication, 2003). Because ASM is broader in scope than that guidance, it will be necessary for the federal agencies to develop guidance to further define the management decision periods that are inherent to ASM.

Full-scale ASM that includes public participation during each decision period should be targeted to the more complex and high-risk sites where projected large costs are at stake. ASM is particularly appropriate for sites with multiple or recalcitrant contaminants and multiple stressors and heterogeneous hydrogeology because progress at such sites is likely to have stalled prior to reaching cleanup goals. Prior to widespread adoption, the Navy should consider pilot testing ASM at a limited number of high-risk, complex sites to allow Navy managers to better understand any transactional costs and delays that may accompany ASM implementation.

REFERENCES

Air Force. 2001. Final remedial process optimization handbook. Prepared for Air Force Center for Environmental Excellence, Technology Transfer Division, Brooks Air Force Base, San Antonio, Texas, and Defense Logistics

BOX 2-7 Trend Toward ASM in the Navy

A Navy site in Pensacola, Florida, was recently evaluated for the potential to reduce cleanup costs while maintaining or enhancing protectiveness by applying some of the principles of ASM (Navy, 2000). A TCE plume beneath a sludge drying bed had been undergoing remediation by pump-and-treat since 1987, using seven recovery wells. In 1995 the monitoring data were reviewed and indicated that the groundwater contamination had been reduced to Maximum Contaminant Levels (MCLs) at most of the site, although several high-concentration plume areas of 3,000–4,000 μg/L were present in the vicinity of monitoring well GM-66 (see Figure 2-8).

Based on recovery well concentrations and in cooperation with the Florida Department of Environmental Protection, the Navy decided to reduce the number of recovery wells to three and to focus on reducing the high TCE concentrations near GM-66. In 1996 the monitoring data were reviewed again, and it was decided to discontinue pump and treat altogether and monitor for natural attenuation. In 1998 a program of *in situ* chemical oxidation was undertaken to address the removal of the high-concentration source areas. Fenton's Reagent was used

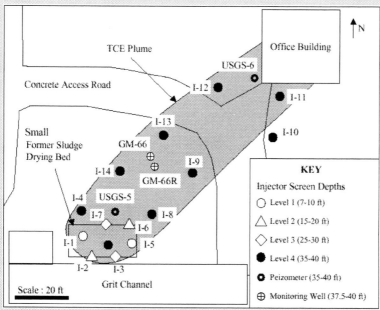

FIGURE 2-8   TCE plume delineation at Pensacola, FL, site.   SOURCE: Navy (2000).

as the oxidant in two phases of implementation, and ultimately a 97 percent reduction in chlorinated VOCs was achieved by early 2000 (see Figure 2-9). The
latter data for well GM-66/66R indicate that the system may be showing a rebound effect in a portion of the domain. This trend should be monitored and assessed over time, with appropriate additional remedial action taken as needed. It
was agreed that monitored natural attenuation would be used to reduce the remaining contaminant levels to MCLs.

The Navy estimated that by implementing this alternative remediation
scheme, a life-cycle cost savings of $2.56 million was achieved in monitoring and
treatment, and this was accompanied by a reduction in cleanup time. This case
study illustrates some facets of ASM because the original remedy was changed
after assessing the results of remediation repeatedly over time. However, it is
not clear what evaluation and experimentation activities were ongoing at the site
during implementation of the original remedy and whether they may have formed
the basis for the suggested changes.

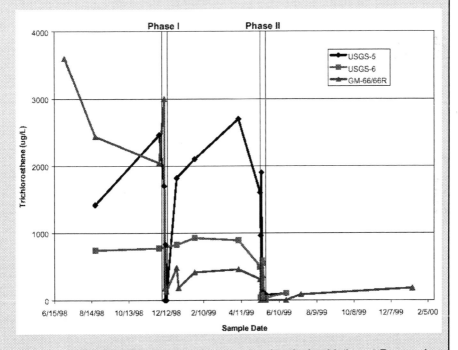

FIGURE 2-9 TCE levels resulting from *in situ* chemical oxidation at Pensacola,
FL, site. SOURCE: Navy (2000).

Agency, Environmental Safety Office, Fort Belvoir, VA. Available at http://www.afcee.brooks.af.mil/er/rpo/rpohandbook.pdf.

Andrews, R. 1997. Risk-based decision making. Pp. 208–230 In: Environmental policy in the 1990s. N. Vig and M. Kraft (eds.). Washington, DC: Congressional Quarterly.

ASTM. 1999. Standards on design, planning and reporting of ground water and vadose zone investigations (2nd Edition). Philadelphia, PA: American Society for Testing and Materials.

Ayres, R. U. 1969. Technological forecasting and long-range planning. New York: McGraw Hill.

Baecher, G. B., and C. C. Ladd. 1997. Formal observational approach to staged loading. Transportation Research Record, Annual Issue.

Brooks, M. C., W. R. Wise, and M. D. Annable. 1999. Fundamental changes in *in situ* air sparging flow patterns. Ground Water Monitoring and Remediation, Spring:105–113.

Buchman, M. F. 1999. NOAA screening quick reference tables. NOAA HAZMAT Report 99-1. Seattle, WA: NOAA Coastal Protection and Restoration Division.

Calabrese, E. J., and L. A. Baldwin. 1993. Performing ecological risk assessments. Chelsea: Lewis Publishers.

Christ, J. A., M. N. Goltz, and J. Huang. 1999. Development and application of an analytical model to aid design and implementation of in situ remediation technologies. J. Contam. Hydrol. 37(3–4):295–317.

Collins, J., and J. Porras. 1997. Built to last. New York: Harper Business Press.

Committee on Environment and Natural Resources. 1999. Ecological risk assessment in the federal government. CENR/5-99/001. Washington, DC: Committee on Environment and Natural Resources of the National Science and Technology Council.

Crognale, G. 1999. Environmental management strategies: the 21st century perspective. Upper Saddle River, NJ: Prentice Hall.

Crumbling, D. M., C. Groenjes, B. Lesnik, K. Lynch, J. Shockley, J. Van Ee, R. Howe, L. Keith, and J. McKenna. 2001. Managing uncertainty in environmental decisions. Environ. Sci. Technol. 35:405–409.

deGeus, A. 1997. The living company. Boston: Harvard Business School Press.

Deming, W. E. 1993. The new economics for industry, government, and education. Cambridge, MA: MIT Press.

Department of Defense (DoD). 1997. Guidance responsibility for additional environmental cleanup after transfer of real property. Washington, DC: DoD Office of the Deputy Under Secretary of Defense (Acquisition and Technology).

DoD. 2001. Management guidance for the DERP. Washington, DC: Office of the Deputy Under Secretary of Defense (Installations and Environment).

Department of Energy (DOE). 1998. Using remedy monitoring plans to ensure remedy effectiveness and appropriate modifications. RCRA/CERCLA Information Brief, DOE/EH-413 9809. Washington, DC: DOE Office of Environmental Policy and Assistance.

DOE. 2001. Adaptive sampling and analysis programs (ASAP). Innovative Technology Summary Report DOE/EM-0592. Washington, DC: U.S. Department of Energy.

Dixit, A., and R. Pindyck. 1994. Investment under uncertainty. Princeton, NJ: Princeton Univ. Press.

Edelstein, M. R. 1988. Contaminated communities: the social and psychological impacts of residential toxic exposure. Boulder, CO: Westview Press.

Elder, C. R., and C. H. Benson. 1999. Air channel formation, size, spacing, and tortuosity during air sparging. Ground Water Monitoring and Remediation, Summer:171–181.

Environmental Protection Agency (EPA). 1989. Methods for evaluating the attainment of cleanup standards. Volume 1: soils and solid media. EPA 230/02-89-042. Washington, DC: EPA.

EPA. 1990. National oil pollution and hazardous substances contingency plan. Federal Register 56:8666–8728.

EPA. 1992. Framework for ecological risk assessment. EPA/630/R-92/001. Washington, DC: EPA Risk Assessment Forum.

EPA. 1993. Guidance for evaluating the technical impracticability of groundwater restoration. OSWER Dir. No. 9234.2-25. Washington, DC: EPA Office of Solid Waste and Emergency Response.

EPA. 1996a. The role of cost in the Superfund remedy selection process. Washington, DC: EPA.

EPA. 1996b. Soil screening guidance: user's guide (2nd edition). Publication 9355.4-23. Washington, DC: EPA.

EPA. 1996c. Memorandum from Stephen D. Luftig, Director, Office of Emergency and Remedial Response, to Regional Directors, Re: National Review Board (September 26, 1996). Available at http://www.epa.gov/superfund/programs/nrrb/9-26-96.htm.

EPA. 1997a. Rules of thumb for Superfund remedy selection. Washington, DC: EPA.

EPA. 1997b. Incidence and severity of sediment contamination in surface waters of the United States. Volume 1, Appendix D. Washington, DC: EPA Office of Water OST.

EPA. 1997c. Ecological risk assessment guidance for Superfund: process for designing and conducting ecological risk assessments. EPA/540-R97-006. Washington, DC: EPA Office of Solid Waste and Emergency Response.

EPA. 1998. The incidence and severity of sediment contamination in surface waters of the United States. Volume 1: National sediment quality survey. EPA 823-R-97-006. Washington, DC: EPA Office of Water OST.

EPA 1999a. Risk assessment guidance for Superfund. Volume 1–Human health evaluation manual, supplement to part A: community involvement in Superfund risk assessments. EPA 540-R-98-042. Washington, DC: EPA

Office of Solid Waste and Emergency Response.

EPA. 1999b. The benefits and costs of the Clean Air Act, 1990 to 2010. EPA-410-R-99-001. Washington, DC: EPA Office of Air and Radiation.

EPA. 2000a. Draft public involvement policy. Federal Register 65:82,335.

EPA. 2000b. Close-out procedures for National Priority Sites. EPA 540-R-98-016. Washington, DC: EPA Office of Emergency and Remedial Response.

EPA. 2001a. EPA National Remedial Review Board. Available at: http://www.epa.gov/superfund/programs/nrrb/index.htm

EPA. 2001b. Comprehensive five-year review guidance. EPA 540-R-01-007. Washington, DC: EPA Office of Solid Waste and Emergency Response.

Federal Remediation Technologies Roundtable (FRTR). 1995, 1997, 1998a, 2000, 2001. Abstracts of Remediation Case Studies, Vol. 1, March, 1995; Vol. 2, July 1997; Vol. 3, Sept 1998; Vol. 4, June 2000; Vol. 5, May 2001. Washington, DC: FRTR.

FRTR. 1998b. Remediation case studies: groundwater pump-and-treat (chlorinated solvents), Volume 9. EPA-542-R-98-013. Washington, DC: FRTR.

FRTR. 1998c. Remediation case studies: groundwater pump-and-treat (non-chlorinated solvents), Volume 10. EPA-542-R-98-014. Washington, DC: FRTR.

General Accounting Office (GAO). 2000. Environmental Protection Agency: use of precautionary assumptions in health risk assessments and benefits estimates. GAO-01-55. Washington, DC: GAO.

Gregory, R., J. Flynn, and P. Slovic. 1995. Technology stigma. American Scientist 83(3):220–223.

Hax, A. C., and D. L. Wilde II. 1999. The Delta model: adaptive management for a changing world. MIT Sloan Management Review 40(2), Reprint 4021.

Holling, C. S. (ed.). 1978. Adaptive environmental assessment and management. New York: John Wiley & Sons.

Javandel, I., and C. F. Tsang. 1986. Capture-zone type curves: a tool for aquifer cleanup. Ground Water 24(5):616–625.

Kaslow, T. W., and R. S. Pindyck. 1994. Valuing flexibility in utility planning. The Electricity Journal 7(2):60–65.

Kasperson, R. E., D. Golding, and S. Tuler. 1992. Social distrust as a factor in siting hazardous facilities and communicating risks. Journal of Social Issues 48:161–187.

Kueper, B. H., J. D. Redman, R. C. Starr, S. Reitsma, and M. Mah. 1993. A field experiment to study the behaviour of tetrachloroethylene below the water table: spatial distribution of residual and pooled DNAPL. Journal of Ground Water 31(5):756–766.

Kulatilaka, N. 1993. The value of flexibility: the case of a dual-fuel industrial steam boiler. Financial Management 22(3):271–280.

Lee, K. 1993. Compass and gyroscope: integrating science and politics for the environment. Covelo, CA: Island Press.

Lee, K. N. 1999. Appraising adaptive management. Conservation Ecology

3(2):3.

Luthy R. G., G. R. Aiken, M. L. Brusseau, S. D. Cunningham, P. M. Gschwend, J. J. Pignatello, M. Reinhard, S. J. Traina, W. J. Weber, and J. C. Westall. 1997. Sequestration of hydrophobic organic contaminants by geosorbents. Environ. Sci. Technol. 31(12):3341–3347.

Mackay, D. M., and J. A. Cherry. 1989. Groundwater contamination: pump-and-treat remediation. Environ. Sci. Tech. 23(6):630–36.

Maughan, J. T. 1993. Ecological assessment of hazardous waste sites. New York: Van Nostrand Reinhold.

Mays, L. W., and Y-K. Tung. 1992. Hydrosystems engineering and management. New York: McGraw-Hill.

McCarthy, J. M., O. F. Canziani, F, Osvaldo, N. A. Leary, D. J. Dokken and K. S. White (eds.). 2001. Climate change 2001: impacts, adaptation, and vulnerability. Report of Working Group II to the Third Assessment Report of the Intergovernmental Panel on Climate Change. Cambridge, England: Cambridge University Press.

Merton, R. 1973. The theory of rational option pricing. Bell Journal of Economics and Management Science 4:141–183.

Montgomery, D. C. 1996. Introduction to statistical quality control. New York: John Wiley & Sons.

Nambi, I., and S. E. Powers. 2000. NAPL dissolution in heterogeneous systems: an experimental investigation in a simple heterogeneous system. J. Contam. Hydrol. 44(2):161–184.

National Research Council (NRC). 1983. Risk assessment in the federal government: managing the process. Washington, DC: National Academy Press.

NRC. 1989. Improving risk communication. Washington, DC: National Academy Press.

NRC. 1994. Alternatives for ground water cleanup. Washington, DC: National Academy Press.

NRC. 1996. Understanding risk: informing decisions in a democratic society. Washington, DC: National Academy Press.

NRC. 1997. Innovations in ground water and soil cleanup: from concept to commercialization. Washington, DC: National Academy Press.

NRC. 1999a. Environmental cleanup at Navy facilities: risk-based methods. Washington, DC: National Academy Press.

NRC. 1999b. Downstream: adaptive management of Glen Canyon Dam and the Colorado River ecosystem. Washington, DC: National Academy Press.

NRC. 2001a. A risk management strategy for PCB-contaminated sediments. Washington, DC: National Academy Press.

NRC. 2001b. Assessing the TMDL approach to water quality management. Washington, DC: National Academy Press.

NRC. 2002. The Missouri River ecosystem: exploring the prospects for recovery. Washington, DC: National Academy Press.

NRC. 2003. Bioavailability of contaminants in soils and sediments: processes, tools, and applications. Washington, DC: National Academy Press.

Navy. 2000. Naval Air Station Pensacola. Optimization of remedial action operation to treat chlorinated hydrocarbons in groundwater. Available at http://enviro.nfesc.navy.mil/erb/erb_a/support/wrk_grp/raoltm/case_studies/rao_pensacola.pdf

Navy. 2001. Memorandum from Goodman, S. W., Re: Policy on land use controls associated with environmental restoration activities. January 17th.

NAVFAC. 2000. Guide to optimal groundwater monitoring. Prepared for the Naval Facilities Engineering Research Center by Radian International. White Rock, NM: Radian International.

NAVFAC. 2001. Guidance for optimizing remedial action operation (RAO). Special Report SR-2101-ENV. Prepared for the Naval Facilities Engineering Service Center. Research Triangle Park, NC: Radian International.

Nonaka, I., and H. Takeuchi. 1995. The knowledge creating company. New York: Oxford Univ. Press.

O'Brien, M. 2000. Making better environmental decisions: an alternative to risk assessment. Cambridge, MA: MIT Press.

Peyton B. M., T. P. Clement, and J. P. Connolly. 2000. Modeling of natural remediation: contaminant fate and transport. Pp. 79–120 In: Natural remediation of environmental contaminants: its role in ecological risk assessment and risk management. M. Swindoll, R. G. Stahl, and S. J. Ells (eds.). Pensacola, FL: SETAC Press.

Pierce, F. J., and W. E. Larson. 1993. Developing criteria to evaluate sustainable land management. Pp. 7–14 In: Proceedings of the eighth international soil management workshop: utilization of soil survey information for sustainable land use, May 3, 1993. J. M. Kimble (ed.). Lincoln, NE: USDA Soil Conservation Service National Soil Survey Center.

Pignatello, J. J., and B. S. Xing. 1996. Mechanisms of slow sorption of organic chemicals to natural particles. Environ. Sci. Technol. 30(1):1–11.

Pirie, R. B. 1999. Memorandum 99-02—Land Use Controls. 25 May 1999.

Pittinger, C. A., R. Bachman, A. L. Barton, J. R. Clark, P. L. deFur, S. J. Ells, M. W. Slimak, R. G. Stahl, and R. S. Wentsel. 1998. A multi-stakeholder framework for ecological risk management: summary from a SETAC Technical Workshop. Environmental Toxicology and Chemistry Supplement 18(2):1–20.

Presidential/Congressional Commission on Risk Assessment and Risk Management. 1997a, b. Framework for environmental health risk management, volumes 1 and 2. Washington, DC: U.S. Government Printing Office.

Renn, O. 1999. A model for an analytic-deliberative process in risk management. Environ. Sci. Technol. 33(18):3049–3055.

Sexton, K. 1999. Setting environmental priorities: is comparative risk assessment the answer? Pp. 195–219 In: Better environmental decisions: strategies for governments, businesses, and communities. K. Sexton, A. A. Marcus, K. W. Easter and T. D. Burkhardt (eds.). Washington, DC: Island Press.

Shindler, B., and K. A. Cheek. 1999. Integrating citizens in adaptive management: a prepositional analysis. Conservation Ecology 3(1):9.

Simpson, R. D., and N. L. J. Christensen. 1997. Ecosystem function and human activities: reconciling economics and ecology. New York: Chapman and Hall.

Slovic, P. 1993. Perceived risk, trust, and democracy. Risk Analysis 13(6):675–681.

Sokol, R. C., C. M. Bethoney, and G. Y. Rhee. 1998. Effects of Aroclor 1248 concentrations on the rate and extent of polychlorinated biphenyl dechlorination. Environ. Toxicol. Chem. 17(10):1922–1926.

Stahl, R. G., Jr., R. A. Bachman, A. L. Barton, J. R. Clark, P. L. deFur, S. J. Ells, C. A. Pittinger, M. W. Slimak, and R. S. Wentsel. 2001. Risk management: ecological risk-based decision-making. Pensacola, FL: SETAC Press.

Stahl, R. G., Jr., C. A. Pittinger, and G. R. Biddinger. 1999. Developing guidance on ecological risk management (guest perspectives). Risk Policy Report May 14:38–40.

Suter, G. W. 1993. Ecological risk assessment. Chelsea: Lewis Publishers.

Travis, C. C., and J. M. Macinnis. 1992. Vapor extraction of organics from subsurface soils: is it effective? Environ. Sci. Technol. 26(10):1885–87.

Trigeorgis, L. 1996. Real options: managerial flexibility and strategy in resource allocation. Cambridge, MA: MIT Press.

U.S. Army Corps of Engineers (USACE). 1996. Risk assessment handbook. Volume II: environmental evaluation. EM 200-1-4. Washington, DC: U.S. Army Corps of Engineers.

Walters, C. 1986. Adaptive management of renewable resources. New York: Macmillan.

Walters, C. 1997. Challenges in adaptive management of riparian and coastal ecosystems. Conservation Ecology 1(2):1.

Walters, C. J., and C. S. Holling. 1990. Large-scale management experiments and learning by doing. Ecology 71: 2060–2068.

Weber, W. J., P. M. McGinley, and L. E. Katz. 1991. Sorption phenomena in subsurface systems: concepts, models and effects on contaminant fate and transport. Water Research 25(5):499–528.

Williams, B. K. 2001. Uncertainty, learning, and the optimal management of wildlife. Environmental and Ecological Statistics 8:269–288.

Woodhouse, E. J. 1995. Can science be more useful in politics? The case of ecological risk assessment. Human and Ecological Risk Assessment 1(4):395–406.

Zeiss, C., and J. Atwater. 1991. Waste disposal facilities and community response: tracing pathways from facility impacts on community attitude. Can. J. Civ. Eng. 18(1):83–96.

3

Monitoring and Data Analysis to Support
Adaptive Site Management

Adaptive site management (ASM) is dependent on the the development of analytical tools to help site managers determine when, and to what degree, a change of remedy will better achieve the goals of cleanup. At the same time, these tools should help demonstrate to diverse stakeholder groups that changes are warranted. It is important to gain support from the affected public and from public or private transferees prior to making changes in remedial strategies, even when an agreement has already been reached between the lead regulatory agency and the responsible party. Consensus can best be achieved if there are objective methods that help evaluate the potential changes.

This chapter considers analytical tools and monitoring techniques that can aid in the assessment of remediation performance and help site managers decide if the current remedy-in-place should be reevaluated. Monitoring programs supply the information required to support the four management decision periods (MDP) described in Chapter 2. For example, analysis of monitoring data is needed to determine whether performance standards and operational expectations have been met, whether remedial goals have been achieved, and ultimately whether site closeout can occur.

ANALYTICAL TOOLS FOR EVALUATING REMEDY
EFFECTIVENESS AND NEED FOR CHANGE

Both graphical and tabular techniques exist to help make decisions about the effectiveness of remedies and the need for change. Tabular methods attempt to characterize the various objectives and attributes of interest for alternative remediation plans and display them on a single table so that they may be considered together. These objectives could include human health and ecosystem risks (or risk reductions), contaminant mass remaining (or removed), projected time and cost to completion of remediation, projected land use and property values at or near the site, and a qualitative indication of the likely extent of support or opposition among different stakeholder groups. This presentation should help illuminate major advantages and disadvantages of each alternative, and indicate the tradeoffs between the desired objectives that occur in switching from one remediation plan to another.

More formal analysis is also possible using various techniques of multiattribute utility theory (Keeney and Raiffa, 1976; Keeney, 1980; Merkofer and Keeney, 1987; Edwards and Barron, 1994; Clemen, 1996; Farber and Griner, 2000). Examples include the assignment of weights to different objectives (both by the site manager and by different stakeholders) to see how sensitive preferred alternatives are to these differing weights. As a hypothetical example, the eight objectives identified in Chapter 2 could be used, with differential weight being given to them to reflect laws and regulations and stakeholder preferences. The outcomes of different remedies can be ranked in an attempt to identify the most promising alternative. Such techniques have been employed to help facilitate stakeholder deliberations and decisions for other environmental management problems (Jennings et al., 1994). Often, such deliberations are best supported with simple and effective graphical presentations for each alternative, as discussed below. One weakness of this approach is that it can be difficult and costly (in terms of time and resources) to obtain quantitative values for all objectives.

In addition to the tabular approaches, a number of graphical options can be developed to illustrate when changes in a remedy might be necessary. For remediation operations based upon contaminant extraction (e.g., pump-and-treat or soil vapor extraction), the most straightforward graph would be one that displays mass removal over time, as shown in Figure 3-1. Indeed, such graphs are already commonly prepared in practice, as discussed in Chapter 2 in the Lawrence Livermore case study (and other case studies described later). Recent Navy guidance (NAVFAC, 2001) advocates preparation of performance plots of monthly operation and cost data similar to Figure 3-1.

Although mass removal is one objective measure of the remediation performance, cleanup goals are normally based upon reduction of total pollutant concentrations to health-based standards. [Such cleanup goals contain an implicit assumption that total concentration levels determine risk, which may or may not be accurate depending on the bioavailability of the contaminant (NRC, 2003)]. Therefore, another way to assess the progress of remediation is to plot the temporal changes in concentration at chosen "sentinel" monitoring wells (e.g., wells located at the down-gradient property boundary or adjacent to critical receptors). Such a plot is represented by Figure 3-2, which shows both hypothetical contaminant concentration over time as well as the *reduction* in contaminant concentration (or reduction in risk) over time. This second measure is more re-liable, because calculation of the baseline risk associated with the initial contaminant level is fraught with uncertainty, whereas there is less uncertainty about the risk *reduction* (as measured by the surrogate concentration reduction).

The hypothetical graphics shown in Figures 3-1 and 3-2 are drawn to represent a single remediation technique (e.g., pump-and-treat, soil vapor extraction). Analogous curves using different measured parameters could also be drawn to describe containment technologies (e.g., sediment capping), which aim to limit the contaminant mass flux through a "compliance" boundary. Of course, a containment technology would have

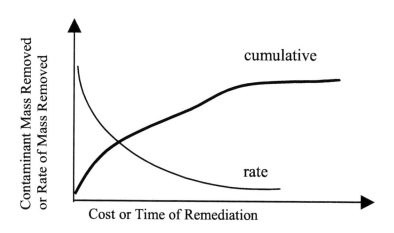

FIGURE 3-1 A hypothetical plot of contaminant mass removed over time or over cost, for a remedy based on extraction of mass. Both the cumulative mass removed and the rate of mass removed are shown.

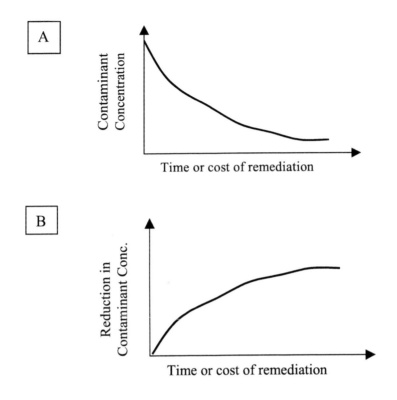

FIGURE 3-2 Hypothetical plot of (A) contaminant concentration over time or over cost and (B) reduction in contaminant concentration over time or cost.

little or no impact on mass removal (Figure 3-1), but would achieve dramatic reduction in risk; this is an example of an exposure reduction ("E") strategy as previously discussed in reference to Figures 2-1 and 2-2.

Ideally, there would be a set of performance curves like those in Figures 3-1 and 3-2 for different remediation methods or management options such that the curves could guide decisions as to which option to select and when to change from one approach to another. As an illustration of such curves, consider Figure 3-3, which shows a family of hypothetical curves for the risk reduction over time for various types of remediation systems. Curves A, B, C, D, E, and F within Figure 3-3 suggest a wide range of potential results from different remedies. Attaching specific strategies to any given curve is not possible without more information on the type of contamination, the predominant exposure pathway, and the affected receptors. However, one can speculate that Curve A represents a mass removal strategy such as *in situ* chemical oxidation of dense nonaqueous phase liquids, where a high percentage of mass must be destroyed before a significant reduction in groundwater concentrations and thus risk is achieved (see Box 5-12 for more explanation of this behavior). Curves B, C, and E could represent any number of strategies where risk is reduced incrementally over time from the source zone, including monitored natural attenuation. Curve F may represent a strategy like containment or a landfill cap where no contaminant mass is reduced, with the dotted line representing the possibility of future catastrophic failure.

The "effectiveness" of any particular remedy could be based on the ratio of risk reduced per unit of time or cost. (Keep in mind that it is difficult to quantify risk, and thus the ordinate axis may actually represent reduction in concentration.) Higher ratios would be desirable, and any remedy that provided the higher ratios may be considered well suited for the particular risk reduction goal. Lower ratios would suggest that either the remedy is not appropriate for meeting the risk reduction goal or the remedy needs to be optimized.

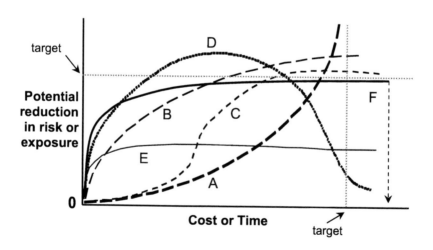

FIGURE 3-3 Hypothetical graphical representations of the change in risk with time or cost for different remedies.

Of these curves, only Curve E suggests a remedy that is totally ineffective in meeting cleanup goals and perhaps would provide the strongest graphical illustration of the need to change or modify the remedy. Clearly, remedies for which costs are increasing without any noticeable reduction in risk should not be continued. The case for change may be less clear for Curve A, which will eventually result in risk reduction albeit at longer time periods and higher costs than Curves B and C. The shape of Curve B is analogous to several case studies (see Box 2-3 and Appendix B) in that there is a relatively large initial reduction in contaminant concentration followed by a long period of relatively small reduction. Curve D represents a unique case in the sense that risk is seemingly being reduced quite effectively, yet as the remedy continues longer, the risk increases. This type of result may occur when source materials are drawn into an area or aquifer as a result of the remedy, increasing the concentrations of the contaminants to such a degree that higher risk results. This may also be the same type of curve that would result when an effective remedy is turned off and a rebound in concentrations occurs as the plume continues to move through the monitoring wells (but only if time, not cost, is the x-axis; if you turned off a remedy, presumably the cost disappears).

In addition to the qualitative assessments of the various curves described above, graphical tools could provide more quantitative guidance, assuming that reliable and accurate values for cost and risk reduction can be measured. For example, if there is a desired target goal for risk reduction, then a horizontal line can be drawn from this target to find the "least cost" remediation scheme. Using the example illustrated in Figure 3-3, Curve B would be conceptually the most desirable over the mid term, although Curve D achieves the target risk reduction at the least cost over the short term. Conversely, if there is a target remediation budget, then a vertical line can be drawn from this target to find the most effective remediation scheme. Using the example illustrated above, Curve A would be selected.

These examples are intended to be illustrative, and more detailed quantitative assessments are possible. For example, the slopes of the curves in Figure 3-3 measure the marginal risk reduction per unit investment, and these can be used in principle to optimally switch from one curve to another. Of course there may be other constraints that preclude such flexibility, and the difficulties in generating the risk reduction estimates must also be appreciated.

As discussed in Chapter 2, risk reduction may not be the sole objective of a site remediation strategy. For example, if both contaminant mass removal *and* risk reduction objectives are sought, then the problem becomes more complicated to visualize; however, graphical tools such as the one illustrated in Figure 2-2 could be developed. In this case each remediation system is represented by two curves, one measuring its performance for the risk reduction objective, and the other for the mass reduction objective. The time horizon for remediation is another objective that is often not considered explicitly during the remedy selection phase. However, short remediation times would be highly desirable in scenarios where the property is to be transferred for economic development. In most cases a single remediation strategy will not be capable of simultaneously satisfying all the objectives. The value of such a multidimensional graphical plot is that tradeoffs among objectives and strategies become evident, thus establishing a framework for stakeholder input and negotiation.

Although development of performance curves is advocated in recent Navy guidance (NAVFAC, 2001), they are not routinely developed at most sites, particularly for soil and groundwater contamination. Rather, the general sequence of events is to determine a remedial goal and then choose a technology that will meet the goal at lowest cost. For sediment contamination, it is more typical to use the type of predictive models that could generate these performance curves in choosing the remedy (e.g., see Figure 3-4). The committee strongly recommends that the Navy make a concerted effort to collect the appropriate performance data so that these curves can be generated for various types of remedial actions and hydrogeologic settings. Indeed, data likely exist from Department of Defense (DoD), Department of Energy (DOE), and Superfund sites, as well as from government demonstration programs like the Environmental Security Technology Certification Program. The goal is to develop a set of models for broad classes of remedies, contaminants, exposure pathways, and receptors that can then be calibrated (most logically during the feasibility study) with site-specific data to generate performance curves applicable to a specific site. Developing the models in the first place will require data collection at sites where remedies are already in place, including data on contaminant concentrations at compliance or receptor locations if risk reduction is a desired metric. The benefits of this exercise are accrued later when the resulting models are calibrated with site-specific information and then used to inform remedy selection. Because such models reflect our current understanding of subsurface processes, which in some cases is limited, the models should be updated as performance monitoring data become available.

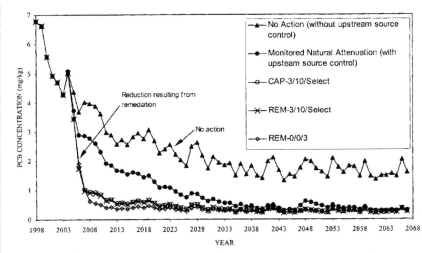

FIGURE 3-4 Model projections for polychlorinated biphenyl (PCB) concentrations in Thompson Island fish from 1998 to 2068 for various remedial alternatives as outlined by EPA Region 2. The noise in the "no action" projection is due to year-to-year variability in the projected flow record, which reflects the statistics of historical flows. SOURCE: Reprinted, with permission, from the National Research Council (2001). © (2001) National Academies Press.

Graphical tools can also be used to make decisions after implementation of a remedy, particularly in conjunction with the specific management decision periods of ASM. In addition to answering the three questions of MDP2 (is the remedy meeting performance standards, is it meeting operational expectations, and is it meeting the remedial goal), graphical analysis of monitoring data can enable identification of asymptotic conditions where concentrations are not low enough at the site to achieve the health-based remedial goal, and operation and maintenance costs have become high enough to raise concerns. Interpretation of the graphs to provide yes or no answers to these questions will be subjective, because there will likely be disagreement about various critical performance criteria (e.g., at what dollar value does the cost per pound removed become cost-inefficient, or at what slope of the concentration versus time curve should the remedy be changed). Nevertheless, the graphs will indicate trends that provide information needed by the remedial project managers (RPMs), regulators, and stakeholders for decision making.

Several case studies already exist demonstrating how graphical tools can aid in making decisions to modify remedies and in evaluating remedial objectives. In almost all these examples, concentration is the measured parameter and is used as a surrogate indicator of risk. The first study is from the set of volumes published by the U. S. Environmental Protection Agency (EPA) under the auspices of the Federal Remediation Technologies Roundtable (EPA, 1998a). At Pope Air Force Base, as much as 75,000 gallons of JP-4 free product are floating on top of the water table; some dissolved volatile organic compounds (VOCs) have also been detected in groundwater samples. The remediation system consists of a free product cut-off trench and a dual pump recovery system. Figure 3-5 shows a decreasing removal rate over time for free product at this site, such that the cumulative recovery curve begins to flatten after April 1995. (Note that the EPA report includes another case study for a different free product removal site at Pope AFB where the cumulative removal continues to increase approximately linearly over time.) As of the last reported date (October 1996), approximately 3,500 gallons had been recovered, less than 5 percent of the estimated spill volume of 75,000 gallons.

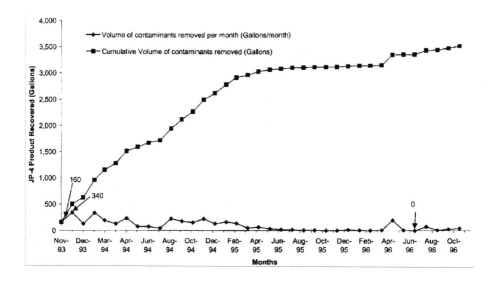

FIGURE 3-5 Monthly and cumulative free product removal at Site SS-07, Pope AFB. SOURCE: EPA (2000a).

An interesting aspect of this case study is that estimates of operation and maintenance (O&M) costs are combined with the above data to produce a graph showing cumulative costs versus cumulative pollutant mass removed. This graph, presented in Figure 3-6, illustrates the economic impact of the "tailing" behavior—as the remediation progresses, it becomes increasingly more costly to remove a given unit of contamination. However, it should be noted that Figure 3-6 was produced by making the simple assumption of constant average monthly O&M costs. Because the monthly costs are constant, and the monthly removal rate decreases over time as shown above, the cost per unit gallon removed will increase. This graph indicates that the performance of the remediation system has declined because little additional mass is being removed as funds continue to be spent, signaling that the system should be reevaluated. A reevaluation may or may not lead to a change in remedy, depending on the expected performance of the technology and other factors.

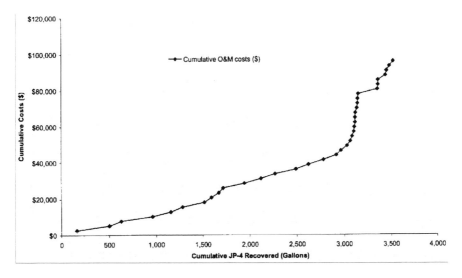

FIGURE 3-6 Free product removal versus cumulative operation and maintenance costs at Pope AFB. SOURCE: EPA (2000a).

A second case study is for the Campbell Street Fuel Farm groundwater pump-and-treat system located at Marine Corps Air Station New River, which is co-located with Marine Corps Base Camp Lejeune, North Carolina (NAVFAC, 2001). The fuel farm is an active storage facility for JP-5, and release of fuels at the site has led to contamination of soil and groundwater by benzene, toluene, ethylbenzene, and xylene (BTEX) and VOCs. Contaminated groundwater is limited to the upper portion of a surficial aquifer with its water table 6 to 7 feet below ground surface. Initial remedial actions at the site were excavation of contaminated soil and removal of measurable free product. A groundwater pump-and-treat system began operation in July 1996; the system includes interceptor trenches and several extraction wells that were installed in plume hot spots. The trenches are downgradient of the contaminant plume, and all intercepted water is directed toward sumps for removal.

Figure 3-7 shows that the VOC mass removal rate has decreased significantly over time; while 3.5 pounds were removed during July 1996 through March 1999, less than 0.5 pounds have been removed since December 1997. Figure 3-8, a plot of the cumulative cost versus cumulative mass removed, dramatically displays the tailing behavior of the system. It can be seen that approximately $175,000 was spent to remove the first 3 pounds of VOCs, but an additional $325,000 was spent to remove the next 0.5 pounds. The graphical data below were used in conjunction with other analyses and assessments to recommend that the trenches be shut down and that monitoring data be collected to evaluate the degree to which the plume was being affected by natural attenuation processes. Figure 3-8 suggests that caution and knowledge of the chosen treatment are needed when interpreting such graphs for the purpose of making changes to the remedial system (as discussed in Chapter 2 with respect to MDP2). It would have been premature to abandon the pump-and-treat system at the first sign of cost inefficiency in late 1996. Fortunately, site managers recognized that such systems generally take years before performance reaches an asymptote; continued operation resulted in a substantially longer period of effective mass removal.

When the graphical tools indicate the remediation system should be reevaluated, changing the remedy can improve the system, as illustrated graphically in Figure 3-9. This figure schematically depicts contaminant concentration versus time when changing from a suboptimal remedy (such as that depicted by Curve E in Figure 3-3) to another remedy (Curve B in Figure 3-3). Changing the remedy should alter the concentration versus time curve such that the target contaminant level is reached sooner.

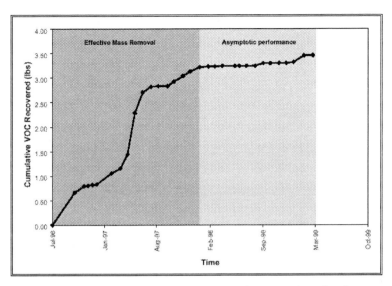

FIGURE 3-7 Cumulative mass recovered versus time for the pump-and-treat system at the Campbell Street Fuel Farm. SOURCE: NAVFAC (2001).

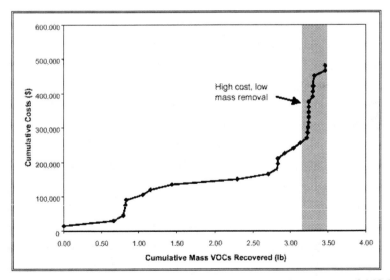

FIGURE 3-8 Cumulative costs versus cumulative mass of VOCs removed for the pump-and-treat system at the Campbell Street Fuel Farm. SOURCE: NAVFAC (2001).

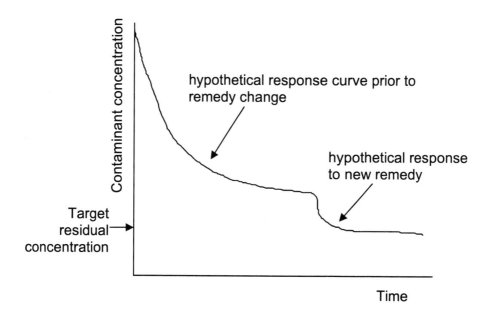

FIGURE 3-9 Hypothetical effect of changing the remedy on the concentration versus time curve.

Despite the problems that may arise if data are highly variable, there is merit in undertaking the graphical approach described above provided that the data reveal trends. In general, the remedy should be revisited when there are reasonable quantitative data showing that the existing remedial action cannot attain the health-based remedial goal selected for the site after being operated for an appropriately long period of time. In order for this to occur, the slope of the line tangent to the concentration versus time curve must be approaching zero (the so-called asymptote), yet the concentration must remain above the site-specific remedial action goal. Second, it should be shown that the cost of treatment is increasing sharply as the incremental mass removed decreases, even though the annual costs may remain constant. Whenever possible, visual interpretations of these data should be supported by statistical analyses to ensure that the inferred value of a trend (or the lack of a trend) is statistically significant. [Overviews of statistical methods for trend analysis are found in Lettenmaier (1976, 1977), Gilbert (1987) and Gibbons and

Coleman (2001)]. Simple methods include cumulative sum charts that accentuate the long-term effects of trends and changes in the mean (Berthouex et al., 1978), and linear regressions of concentration versus time from which it can be determined whether the slope of the fitted regression model is significantly different from zero. Methods for data quality control can also be adapted for trend analysis (Starks and Flatman, 1991). With large datasets, methods for time-series analysis can be used to identify and remove trends and statistical periodicities in the data over different time scales (Box and Jenkins, 1994). Nonparametric methods may also be employed for trend detection; these are especially appropriate when, as often occurs with environmental data, the variations in the measurements are not normally distributed.

When cost and concentration data analyses reveal declines in remedy performance prior to reaching the cleanup goal, the responsible party should undertake reconsideration of the remedy with the same public participation steps that are utilized in the original remedy selection process. In order for these exercises to be effective, the Navy, in consultation with stakeholders, should select a unit cost for the continued operation of the remedial action at the site under investigation, above which the existing remedy is no longer considered a tenable option. This value will necessarily vary from site to site to reflect the type of technologies used, site conditions including the existing contaminant concentrations compared to the cleanup goal, the toxicity of the contaminants, the likelihood of future exposure, and other factors. It is possible that there will be some regulatory and stakeholder reluctance to using a metric such as "cost per pound of contaminant removed" for decision making. Members of the public are often suspicious of risk assessment in general (see Box 2-2), particularly attempts to place a monetary value on individual lives and public health. Nevertheless, in the committee's experience, most community activists react constructively when given pertinent technical and financial information and the chance to fully participate in decision making. Typically, if incrementally more cleanup can be demonstrated to make the local environment significantly safer, most stakeholders will insist upon the higher-level response. If additional actions will only marginally improve safety, and this can be conveyed using the types of graphical presentations discussed previously, stakeholders will give it due consideration (as was evidenced by community activist sentiment regarding a mercury-contaminated site in Oak Ridge, TN—NRC, 2003). Graphs showing predicted and real performance curves (and other evidence that responsible parties and regulators have the public's interests in mind) are also more likely to make the public receptive to limited cleanup at certain locations if other sites receive *greater* attention as a result of the same type of analysis.

Consideration of Uncertainty

The discussion of tabular and graphical tools above neglects the inherent uncertainties that are present in risk assessment and performance assessment of groundwater remediation. Uncertainty is a significant reality associated with all environmental monitoring programs and is the result of (among other things) limited spatial and temporal data from which inferences must be drawn. Uncertainty is particularly prevalent in our understanding of subsurface properties, including stratigraphy, presence or absence of preferential flow paths or fractures, porosity, hydraulic conductivity, and boundary conditions. There is also substantial uncertainty at a given site regarding the nature and extent of contamination, the type of biological and geochemical processes that might be taking place that affect contaminant fate and transport, and the exposure mechanisms that translate into deleterious effects (NRC, 1999). As a result, there is significant uncertainty associated with any estimated contour map of a contaminant plume as well as the total contaminant mass. There may also be significant uncertainty about whether the measured total mass of a contaminant in the subsurface is directly correlated with exposure or risk. Because of these uncertainties, it is not possible to assign a single value to either the baseline risk, or to the risk reduction that could be achieved by a given remediation technology.

The extent of uncertainty about site conditions and remedial performance has implications for decision making throughout ASM. For example, at MDP2 the uncertainty in performance monitoring data plays a significant role in determining whether cleanup goals are being met. Mass removal achieved by ongoing remediation (Figure 3-1) is generally known to a high degree of certainty, but the critical factor is how close the asymptotic cumulative value is to the total pollutant mass at the site. In many cases, it may not be known whether the curve is leveling off at 5 percent, 50 percent, or 95 percent of the total (but unknown amount of) onsite contamination. The uncertainty may be particularly high at complex sites with a high degree of heterogeneity, multiple aquifer layers, fractured rock, and/or the presence of nonaqueous phase liquids (NAPLs) that can move in unusual ways from source zones, or remain entrapped at disparate locations on- or off-site. Similarly, river or coastal sediment beds with unusual hydrologic and sediment transport and deposition/resuspension regimes can lead to a high degree of uncertainty in the quantity and location of remaining contamination following cleanup efforts.

The overall uncertainty in the total mass onsite is shown schematically in Figure 3-10. There, M_t denotes the unknown total contaminant mass in the system, and the double-headed arrow is meant to convey uncertainty in that value. (Note that there could also be uncertainty in the asymptotic value of the cumulative mass removed.)

Uncertainty can also be represented on graphs that plot the reduction in contaminant concentration as a function of time or cost of remediation (such as Figure 3-2). These data could be generated from monitoring at chosen compliance or sentinel wells. However, because of the inherent spatial variability of contaminant fate-and-transport processes, there will always be uncertainty about the contaminant levels in portions of the site that are not monitored. Moreover, there are uncertainties that arise in computing human or ecological risk from ambient groundwater concentration values, given a lack of knowledge about how much of the total contaminant concentration is actually bioavailable. Incorporating this uncertainty into the graphical representations of concentration and risk reduction in Figure 3-2 and 3-3 is even more challenging.

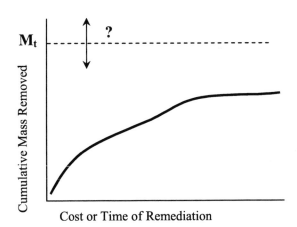

FIGURE 3-10 Hypothetical graph of cumulative mass removed over cost or time, showing uncertainty in the value of the total mass present (M_t).

One possible graphical technique is the concept of statistical confidence bands. The concept assumes that the risk (or concentration) reduction achievable for any given cost is a random variable with a certain probability density function. The random nature results from all the uncertainties in the system—for example, uncertain initial contaminant mass, uncertain remaining contaminant mass for a given remediation cost, uncertainty in groundwater fate-and-transport models, and uncertainty in dose–response models. The use of confidence bands is demonstrated conceptually in Figure 3-11. For any given remediation cost, the confidence bands could represent, for example, the 5 percent and 95 percent probability levels. That is, there is a 95 percent probability that the risk reduction is less than the upper curve, a 5 percent probability that the risk reduction is less than the lower curve, and thus a 90 percent probability that the risk reduction is between the upper and lower curves. The solid center curve might represent the mean or "best estimate."

The figures discussed above are for treatment-based remediation strategies where there is a direct correlation between performance and time. Strategies based upon exposure pathway intervention, such as sediment capping, onsite containment, or institutional controls (see the "E"-type strategies discussed in relation to Figures 2-1 and 2-2), perform in either a success or failure mode. Thus, performance uncertainty involves mainly the time to potential failure and, to a lesser extent, the nature of the failure (e.g., catastrophic or gradual). This is illustrated schematically in Figure 3-12.

There is an increasing body of literature that presents ideas along the lines discussed here, especially the concept of formally incorporating uncertainty into remediation design (mostly pump-and-treat). The main types of uncertainty considered are related to site hydrogeology. A typical statement of the design problem is as follows: design a remediation system that is guaranteed to work with a probability of at least X percent. Tradeoffs between reliability and cost are developed by varying the success probability level. A recent review of this work is given by Freeze and Gorelick (1999). Some more recent work (Minsker and Smalley, 1999) is extending these design concepts to be based more directly upon human health risk. Although most published work emphasizes development of the methodology with application only to hypothetical scenarios, Russell and Rabideau (2000) present an application to an actual site near Buffalo, New York.

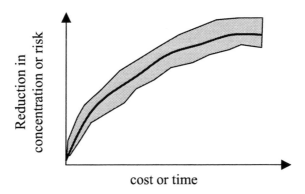

FIGURE 3-11 Statistical confidence limits around the curve of concentration reduction over cost or time. The upper and lower curves correspond to the 5 and 95 percent probability levels.

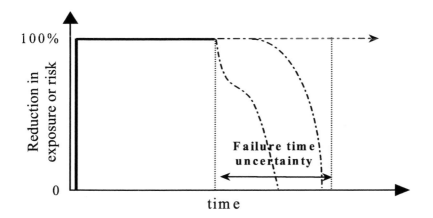

FIGURE 3-12 Statistical confidence limits around the time of potential remedy failure.

The work of Maxwell et al. (1998) (initially discussed in NRC, 1999) further illustrates formal uncertainty analysis concepts. The significant feature of Maxwell's work is that it combines uncertainty about groundwater fate and transport with variability in human receptors due to factors such as body weight and daily habits of water consumption and vapor inhalation. Typical results show the probability of increased cancer risk for different fractiles of variability in the receptor population given an exposure pathway of drinking contaminated groundwater. In more recent work, Maxwell et al. (2000) extend these concepts to evaluate the impact of different pump-and-treat remediation systems on reducing risk for a hypothetical contamination scenario. Their results do show that remediation reduces risk but, interestingly, there are differing amounts of risk reduction to different segments of the population. Their results are presented in the form of Figure 3-11, with confidence bands added to reflect fate-and-transport uncertainty. In order to consider variability among receptors, different curves (each with different confidence bands to reflect uncertainty) are drawn for different members of the population. Figure 3-13 provides an example, in which the curve corresponds to one segment of the receptor population, and the vertical bars indicate the uncertainty (approximate confidence limits) that is due to geological variability for the two different pumping rates that were studied.

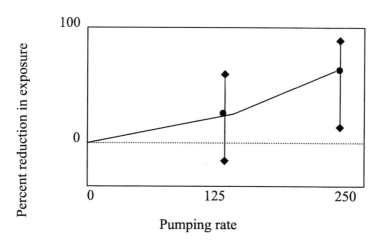

FIGURE 3-13 Uncertainty in the percent reduction in exposure achieved as remediation efforts are changed by varying the pumping rate. SOURCE: Maxwell et al. (2000).

In another interesting recent contribution related to contaminated sediments, Stansbury et al. (1999) present a methodology for accounting for uncertainty in human health risks, ecological impacts, and remediation costs of different strategies for disposal of dredged material. The method is demonstrated using an example of contaminated sediment disposal at Elliott Bay, near Seattle, where the possible remedial alternatives are (1) unconfined aquatic disposal (UAD), (2) capped aquatic disposal (CAD), (3) near-shore confined disposal facility (CDF), (4) upland disposal (UPL), and (5) upland secure disposal (UPS). Unconfined aquatic disposal is open water discharge of the dredged material. Capped, or confined, aquatic disposal is open water discharge of the dredged material into a prepared or existing depression in the sediment and capping with clean sediment. A near-shore confined disposal facility is an in-water landfill, generally with only primary sedimentation as treatment during placement. Upland disposal and upland secure disposal are both conventional landfills, the first with simply primary treatment and the second with more elaborate containment. These remediation alternatives achieve their effectiveness at the time they are implemented, whereas most groundwater remediation alternatives need to operate over extended time periods.

Stansbury et al. (1999) do not use probabilistic techniques but rather adopt the formalism of "fuzzy set" methods to incorporate uncertainty. In this approach, uncertainty is represented by a range of "plausible" and "most likely" parameter values; this range can be established using a variety of information sources including measured data and engineering judgment. An example result is shown below in Figure 3-14, which shows tradeoffs among human health risk, uncertainty, and cost for the five remedial alternatives described above. For each alternative, the inner rectangle represents "high confidence" while the outer rectangle is still plausible but with lower confidence. The results show that upland secure disposal provides the greatest human health benefit, but at a very high cost. There is also a relatively large degree of uncertainty in the human health risk estimate; i.e., a given disposal strategy may provide relatively low cancer risk under one set of assumptions, yet it may also result in a high risk under a different, yet plausible, set of assumptions.

***

NRC (1999) identified various ways to approach uncertainty in hazardous waste cleanup. In the face of limited information that typifies many sites, the use of conservative cleanup goals has been prevalent.

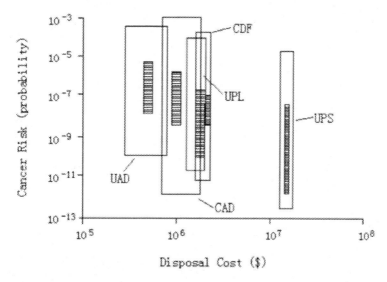

FIGURE 3-14 Tradeoffs between lifetime cancer risk and disposal costs for five disposal alternatives. For each alternative, the outer box denotes the range of plausible values, while the inner hatched box denotes the range of most likely values. UAD = unconfined aquatic disposal, CAD = capped aquatic disposal, CDF = near-shore confined disposal facility, UPL = upland disposal, and UPS = upland secure disposal. SOURCE: Reprinted, with permission, from Stansbury et al. (1999). © (1999) Journal of Water Resources Planning and Management.

Alternatively, attempts have been made to develop more comprehensive programs of site monitoring and characterization by, e.g., increasing the number of monitoring wells. Clearly there are distinct tradeoffs between these two approaches, both in terms of the information gathered and cost. For example, the use of only one monitoring well located within a contaminated area would require significant extrapolation as to what is occurring at more distant edges of the plume. This may result in a conservative approach to operating a pump-and-treat system even if the well yields consistent results below cleanup criteria. In contrast, a monitoring system that plasters an affected plume with sampling points would presumably be much more precise in its determination of when a system is in compliance and hence when remediation can stop, although it will also be more costly.

Because many remediation systems are overdesigned to account for uncertainties (i.e., via an engineering "safety factor"), there may be significant economic value in collecting data to assess and reduce uncertainties in remedial performance. It was for this reason that NRC (1999) favored the more rigorous data collection approach over the use of conservative goals. The formal uncertainty analyses presented above provide substantial benefits to such data collection efforts. First, such analyses are valuable aids for site decision making because they provide a graphical display of the variability and uncertainty that are inherent features of any remediation problem, which can be used, among other things, for communicating information to stakeholders (including the extent of confidence in predicted and actual remedial performance). Moreover, by analyzing and ranking the various factors that contribute the greatest to overall uncertainty, it is possible to direct data collection activities that might reduce uncertainty toward the most critical parameters. In Stansbury et al. (1999), it was found that the rather large range in cancer risk shown in Figure 3-14 for all disposal alternatives was due mainly to the uncertainty in the dose–response relationship for the contaminants rather than to uncertainty in exposure pathways. This suggests that reducing uncertainty could be better accomplished by investing in additional research on dose–response relationships rather than by exploring other remedial options. Similarly, Russell and Rabideau (2000) used sensitivity analysis to examine the impact of various modeling assumptions upon management decisions. Several authors have studied how information obtained from specific data collection programs can be used to most effectively reduce uncertainty in contaminant fate-and-transport predictions and also in resulting site management decisions (e.g., James and Gorelick, 1994; Wagner, 1999; Sohn et al., 2000). It should be noted that such uncertainty analyses may only be feasible for larger, more complex sites where a fate-and-transport model is already available.

MONITORING

Monitoring plays a pivotal role at all stages in adaptive site management—from initial site discovery to site closeout. A cursory examination of Figure 2-7 might suggest that monitoring is needed only to answer the three questions posed during MDP2. However, monitoring programs are essential to facilitate site characterization and risk assessment (Step 1), to adequately conduct experimentation and evaluation, to produce the data necessary for constructing the performance evaluation graphs described earlier (which would be used during remedy selection or MDP3), and to determine whether residual contamination exists that will prevent site closeout during MDP4. The focus of the monitoring programs is necessarily site- and time-specific. For example, a soil remedial action may primarily require sampling during excavation (performance monitoring) and immediately after remediation work is complete (site closeout). For sediment and groundwater remedial actions, much longer-term monitoring programs might be developed that have their roots in initial site characterization activities, continue through remediation, and extend for significant periods of time beyond the termination of active remediation. In the case of groundwater, most sites begin with an inherited set of monitoring points already established, and so part of the monitoring design process also includes determining to what extent this existing network can be used or must be abandoned or expanded. Depending on the chosen remedial actions, monitoring programs may represent the majority of remedial action costs (such as for monitored natural attenuation) or only a small percentage.

The design of a remedial action performance monitoring network requires determining the parameters of interest, identifying the numbers and locations of monitoring points, specifying sampling protocols, frequencies, and analytics, and, finally, developing the data analysis methods that will support the decisions that have to be made. Traditional characterization and monitoring programs tend to pre-specify sample numbers, locations, sampling frequency, and analytics, where the emphasis for analytics has been on offsite laboratory analyses. This traditional type of data collection presents several limitations, particularly in the context of subsurface characterization and monitoring. The costs are sometimes prohibitive, driven both by sample analytical costs and the capital investment required for monitoring wells. High monitoring costs, particularly for monitoring programs that extend over time, result in pressures to limit data collection. Limited data collection, in turn, results in decision making that relies on datasets too sparse to adequately address the inherent heterogeneities and uncertainties associated with subsurface systems. Finally, by pre-specifying sample numbers and locations and relying on offsite laboratory analyses with long turnaround times for analytical results, traditional characterization and monitoring programs are ill equipped to handle unexpected results when they are encountered. Fortunately, in the last several years there have been technological advances in sensors, field analytics, and sample collection technologies that can help to lower costs and/or increase the effectiveness of monitoring programs. New approaches for designing and implementing environmental data collection programs have also been developed.

The following section discusses several different aspects of monitoring, starting with the parameters that are commonly used to measure remedy performance (relevant during MDP2, evaluation and experimentation, and long-term stewardship). The focus then shifts toward innovative monitoring network design that will facilitate use of ASM by allowing the entire remedial implementation period to be more adaptive. This includes discussion of new sampling technologies as well as ways to enhance existing sampling networks. The former is applicable to all stages of cleanup, from site characterization to long-term monitoring, while the enhancement of existing networks pertains primarily to long-term monitoring of contaminated groundwater. Thus, the case studies presented span various stages of cleanup, from initial characterization of a contaminated sediment site to optimization of groundwater monitoring. Indeed, because "site characterization" and "long-term monitoring" describe the same general activity—data collection with the purpose of understanding surface/sediment/subsurface contamination events at particular points in time—it should not be surprising that the same sampling technologies are appropriate for both characterization and later monitoring activities.

Monitored Performance Parameters

The performance evaluation graphs presented earlier focus on several key parameters measured over time. These include risk and risk reduction, contaminant concentration, contaminant mass removal, and cost. Aside from these primary parameters, there can also be a host of secondary, tertiary, and technology-specific parameters that might be included in a monitoring program. The section below discusses many of the most common performance parameters used for assessing remedy performance in contaminated soil, sediment and groundwater scenarios. To be rigorous, the monitoring system should provide multiple lines of evidence (as manifested by a variety of measured parameters) that a remedy is or is not effective.

Risk Parameters

Most cleanup goals in RODs are expressed as contaminant concentrations that correspond with a risk falling in the range of 10^{-6} to 10^{-4} for carcinogenic compounds. There are performance parameters that directly address risk without measuring mass or concentration reduction, such as growth and/or mortality of a target organism. Such parameters are measured in effects-based toxicity tests, and they are used primarily where cleanup is driven by ecological concerns because of the acceptability of performing these types of tests on plants and animals. For example, a suite of methods is available to assess toxicity of contaminants in soils and in freshwater and marine sediments to invertebrates and other animals, and newer methods that harness molecular biological techniques are being developed for high throughput toxicity testing of sediments (EPA, 2000b; NRC, 2002). Toxicity test results from a study area can be compared to those of samples taken in a reference area where the contaminants are absent or are present at reduced levels to determine whether toxicity in the study area is elevated above a level considered acceptable or shown to cause negative effects.

The use of such effects-based parameters raises two implementation issues that must be addressed. First, because of the time required for substantive results from remedial actions to be reflected by such measures, short-term measurements such as contaminant volume, mass, or concentration reductions will almost certainly be needed to supplement the long-term monitoring of toxicity. Second, there can be substances in the sediment or soil that cause a toxic response other than the contaminants of concern, making interpretation of results difficult. As a result, it is important to be familiar with the conduct of these tests, with the types of spurious results that might result in some sample types or matrices, and with how to interpret the data appropriately so that inaccurate conclusions are not made.

Indicators of Exposure and Risk

One of the main elements of risk is exposure, for which a variety of monitored parameters are indicative. Contaminant concentrations at key locations or in key media (e.g., in sentinel monitoring wells for groundwater, or in the overlying water column in the case of sediments) are commonly used and often codified in RODS, as mentioned above. It is important to differentiate concentration measurements that are direct indicators of exposure, such as water column, plant, invertebrate, or fish tissue concentrations, from total concentrations in soil and sediment, which, depending on the receptor and exposure pathway, may be more indirect indicators of exposure.

Remedial action performance monitoring programs almost always include *in situ* concentration monitoring as a significant component. Examples of this kind of monitoring include monitoring wells for groundwater and sediment sampling for contaminated sediments. The results are used to compare to concentration-based remedial goals, to develop spatially averaged concentration values, and to construct concentration isopleths. *In situ* spatially averaged concentration values, when combined with mass removal measurements (discussed later), allow both for a comparison with compliance requirements and for estimation of when these compliance requirements might be achieved. Concentration isopleths can be used to identify areas that are or are not in compliance with cleanup requirements.

In contaminated sediments not subject to physical disturbance like erosion, bioturbation—the mixing associated with the normal life-cycle activities of benthic organisms—is typically the most important mechanism for transporting contaminants to the sediment–water interface (Reible et al., 1991). Because more than 90 percent of the 240 observations of bioturbation mixing depths in both fresh and salt water were 15 cm or less and more than 80 percent were 10 cm or less (Thoms et al., 1995), surficial sediments are thought to be most important in contributing to exposure of (1) organisms in the sediment or overlying water and (2) animals that may feed off of these organisms. Isolated deeper penetrations by individual organisms apparently have limited impact on a population-wide basis. If only this surface layer contributes to exposure, then the surface area weighted average concentration (SWAC) in sediments presents a convenient monitoring metric. This metric has been employed as a measure of exposure and risk at several contaminated sediment sites—for example, within the ROD for the Sheboygan Superfund site and for the remedial investigation and feasibility study (RI/FS) at the Fox River site (Wisconsin DNR, 2001). It should be emphasized, however, that the biologically active layer is not necessarily static, and erosion can expose deeper sediments or deposition can bury surficial sediments with time.

Using sediment contamination as an example, a variety of direct and indirect concentration metrics can be used during MDP2. Thus, MDP2a (compliance monitoring) might seek to ensure that water quality standards are not violated during implementation of a remedial approach. MDP2b (monitoring to ensure that operational expectations have been met) could employ surficial sediment concentrations such as SWACs. MDP2c (monitoring to ensure achievement of remedial goals) might involve fish tissue concentration measurements.

In groundwater extraction systems, changes in contaminant concentration in produced fluid over time are a typical metric. The primary issue with this metric is that although extracted fluid contaminant concentrations are easy to measure, they are difficult to interpret from a performance perspective. For example, steady values of measured concentrations may mean that the system is performing well (particularly if these measurements can be linked to large mass extractions as planned). However, the same values may indicate a poorly performing system if levels are higher than cleanup goals. As with almost all of the metrics discussed in this section, contaminant concentrations need to be interpreted in conjunction with other remedial performance measurements.

Mass Removal

Although closure requirements are traditionally posed as either concentration or risk-based standards, in some cases cleanup is stated in terms of mass removal. Even in cases where mass removal does not necessarily translate into cleanup requirement compliance, it is obviously linked to attaining such standards. Thus, for remedial systems that physically extract and then remove or destroy contaminants, mass removal can function as a directly measurable performance parameter. Although this metric is less related to risk than concentration, mass removal is easy to measure and is not subject to spatial variability to the same extent as concentration. Mass removal is commonly measured for pump-and-treat systems and vapor extraction systems for groundwater and vadose zone contamination, respectively, and for excavation/dredging and disposal for soil and sediment contamination. Mass removal measurements are much more difficult for systems that rely on *in situ* processes to degrade or destroy contamination, such as *in situ* bioremediation or natural attenuation. The issues are twofold. Changes in concentrations at fixed monitoring points over time can be indicative of either degradation or simple transport and contaminant redistribution. Estimates of total mass degradation rely on interpolating from relatively sparse monitoring datasets to the system as a whole.

Specific metrics related to mass removal include the rate of contaminant mass removal. This rate could be measured in an instantaneous sense (i.e., the current rate of removal), or it could be measured in an aggregate sense (i.e., the rate of removal over the last quarter or over the last year). The latter, in particular, may be important for identifying a decline in performance over time. For systems where contaminant mass is physically removed and can be measured, implementing this type of metric is straightforward. For *in situ* systems, the challenge is obtaining accurate estimates of contaminant mass removal or destruction.

The percentage of total mass removed may also serve as a performance metric. The problem in implementing this type of metric is having an accurate estimate of the original contaminant mass; such information is frequently unknown. Sampling programs are discrete events in time and space, requiring inferences regarding spatial and temporal trends, often based on very limited datasets. For example, estimates of total *in situ* contaminant mass based on relatively large RI/FS datasets can be grossly in error, largely because the data gathering performed for an RI/FS is not intended and should not be assumed to be adequate to design the remedy. A site near Tonawanda, New York, had an estimated 14,000 cubic yards of contaminated soils. This estimate was based on 341 soil samples collected from 116 soil cores over a five-acre site during the RI/FS. By the time remediation was complete, 45,000 cubic yards of contaminated soils had been removed (Durham et al., 1999). Thus, it should be recognized by regulators, the Navy, and the public alike that additional sampling data will almost always be required after the RI/FS.

Secondary, Tertiary, and Technology-Specific Performance Parameters

Besides mass removal and *in situ* concentration, there can be a host of secondary, tertiary, and technology-specific performance parameters that might be included in a monitoring program. Examples of secondary and tertiary parameters include daughter products from bioremediation processes, pH, dissolved oxygen, redox potential, dissolved carbon content, and depth to the water table. Examples of technology-specific performance parameters include drawdown for extraction wells, tracers for enhanced *in situ* bioremediation, and airflow rates for vapor extraction systems. Secondary, tertiary, and technology-specific performance parameters are used in combination with primary metrics to evaluate the efficacy of a remedial system. Circumstantial evidence provided by these types of performance parameters is significant and may be crucial to making the correct ASM decisions. Examples of the use of such data to draw inferences about the performance of a remediation plan are provided in Kampbell et al. (1998), EPA (1998b), Stiber et al. (1999), Wiedemeier et al. (1999), and NRC (2000). These protocols place a special emphasis on data to support the suitability for, and success of, natural attenuation because monitoring is central to implementation of this remedial strategy. However, similar data analysis methods can and should be developed and applied to evaluate the progress of other remediation methods.

Adaptive Monitoring Network Design

The design and implementation of monitoring programs can be made more adaptive to keep data collection activities, as well as the remedial action, as focused and cost-efficient as possible. Drivers for adjusting monitoring programs include changes in site understanding that lead to improved site conceptual models, unexpected monitoring results, alterations in remedial actions, improvements in monitoring technology, and changes in the type of information required by regulations.

In the last several years there have been significant technological advances in decision analysis, field analytics, and data collection technologies for characterization and monitoring work. These present several opportunities for making the characterization and monitoring process more adaptive and more supportive of an ASM approach. They include (in order of maturity and acceptance) (1) enhancing or optimizing existing monitoring networks, (2) incorporating sensors and field analytics in monitoring design, (3) using new technologies for collecting samples such as direct push systems and passive diffusion samplers, and (4) replacing static sampling and analysis plans with dynamic work plans. The following sections discuss each of these potential enhancements to remedial action monitoring programs, providing details on technology maturity and case studies.

Enhancing Existing Monitoring Networks

The first opportunity for adaptive sampling and analysis as remediation proceeds is to allow monitoring locations to be dropped or sampling intervals lengthened in response to monitoring data that show a system performing well. In the same vein, enhancements could involve adding monitoring locations or increasing the sampling frequency for existing locations for a remedial system that shows signs of deteriorating performance. There is often significant financial incentive to use as many existing groundwater wells as possible because of the costs associated with implementing new wells. Monitoring costs come in two forms—the capital cost of installing monitoring systems and the longer-term cost of sampling and maintaining the system. For deep vadose zone systems, installation costs can range into the hundreds of thousands of dollars per well (DOE, 1998). For shallow groundwater systems, these costs may be on the order of tens of thousands of dollars per installation. In any case, capital costs typically dwarf annual sampling costs.

The most widely used method for improving remedial action monitoring network performance is to determine whether monitoring locations need to be changed (i.e., old monitoring locations abandoned or new locations added) or sampling intervals adjusted. A variety of techniques have been suggested for assisting in this process. These techniques include relatively sophisticated fate-and-transport models, geostatistical and time series analyses, and mathematical optimization methodologies as well as relatively simple "rule-of-thumb" techniques.

The optimal design of monitoring networks in surface and subsurface hydrology is a classic problem that has received extensive attention in the scientific literature. Most of the previous work in the groundwater field falls into two categories: networks for site and plume characterization (e.g., Loaiciga et al., 1992) and networks for plume detection at landfills and hazardous waste sites (e.g., Meyer et al., 1994). There has been significantly less work to address questions of remedial action performance evaluation and long-term monitoring—questions that are directly relevant to MDP1 and MDP2 in the ASM process.

Long-term monitoring networks. With the realization that many contaminated sites will not be quickly closed and will thus require long-term monitoring and management, research in monitoring network optimization has shifted toward the objective of reducing long-term sampling costs without sacrificing information gained or protectiveness. The goal of the research published to date is to eliminate data redundancy by identifying a subset of monitoring wells and a reduced sampling schedule that effectively capture a groundwater plume's evolution. Temporal redundancy refers to whether wells are being sampled too frequently, and spatial redundancy refers to whether too many wells are being sampled. An early example that focused on temporal redundancy is the work of Johnson et al. (1996), who were motivated by the observation that in 1993, the laboratory fees alone required for analyzing groundwater samples at the Savannah River Site amounted to nearly $10 million. These researchers developed a simple technique to reduce sampling schedules through analysis of the time series at individual wells. A trial application of their method resulted in an estimated cost savings of $1.8 million at the Savannah River Site.

Several recent studies have combined methods such as fate-and-transport modeling, geostatistics, and optimization to investigate the temporal and spatial redundancy of existing sampling networks (e.g., Cameron and Hunter, 2000; Rizzo et al., 2000). An example of a study that focuses on identifying spatial redundancy in monitoring networks is Reed et al. (2000), which describes a method that combines groundwater fate-and-transport simulation, kriging, and optimization. This method can be used to identify subsets of monitoring wells to sample for producing an estimate of the total mass of the plume mass that is "acceptably close" to that which would result from sampling all of the available monitoring wells. As discussed in Box 3-1, application of this methodology to the Hill Air Force Base indicated that sampling costs could be reduced by nearly 60 percent.

In recognition of the importance of long-term monitoring optimization, several agencies have developed useful formal decision support tools for network design (see the Federal Remediation Technologies Roundtable (FRTR) web site at http://www.frtr.gov/optimization/monitoring/). An example is the MAROS software developed for the Air Force Center for Environmental Excellence, described in Box 3-2 (Aziz et al., 2000). This software package includes (1) parametric and nonparametric statistical analysis of concentration time series, (2) a sampling frequency determination algorithm based upon the "cost effective sampling" method of Ridley and MacQueen (1995), (3) a plume-mapping method, based on Thiessen polygons, that computes the relative importance of each well in estimating the overall average concentration of the plume, and (4) a stepwise optimization that sequentially removes wells that are "redundant" for computing the average plume concentration.

The Navy is clearly interested in optimizing its long-term monitoring systems, as evidenced by the recent development of guidance for the design and evaluation of groundwater monitoring programs (NAVFAC, 2000). This guidance is fairly general in nature, but it does emphasize the importance of annual reviews for monitoring programs, and the potential need for revisiting both remedial strategies and monitoring program design based on the results of those reviews. The guidance suggests various techniques that might be useful in improving monitoring system performance, including basic statistical comparisons, geostatistics, groundwater modeling, and data presentation using geographic information systems (GIS), but it provides little supporting detail.

BOX 3-1
Groundwater Monitoring Optimization at Hill Air Force Base

A BTEX plume previously studied at Hill Air Force Base in Utah was numerically simulated for the purpose of demonstrating the methodology of Reed et al. (2000) for optimizing the choice of monitoring well locations. The areal extent of the two-dimensional, steady-state simulated plume (21,000 m^3) and the locations of 30 potential monitoring wells are shown in Figure 3-15. (Two-dimensional modeling was justified based on the presumed full vertical extent of the plume over the 0.9-m saturated zone.) The total mass of BTEX within a defined subdomain as shown in Figure 3-15 was calculated to be 37.6 kg. Contaminant plume simulation is used to project the migration and mass of BTEX.

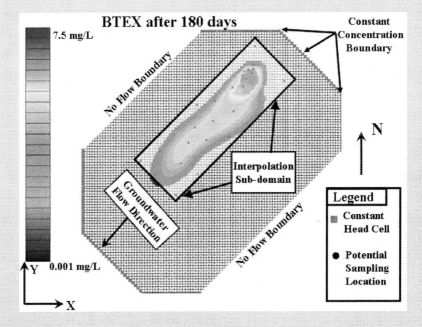

FIGURE 3-15 Simulated BTEX plume and potential monitoring well locations enclosed in the subdomain used for total mass calculations. SOURCE: Reprinted, with permission, from Reed et al. (2000). © (2000) American Geophysical Union.

Continued

BOX 3-1 Continued

Formal mathematical optimization (i.e., a generic algorithm) was used to identify optimal solutions in which a reduced number of sampling points provided accurate mass estimates. Mass estimates were computed using three different approaches for plume interpolation, including kriging, inverse distance weighting, and a hybrid heuristic that uses both of these methods in combination.

The inverse distance weighting scheme, which is extremely fast computationally, chose an optimal sampling network consisting of 15 wells (Figure 3-16A) and yielded a mass estimate of 46.4 kg. This mass estimate provides nearly the same mass estimate as if 30 wells had been chosen (46.7 kg). Both of these estimates have about 24 percent error compared with the known mass, but the optimal solution would reduce operating and maintenance costs by 50 percent.

The kriging-based optimization solution chose 12 monitoring wells (Figure 3-16B) and estimated the mass of the plume using these wells to be 35 kg, which was also the mass estimated using this method for all 30 wells. This mass estimate was within 7 percent of the known mass, and the total costs were reduced by 60 percent by eliminating 18 monitoring wells. This scheme is more accurate but computationally more expensive and requiring greater technical skill and effort than the inverse distance weighting scheme.

A hybrid solution algorithm combines the best features of the above two approaches. This solution approach chose 13 monitoring wells (Figure 3-16C) and a mass estimate of 35 kg, with a computational time reduced by 67 percent compared to the kriging approach. The mass error is the same as the kriging approach, but the identified solution is not as optimal as that found by the kriging approach because one additional well is required, and therefore the approach represents a tradeoff between computational efficiency and solution cost.

This case study illustrates the beneficial information that can be gleaned from applying mathematical optimization techniques to design a monitoring well system, or to adjust a monitoring well system that is already in place by adding or removing wells. The level of sophistication of a user would be expected to be relatively high owing to the required use of mathematical optimization.

FIGURE 3-16   Optimization of well monitoring networks using three approaches: (A) inverse distance, (B) kriging, and (C) hybrid solution. Note: The highest BTEX concentrations, which are present in the center of the plume, correspond to the top of the concentration scale bars to the left of each figure. SOURCE: Reprinted, with permission, from Reed et al. (2000). © (2000) American Geophysical Union.

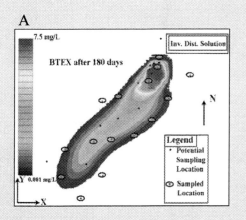

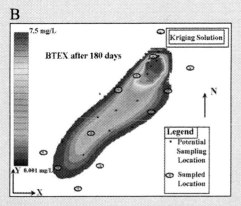

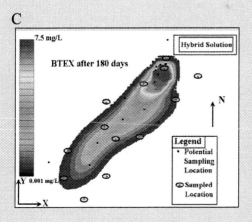

BOX 3-2
MAROS—The Monitoring and Remediation Optimization System

In recognition of the importance of long-term monitoring optimization, several agencies in the Federal Remediation Technologies Roundtable have developed formal decision support tools for network design (http://www.frtr.gov/optimization/monitoring/). An example is the Monitoring and Remediation Optimization System (MAROS) developed for the Air Force Center for Environmental Excellence (AFCEE) by Groundwater Services, Inc. (Aziz et al., 2000; http://www.afcee.brooks.af.mil/er/rpo.htm).

MAROS is a simple and flexible tool that aims to "optimize" long-term monitoring by adjusting the temporal frequency of sampling and identifying spatially redundant wells. The main information used is the concentration versus time data from the existing monitoring wells for up to five constituents of concern (COCs); these data comprise the so-called "primary lines-of-evidence." Parametric and nonparametric statistical analyses of the time-series trends are used to classify each well and each COC into one of the following categories: decreasing, probably decreasing, stable, increasing, probably increasing, and no trend. MAROS also allows the results of groundwater models and empirical information from various "plume-a-thon" studies to comprise "secondary lines-of-evidence." For example, groundwater models can be calibrated and then used to predict future plume growth. Primary and secondary lines-of-evidence are combined and each monitoring well is classified as to whether it requires "extensive" (i.e., quarterly), "moderate" (i.e., biannually or annually), or "limited" (i.e., annually or biennially) monitoring. For example, if a plume shows a highly confident decreasing trend, then it would be in the "limited" category. A more sophisticated approach to determining sampling frequency is developed in the more advanced MAROS modules. This approach is based upon the so-called "cost effective sampling" method developed by Lawrence Livermore National Laboratory. This approach uses regression to determine the rate of concentration change for individual wells; sampling frequency is based upon this rate, with adjustments for overall long-term trends and compound risk.

A plume-mapping method is used to assess the spatial redundancy of a well. For each well in the network, the concentration is estimated by interpolation with the measured values at nearby wells. (Delaunay triangularization is used as the interpolation method.) Comparison of this estimated value with the true measured value yields a quantitative measure of the importance of the nearby wells. This measure is used in a heuristic optimization step that eliminates well locations that do not contribute significant information about the plume.

Attractive features of the MAROS software include its relatively simple construction and analysis, streamlined data entry and the ability to update data and develop new modules, its use of different levels of reporting ranging from a one-page system optimization summary to individual well trends and statistics, and ability to download the software for free and thus be highly accessible to RPMs. MAROS is best for small to medium sites with fewer than 100 wells. Because MAROS is currently being applied at only a select number of sites, case studies not available at this time.

Boxes 3-3 and 3-4 present case studies of Navy sites where an explicit analysis of monitoring program effectiveness has been performed. Most of the reported case studies (including those in NAVFAC, 2000) focus on the cost savings gained from existing groundwater systems by eliminating redundant monitoring points, reducing sampling frequency, and/or refining analytical lists of contaminants of concern. These cost reductions can be significant, with savings greater than 50 percent over baseline being common. In the context of these case studies, "enhancing" or "optimizing" existing monitoring systems is synonymous with cost reduction. It is important to note, however, that in ASM the review of monitoring information and monitoring system performance and/or the modification of a remedial strategy could lead to *increased* monitoring requirements and associated costs. The obvious example of this is when a pump-and-treat system is converted to a strategy that relies on monitored natural attenuation.

BOX 3-3
Optimizing Monitoring of the Eastern Plume, NAS Brunswick, Maine

The "Eastern Plume" at NAS Brunswick has resulted from past solvent disposal practices and contains primarily TCE, PCE, 1,1,1-TCA, and limited degradation products. An interim record of decision (ROD) for extraction and treatment was signed in June 1992, and a final ROD for No Further Action for soils and continued pump-and-treat for groundwater was signed in February 1998. The initial groundwater monitoring program included 36 monitoring wells (30 within the plume and six sentinel wells). The monitoring wells were sampled on a triannual basis for VOCs and other compounds. The annual cost for long-term monitoring in 1996 and 1997 was approximately $550,000.

Reviews of the monitoring data showed that the plume was relatively stable. This prompted the Navy to conduct a geostatistical analysis, which revealed some data redundancy as well as data gaps. The Navy met with federal and state regulators and reviewed the records for each sampling location. This resulted in the following key decisions: (1) installation of five new monitoring wells in regions where the data are sparse, (2) reduction of the total number of wells to be sampled from 36 to 22, with 13 in-plume wells and nine sentinel wells, and (3) reduction in the sampling frequency from three to two times per year. Additional cost savings could be realized by modifying the reporting procedures. The annual cost of the monitoring program is anticipated to be approximately $250,000, a savings of over 50 percent.

BOX 3-4
Reducing Sampling Costs in Long-Term Monitoring
at NAS Fort Worth (Former Carswell AFB)
Source: HydroGeoLogic, Inc. (2000).

In 1993 Carswell Air Force Based officially closed, and a large portion was transferred to the Navy and renamed the NAS Fort Worth Joint Reserve Base. Activities at the site resulted in the generation of a variety of wastes that have contaminated soil and groundwater. The Air Force, under its Installation Restoration Program, is responsible for cleanup of contamination that occurred prior to October 1, 1993. Most of the effort has focused upon a chlorinated solvent plume, and a pump-and-treat system has been operating to prevent migration of the solvents beyond the eastern boundary of the site. Over 260 monitoring wells have been sampled quarterly at a cost of over $300,000 per year. The plume appears to be relatively stable over time, being effectively contained by the pump-and-treat system.

As part of the Year 2000 Groundwater Sampling and Analysis Plan, Hydro-GeoLogic, Inc., proposed to apply advanced geostatistical techniques to optimize the selection of a subset of wells to be sampled. The application developed by HydroGeoLogic, called the Long-Term Monitoring Optimization (LTMO) Tool Kit, utilizes geostatistical and temporal trending methods to develop sampling plans that eliminate spatial and temporal redundancy. For the NAS Fort Worth site, the objective was to minimize monitoring costs by eliminating sampling locations that do not contribute to characterization of the plume along the eastern boundary of the site. The geostatistical technique known as kriging not only yields a contour map of the contaminant plume, but also gives an estimate of contaminant uncertainty. An "importance factor" for each monitoring well can be calculated based upon its contribution to the overall uncertainty over a region of interest. Monitoring locations with small importance factors are candidates for elimination. In this case, application of the LTMO Tool Kit identified more than 60 percent of the wells as spatially redundant. Because of certain fixed expenses, the overall cost savings realized was somewhat less—the 1999 cost for sampling 193 locations was $447,712, and the 2000 cost for sampling 72 locations was $310,794, resulting in an overall savings of $136,918.

Continued

Incorporation of Sensors and Field Analytics into Monitoring

Within the last ten years there have been significant advances in the quality of field analytical techniques, the number of technologies available, and their regulatory acceptability. Thus, a second opportunity for a more adaptive approach to monitoring within a traditional fixed point monitoring system is to build sensors and/or field analytical methods into the characterization or monitoring process. Field analytics such as test kits or portable instrumentation can be used as a complete substitute for

BOX 3-4 Continued

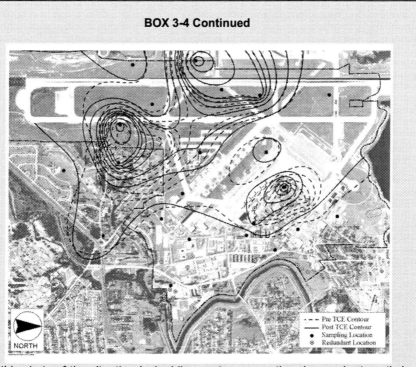

In this photo of the site, the dashed line contours map the plume prior to optimization and are based upon using all the monitoring locations (the black dots plus the white dots). The solid line contours are post optimization and are based only on the black dots. They demonstrate that the use of fewer wells still maintains a good map of the plume contours along the eastern boundary.

laboratory analyses, or they can augment a laboratory-based program by providing on-the-spot analyses to justify the collection and submittal of samples for more traditional laboratory analyses.

Sensors and field analytics can (1) reduce overall characterization and monitoring costs, (2) provide more complete datasets spatially and over time, and (3) produce more timely results than reliance solely on offsite laboratory analyses. Field analytics and sensors reduce overall characterization and monitoring costs because, in general, the per-sample cost associated with a field analytical analysis is much less than that of the corresponding laboratory analysis. As an example, a field-deployable GC/MS tuned for explosives work was used to support characterization of TNT- and DNT-contaminated soils at Joliet Army Ammunition Plant.

Estimated analytical costs were under $60 per sample for the work, compared to the per-sample cost of approximately $250 for standard offsite laboratory analysis (Johnson et al., 1997).

The use of continuous, or nearly continuous, data collection technologies at fixed monitoring points can provide a much more complete set of data upon which to base performance evaluation decisions. This is partially because the lower costs associated with field analytics for *ex situ* sample analyses can allow a larger number of samples to be collected within the same budget as compared to a traditional monitoring program, providing much more complete coverage spatially and temporally. Off-the-shelf, commercially available, continuous depth-to-water-table measurement systems and data loggers are mature examples of these types of technologies. Technologies for providing continual recording of basic parameters such as temperature and pH have also been available for some time. Advances in sensor miniaturization have recently led to commercially available multiparameter sensors that can simultaneously measure dissolved oxygen, conductivity, and resistivity, along with depth. It is only a matter of time before the range of parameters amenable to *in situ* monitoring expands to include at least some common groundwater contaminants of concern.

The use of either dedicated *in situ* measurement systems or field analytics for rapid in-field sample analyses also provides the opportunity to more quickly identify and respond to potential performance issues with a remediation system. In some cases, such as with natural attenuation of groundwater contamination, system evolution occurs at time scales where rapid identification of changing subsurface conditions is not important. However, for engineered barrier systems and some of the more dramatic intrusive subsurface interventions (e.g., thermal heating, Fenton's Reagent, etc.), quickly identifying unexpected contaminant mobilization or other key potential system failures can be crucial to overall remediation success.

Federal agency research and development programs have heavily invested in the last decade in field analytics and sensor technologies that can be applied to hazardous waste site characterization, remediation, and monitoring activities. For example, DOE's Environmental Management Science Program (http://emsp.em.doe.gov/portfolio/multisearch.asp) currently lists more than 70 research and development projects that address data collection or sample analysis issues. Techniques as diverse as antibody methods, *in situ* microsensors, spectroelectrochemical sensors, spectrometric DNA diagnostics, dielectrics and nuclear magnetic resonance, partitioning tracers, electromagnetic imaging, seismic technologies, acoustic probes, conductive luminescent polymers, cavity ringdown spectroscopy, gamma ray imaging, optical array sensors, noble gas detectors, and BioCOM sensors are mentioned. Likewise, DoD's Strategic Environmental Research and Development Program has funded more than 20 research and development activities focused on characterization and monitoring technologies in its cleanup area. Researchers with the Navy's Space and Naval Warfare Systems Command (SPAWAR) have focused specifically on technologies applicable to the more specialized needs of sediments (see Box 3-5).

In response to these advances, there have been regular modifications to recommended EPA analytical protocols, including SW-846. Within the Resource Conservation and Recovery Act (RCRA) program, EPA's SW-846 contains guidance on acceptable analytical techniques for RCRA-related activities. The latest is Draft IVB (EPA, 2000c), which includes several additions pertinent to Navy contaminants of concern. The EPA Technology Innovation Office (TIO) maintains an encyclopedia of field analytical technologies (http://fate.clu-in.org). The FRTR also maintains a table that provides summary performance information for a wide range of analytical techniques, categorized by contaminant class and media. Table 3-1 provides a summary of field analytical techniques based on the information maintained by the FRTR. In addition, EPA's Environmental Technology Verification program (www.epa.gov/etv)— designed to accelerate the use of innovative technology—has issued reports verifying the validity of over 39 monitoring and characterization technologies. These include, for example, cone penetrometer-deployed sensor technologies, groundwater sampling technologies, PCB field analytical measurement techniques, and portable GC/MS.

Alternative Sample Collection Technologies

Subsurface characterization and monitoring programs have traditionally relied on drilling techniques to obtain soil samples at depth and on permanent, screened and developed monitoring wells for acquiring groundwater samples. Just as there have been advances in field analytical techniques, so too there has been progress made in soil, sediment, and groundwater sample collection technologies. The advantages of these advancements include a reduction in sample collection costs, greater sample production rates, and in some cases more representative samples. In addition, when coupled with field analytical methods, these alternative

BOX 3-5
Rapid Field Characterization of Sediments

Rapid field characterization techniques have been developed to speed assessment and reduce costs. These are field-transportable screening tools that provide measurements of chemical, biological, or physical parameters on a real-time or near real-time basis. Specific advantages include the ability to get rapid results to guide sampling locations, the potential for high data mapping density, and a reduced cost per sample. The approaches do have limitations including the nonspecific nature of some tests, sensitivity to sample matrix effects, and some loss in accuracy over conventional laboratory analyses. A variety of tools have been suggested for the rapid characterization of sediments, as shown in the table below.

Screening-Level Analyses Recommended by the Assessment and Remediation of Contaminated Sediments Program for Freshwater Sediments

Analytical Technique	Parameter(s)
X-ray Fluorescence Spectrometry (XRF)	Metals
UV Fluorescence Spectroscopy (UVF)	Polycyclic Aromatic Hydrocarbons (PAHs)
Immunoassays	PCBs, Pesticides, PAHs
Microtox®	Acute Toxicity
SOURCE: EPA (1994).	

The Sediment Management Laboratory of the Space and Naval Warfare Systems Command (SPAWAR), San Diego, CA, has tested the applicability of these characterization technologies for use with sediment, particularly the use of portable XRF to determine metal concentrations (Kirtay et al., 1998; Stallard et al., 1995). The additional spatial resolution afforded by the inexpensive rapid assessment techniques allows a much more thorough characterization of spatial variability at sediment sites and could provide the detailed information necessary for ASM.

sample collection technologies can enable dynamic work plans and adaptive sampling and analysis programs, concepts discussed in the following section.

One example of this innovation is the use of direct push technologies for obtaining subsurface soil, sediment, and groundwater samples. These technologies drive, push, or vibrate small-diameter steel tubes into the ground, up to depths of approximately 100 feet depending on rig type and subsurface lithology. Direct push technologies generally retrieve intact soil cores for *ex situ* sample analysis. With appropriate attachments and modifications, they can also be used to retrieve groundwater and soil vapor samples. Direct push equipment ranges from small, readily transportable units that can be used through floors of buildings, to large dedicated rigs. Box 3-6 describes the adaptation of a direct push technology for use in an estuary environment for rapidly and efficiently retrieving sediment cores.

Direct push technologies can be coupled with field analytics and sensors in a variety of ways to generate pertinent characterization and monitoring data. Properly instrumented direct push rigs can provide information on subsurface lithology through resistivity and stress/strain readings generated by rod advancement. With specialized tips or rod sections, soil, groundwater, and soil vapor samples can be retrieved for *ex situ* analyses. The membrane interface probe (MIP) is an example of a specialized direct push stem design that allows for the near real-time evaluation of subsurface VOC contamination in soils and groundwater when combined with an above-ground detection system such as a photoionization detector or gas chromatograph. Although its detection limits are not sufficient to meet typical groundwater cleanup standards, they are low enough to allow the system to detect the presence of potential subsurface source areas. This type of capability can be extremely useful in refining remedial interventions that target source removal or source degradation.

Specialized direct push tips have been instrumented to support the *in situ* use of x-ray fluorescence (XRF), laser-induced fluorescence (LIF), gamma spectroscopy, and laser-induced breakdown spectroscopy (LIBS) (DOE, 2002). These systems and the data they generate have gained various levels of acceptance by the user and regulatory communities; it is clear from the technical progress to date that they will be widely used in the future. Most work in this area has focused on the generation of pre-remediation characterization information via the DOE and DoD's Site Characterization and Analysis Penetrometer System (SCAPS) programs (EPA, 1995; USAEC, 2000). SCAPS makes use of a cone penetrometer truck to push instrumented tips into the subsurface.

The possibility of rapidly and inexpensively gathering detailed subsurface information in near real time via direct push technologies can change the way remedial action monitoring work is conducted for those settings amenable to direct push technologies. Direct push technologies such as SCAPS can be used to install relatively low-cost *temporary* monitoring points. The combination of direct push with temporary monitoring points allows monitoring to be adjusted cost effectively across space as well as over time in response to data. An obvious example is the temporal tracking of some critical concentration isopleth over time (i.e., the concentration associated with closure guidelines), something that currently is almost impossible to do at most sites using spatially lim-

TABLE 3-1 Summary of Sensor and Field Analytical Techniques

Technique	Analytes	Media			Performance						Applicable to				
		Soil/Sediment	Water	Gas/Air	Selectivity	Susceptibility to Interference	Detection Limits	Turnaround Time Per Sample	Quantitative Data Capability	Technology Status	Relative Cost Per Analysis	Screen Identify	Characterize Quantify	Cleanup Performance	Long-Term Monitoring
VOC, SVOC, TPH and PCB (*in situ* analysis)															
Solid / Porous Fiber Optic	11	E	A	B	B	A	B	A	B	I	A	A	B	A	B
Laser Induced Fluorescence	5, 11	B	A	NA	B	B	B	A	B	III	A	A	B	A	B
VOC, SVOC, TPH and PCB (*ex situ* analysis)															
Photo-Ionization Detector	1, 3	E	E	A	B	C	B	A	C	III	A	A	C	A	C
Flame-Ionization Detector	1-3	E	E	A	B	C	A	A	C	III	A	A	C	A	C
Explosimeter	1	E	E	A	C	C	B	A	C	III	A	A	C	A	C
Gas chromatography (GC) plus detector	1-6, 11	E	E	A	A	A	A	B	A	III	B	A	A	A	A
Catalytic Surface Oxidation	1,3	E	E	A	B	B	B	A	C	III	A	A	B	A	A
Detector Tubes	1,3	E	E	A	B	B	B	A	C	III	A	A	C	A	C
Mass Spectrometry (MS)	1-6	E	E	A	A	B	B	B	A	II	C	B	A	A	B
GC / MS	1-6	E	E	A	A	A	A	C	A	III	C	A	A	A	B
GC/Ion Trap MS	1-6	E	E	A	A	A	A	B	A	II	C	B	A	A	B
Ion Trap MS	1-6	E	E	A	A	B	A	B	A	II	C	A	A	A	A
Ion Mobility Spectrometer	1-4, 6	B	E	B	C	B	A	A	A	II	B	A	B	A	B
Ultraviolet (UV) Fluorescence	1, 3, 5	B	A	B	B	B	A	B	B	II	B	A	B	A	A
Synchronous Luminescence/ Fluorescence	1-4	E	A	A	B	B	A	B	B	I	B	A	B	A	A

Technique	Code														
UV-Visible Spectrophotometry	1, 3, 5	E	A	B	C	C	A	B	A	I	B	A	B	A	A
Infrared Spectroscopy	1-4	E	E	A	B	B	A	A	B	II	B	A	A	A	A
Fourier Transform Infrared (FTIR) Spectroscopy	1, 3, 11	E	E	A	A	B	A	A	B	II	B	A	A	A	A
Scattering / Absorption LIDAR	1, 3	E	E	A	C	C	A	C	B	I	B	A	A	A	A
Raman Spectroscopy/ Surface Enhanced Raman Scattering (SERS)	1-5, 11	E	A	E	C	C	A	A	B	II	B	A	A	A	A
Near IR Reflectance/ Transmittance Spectroscopy	1, 3	A	NA	NA	C	C	A	C	B	I	B	A	A	A	A
Immunoassay Colorimetric Kits	1-6, 11	A	A	NA	B	B	A	B	B	II	B	A	A	A	B
Amperometric and Galvanic Cell Sensor	1, 3	E	NA	A	A	A	A	A	B	II	B	A	A	A	A
Semiconductor Sensors	1, 3	E	A	A	B	B	A	A	B	I	B	A	A	A	A
Piezoelectric Sensors	1, 3	E	E	A	A	A	A	A	C	I	B	A	A	A	B
Field Bioassessment	1-6	A	A	A	C	C	NA	C	C	II	C	A	C	A	B
Toxicity Tests	1-6	A	A	A	C	C	NA	C	B	II	B	A	C	A	A
Room-Temperature Phosphorimetry	4, 5, 6, 12 (PCBs)	B	A	B	A	C	A	B	B	I	B	A	B	A	A
Chemical Colorimetric Kits	2, 4, 5, 11	B	A	NA	B	B	A	B	A	II	B	A	A	B	A
Free Product Sensors	11	NA	A	NA	C	A	C	A	A	III	C	A	A	A	A
Ground Penetration Radar	11	B	C	NA	C	C	C	B	C	I	B	B	B	B	B
Thin-Layer Chromatography	2	E	A	NA	B	B	A	B	B	II	A	C	A	A	A
Metals (ex situ analysis)															
Atomic Absorption (AA) Spectroscopy	7	E	E	A	A	A	C	A	A	I	A	C	C	C	B

150

Technique	Analytes	Media			Performance							Applicable to			
		Soil/Sediment	Water	Gas/Air	Selectivity	Susceptibility to Interference	Detection Limits	Turnaround Time Per Sample	Quantitative Data Capability	Technology Status	Relative Cost Per Analysis	Screen Identify	Characterize Quantify	Cleanup Performance	Long-Term Monitoring
Metals (ex situ analysis)															
Inductively Coupled Plasma-Atomic Emission Spectroscopy (ICP-AES)	7	E	E	A	A	A	A	B	A	I	C	C	A	C	B
X-Ray Fluorescence	7	A	A	E	A	A	B	A	A	III	A	A	B	A	B
Chemical Colorimetric Kits	7, 9	B	A	NA	A	B	B	B	B	II	A	A	B	A	A
Titrimetry Kits	7, 9	B	A	NA	A	B	B	B	B	III	A	A	B	A	A
Immunoassay Colorimetric Kits	7, 12 (Hg)	A	A	NA	B	B	A	A	B	II	A	A	B	A	A
Anodic Stripping Voltammetry	7	A	A	NA	A	B	A	A	A	II	B	B	A	A	B
Fluorescence Spectrophotometry	7, 12 (Hg)	E	E	A	A	B	A	A	A	II	B	A	A	A	A
Amperometric and Galvanic Cell Sensor	7	E	A	NA	A	B	B	A	B	II	A	A	B	A	A
Field Bioassessment	7, 9	A	A	A	C	C	NA	C	C	II	C	A	A	C	B
Toxicity Tests	7, 9	A	A	A	C	C	NA	B	B	II	A	A	A	C	A
Ion Chromatography	7	E	A	NA	B	B	A	B	A	I	B	A	A	A	A
Explosives (ex situ analysis)															
Gas chromatography (GC) plus detector	10	E	E	B	A	A	B	B	A	II	C	B	B	B	B
Mass Spectrometry	10	E	E	B	B	C	B	A	B	II	C	B	B	B	B

Method												
GC / MS	10	E	A	A	A	B	B	A	II	C	B	B	B
Ion Mobility Spectrometer	10	E	A	A	C	A	B	A	I	C	C	C	C
Field Bioassessment	10	A	A	C	C	NA	C	C	II	A	A	A	B
Toxicity Tests	10	A	A	C	C	NA	A	B	II	A	A	A	A
Chemical Colorimetric Kits	10	E	A	NA	B	B	A	B	III	A	A	B	B
Immunoassay Colorimetric Kits	10	E	A	NA	B	B	A	B	III	A	B	B	A

Legend:

Attribute						
Media and/or Applicable To	A	Better	B	Adequate	C	Serviceable
	NA	Not applicable	E	Requires selection of extraction procedure		
Selectivity	A	Measures the specific contaminant directly	B	Measures the contaminant indirectly	C	Measures a part of the compound
Susceptibility to Interference	A	Low	B	Medium	C	High
Detection Limits	A	Low: 100-1000 ppb (soil); 1-50 ppb (water)	B	Midrange: 10-100 ppm (soil); 0.5-10 ppm (water)	C	High: 500+ ppm (soil); 100+ ppm (water)
	NA	Not applicable				
Turnaround Time per Sample	A	Minutes	B	Hours	C	More than a day
Quantitative Data Capability	A	Produces quantitative data	B	Data is quantitative with additional effort	C	Does not produce quantitative data
Technology Status	III	Commercially available and routinely used field technology	II	Commercially available technology with moderate field experience	I	Commercially available technology with limited field experience
Relative Cost per Analysis	A	Least expensive	B	Mid-range expensive	C	Most expensive

Analytes

1- Non-halogenated volatile organics
2- Non-halogenated semivolatile organics
3- Halogenated volatile organics
4- Halogenated semivolatile organics
5- Polynuclear aromatic hydrocarbons (PAHs)
6- Pesticides / herbicides
7- Metals
8- Radionuclides
9- Other inorganics (asbestos, cyanide, fluorine)
10- Explosives
11- Total petroleum hydrocarbons
12- Specific analyte (named in matrix)

SOURCE: Adapted from FRTR (http://www.frtr.gov/site/analysismatrix.html).

BOX 3-6
Hoverprobe Sediment Coring and Water-Quality Profiling

Many Navy and other DoD facilities are located adjacent to surface-water bodies where plumes may discharge to locations such as wetlands and estuaries that are relatively difficult to access and that contain habitat sensitive to the disturbances caused by traditional drill rigs. These difficulties limit the technologies available to obtain necessary hydrogeologic and water-quality information for site characterization and optimization of groundwater monitoring networks. In response to these needs, a unique drilling and water-quality profiling system, mounted on a hovercraft and called the "Hoverprobe 2000," was developed by the U.S. Geological Survey in cooperation with Hovertechnics, Inc., of Benton Harbor, Michigan, and MPI Drilling, Inc., of Picton, Ontario (Phelan et al., 2001). A hovercraft is a versatile vehicle that can be propelled over the surface of land, water, mud, snow, or ice by a cushion of air produced by downwardly directed fans. It can also be landed on the surface of these difficult terrains and proceed to or from a submerged site even if insufficient water is present to float it. A segmented skirt constructed of rubber-coated fabric surrounds the base of the craft and traps most of the pressurized air under the craft. At rest, the Hoverprobe exerts a pressure that is about 10 percent of the pressure exerted on the ground by a standing person, allowing drilling and sampling in wetlands and tidal flats with minimal surface disturbance (Phelan et al., 2001). The vibracore drill on the Hoverprobe uses hydraulically driven cams to generate high-frequency vibrations to drive casing into the subsurface without use of drilling fluids and with almost no cuttings resulting at the surface. The Hoverprobe can be used for the collection of sediment cores, for drive-point water-quality profiling similar to direct push sampling technologies, or for installation of monitoring wells. Continuous sediment cores can be obtained to a depth of about 100 ft from saturated unconsolidated sediments. Drilling and sampling can occur while the craft is on mud or on solid ground or is floating on water, and can continue as water levels or tides shift.

The first use of the Hoverprobe in a groundwater contamination investigation was as part of an evaluation of natural attenuation of chlorinated solvents discharging to freshwater tidal wetlands and creeks at Aberdeen Proving Ground, Maryland. Although monitored natural attenuation has been shown to be a feasible groundwater remediation method for chlorinated solvents discharging to the tidal wetland and creek (Lorah et al., 1997, 1999a,b), the acceptance of a remediation strategy was delayed by the lack of definition of the southern extent of the plumes discharging to the tidal creeks and of the hydrogeology of the creek channel. Regulators were concerned that subsurface migration of contaminants could occur downstream beneath the creek channel, transporting contaminants to an estuary of the Chesapeake Bay without discharge through wetland sediments where biodegradation of the chlorinated solvents occurs. The Hoverprobe allowed investigation at 13 sites along the stream channel that were previously inaccessible because of mud and shallow water (Phelan et al., 2001). Continuous sediment coring and water-quality profiling for chlorinated volatile organic compounds and redox-sensitive constituents were conducted without installation of wells, providing data to define plume boundaries and to refine the hydrogeologic parameters in a groundwater flow model used to assist in evaluating remedial alternatives.

The Hoverprobe and a support hovercraft during drilling and water-quality profiling along the West Branch, Canal Creek, Aberdeen Proving Ground, MD. The support hovercraft was used for transport of samples to nearby laboratory facility for immediate analysis and in case emergency exit was needed.

ited monitoring well information. Although these types of technologies may never be appropriate for deep vadose zone sites or sites with fractured rock flow systems, they would be appropriate for the majority of coastal Navy facilities with relatively near-surface saturated zones and contamination events.

In the case of traditional monitoring wells, techniques for obtaining less expensive and more representative groundwater samples have also been developed. These include low purge technologies and passive diffusion samplers. Passive diffusion samplers can eliminate altogether the need for purging monitoring wells before sampling. Diffusion samplers are a class of samplers, developed by Don Vroblesky at the U. S. Geological Survey, that are based on the laboratory and field confirmation that VOCs can diffuse through low-density polyethylene films and reach equilibrium concentrations that correlate well with actual subsurface contaminant concentrations (USGS, 2001). Types of diffusion samplers include water-to-water samplers and vapor-to-vapor samplers. Both types are applicable to the sampling of groundwater (via wells), the groundwater/surface water interface, pore water in sediments, surface water, and water from treatment systems. Vapor-to-vapor samplers are also effective for measuring *in situ* soil gas and vapor phase concentrations in confined spaces.

The effectiveness of diffusion samplers is dependent upon the samplers being in direct contact with volatile organic compounds. Diffusion samplers should not be deployed in monitoring wells where sand packs are less permeable than the surrounding formation. In addition, diffusion samplers are not recommended for the quantitative measurement of methyl-tertiary butyl ether (MTBE) or acetone.

Multiple diffusion samplers deployed in a vertical array can provide an effective method of vertical contaminant profiling in monitoring wells. Optimal conditions would consist of the diffusion sampler or groundwater monitoring well screen being in direct contact with the surrounding formation, but correctly designed monitoring well sand packs are also appropriate. The presence of vertical gradients across the sampling interval will compromise the resolution of vertical contaminant profiling.

The most promising application for diffusion samplers appears to be for long-term groundwater monitoring in wells, with the potential to reduce long-term monitoring costs by 20 percent to 50 percent. Detailed information regarding the appropriateness, construction, deployment, handling, and analysis of diffusion samplers can be found in USGS (2001).

Dynamic Work Plans

The last opportunity for developing a more flexible and adaptive approach to subsurface performance monitoring is to base a characterization or monitoring program on dynamic work plans. Dynamic work plans differ from more traditional sampling and analysis plans in that they identify the decision logic that will be used for determining the appropriate analytical techniques and sample numbers, locations, and frequency *as work proceeds*, rather than pre-specifying those data collection characteristics. As alluded to above, dynamic work plans rely at least in part on direct push technologies and field analytic techniques. With these technologies, data collection can be adapted in response to the changing information needs of a remedial action, and the remedial action itself can be adjusted or adapted based on feedback from the data collection.

The concept of developing hazardous waste site characterization programs based on dynamic work plans has been implemented under a variety of names, including expedited site characterization (DOE, 1998) and adaptive sampling and analysis programs (DOE, 2001). The EPA TIO has been advocating the Triad approach (EPA, 2001) to environmental data collection, which adds systematic planning to the dynamic work plan/field analytic mix. The EPA Superfund program is currently preparing draft guidance on the development of dynamic work plans. Case studies that document characterization cost reductions associated with these types of approaches usually report savings on the order of 50 percent or more. These savings are derived from reductions in per-unit analytical costs and in the overall number of samples collected.

Although the emphasis has historically been on site characterization, dynamic work plan concepts and associated technologies (field analytics, sensors, direct push, etc.) are equally applicable to the remediation phase of site restoration. In fact, the potential impacts on overall costs and remediation performance are greater during remediation than they are during characterization because savings can be realized both from reductions in data collection costs and from improved remedial action performance. In this context, dynamic work plans are a natural component of ASM.

Dynamic work plans are particularly applicable to contaminated soil excavations or contaminated sediment dredging operations. Box 3-7 describes the adaptive nature of a removal project for soils contaminated with radionuclides. A similar example, but in the context of pesticide-contaminated soils, was reported in USACE (2000). In this example,

---

BOX 3-7
Precise Excavation at the Ashland 2 Site

The U.S. Army Corps of Engineers (USACE) is conducting cleanup of radiologically contaminated properties as part of the Formerly Utilized Sites Remedial Action Program (FUSRAP). The largest cost element for most of the FUSRAP sites is the excavation and disposal of contaminated soil. Conventional approaches to the design of soil excavation/disposal programs delineate excavation boundaries based on existing characterization data. Excavation then proceeds using these design drawings as the basis for determining which soil must be excavated and which can remain. There is considerable evidence that in fact most pre-remediation characterization datasets are inadequate for precisely delineating contamination footprints. The result can be overexcavation of clean soil at considerable unnecessary expense.

A precise excavation approach was implemented at the Ashland 2 FUSRAP site. Data collection was embedded into the excavation program, with data collection consisting of real-time *in situ* sensors, global positioning system units, and an onsite laboratory. Excavation work proceeded in lifts that ranged from 0.5 to 2 feet in depth, with dig-face screening occurring before excavation continued. A pre-excavation estimate of contaminated soil volumes based on RI/FS data placed the total at 14,000 cubic yards. By the time the work was completed, approximately 45,000 cubic yards of soil were identified as being contaminated at levels that were above the cleanup criteria and were excavated for offsite disposal.

A post-excavation analysis specifically of the initial surficial lift showed that if excavation of surficial soil had been based solely on pre-existing data, it would have removed 4,000 cubic yards of minimally contaminated soil (i.e., where soil contaminant concentrations were below the cleanup criteria), and it would have missed 8,000 cubic yards of soil that had contamination is excess of the cleanup

---

immunoassay kits were used to better define excavation footprints and verify dig-face cleanup guideline compliance at the Wenatchee site. In its cost and performance report, the USACE indicated that overall remediation costs were half of what would have been incurred if excavation had proceeded on the basis of existing historical datasets alone.

There is also a place for dynamic work plans within groundwater remedial action monitoring. A simple example is a plan that samples a traditional network of monitoring wells. In this instance a dynamic work plan might rely on passive diffusion samplers for generating samples and on field analytics for screening those samples. Based on the results, a decision might be made to replicate analyses using an offsite laboratory, to expand sampling to adjacent wells that would not have otherwise been

criteria. Preliminary cost estimation work indicated that the additional cost of the excavation support data collection was approximately $168,000 over six months of excavation. Over $1.5 million in cost savings were achieved by avoiding unnecessary offsite disposal costs for just the initial surficial lift (Durham et al., 1999).

sampled in that round, or to increase sampling frequency in the short term. In the situation where a technology such as direct push was available for quickly acquiring groundwater samples from new locations, or for installing temporary monitoring points, the decision might be to expand the network in the short term to address unexpected trends or results in datasets.

Alternatively, a monitoring system might include real-time data acquisition from dedicated *in situ* sensors. A dynamic work plan would identify the types of result scenarios that would require a response, either by requiring additional data collection or by revisiting the remedial system. An example would be real-time monitoring of a leachate collection system for parameters that might indicate a containment cell failure. A second example would be continuous depth-to-water-table sensors positioned around a barrier wall whose relative potentiometric results might indicate loss of groundwater capture. These latter examples do not represent current practices for monitoring system design, but they do suggest ways that dynamic work plans and adaptive sampling techniques could be used to facilitate an ASM approach to remedial action performance evaluation.

MAJOR CONCLUSIONS AND RECOMMENDATIONS

This chapter was meant to provide general guidance on how to assess remedial performance monitoring with graphical tools and on some of the new monitoring tools available to do so. A major challenge in implementing adaptive site management will be to design the information-gathering efforts to support the management decision points fleshed out in Chapter 2. Thus, monitoring plans should be developed from clearly articulated objectives (such risk reduction, reduction in some indicator of risk, or mass removal), they should support the evaluation of remedial operations performance (MDP2), and they should validate or refine site conceptual models. More specific recommendations that link monitoring to the ASM process are provided below.

Plots of mass removal or concentration versus time or cost (or other metrics depending on the remedy) are objective and transparent tools for illustrating remedial effectiveness that should trigger when to either modify or optimize the existing remedy or to change the remedy. Such graphs should be used after remedy selection to address management decision periods 2 and 3 of ASM. Graphical representations should serve both to enhance stakeholder understanding of the options and to make better decisions about implementing or modifying remedies. At individual sites under investigation, the Navy, in consultation with all stakeholders, should select a unit cost for the continued operation of the remedial action, above which the existing remedy is no longer considered a tenable option.

The Navy should collect and analyze data to develop and validate predictive models of remedy performance. The remedy selection process could be made more quantitative and transparent with the provision of design guidance, charts, and models that summarize technology applications and predict their performance in different environmental settings.

Uncertainties in hydrogeologic data, contaminant concentrations, and rates of remediation should be explicitly recognized in the development and application of performance plots. There are many sources of uncertainty in estimating the mass or risk reduction achieved by any remediation scheme. When sufficient site data are available, statistical methods can be used to estimate error or confidence bands on the performance plots. Site monitoring plans should be developed to ensure that the collected data serve to reduce uncertainty.

A concerted effort should be made to increase monitoring program effectiveness (and to reduce costs) by optimizing the selection of monitoring points, incorporating field analytics and innovative data collection technologies such as direct push, and adopting dynamic work plans and adaptive sampling and analysis techniques. Real-time *in situ* monitoring technologies should also be considered as they mature. These techniques enhance the collection of information upon which ASM decision making is based. DoD should continue to support and foster research in chemical, physical, and biological techniques that would provide more rapid and adaptive approaches for monitoring remedy effectiveness.

REFERENCES

Aziz, J. J., C. J. Newell, H. S. Rifai, M. Ling, and J. R. Gonzales. 2000. Monitoring and remediation optimization system (MAROS) software user's guide. Brooks Air Force Base, TX: Air Force Center for Environmental Excellence.

Berthouex, P. M., W. G. Hunter, and L. Pallesen. 1978. Monitoring sewage treatment plants: some quality control aspects. Journal of Quality Technology 10:139–149.

Box, G. E. P., and G. M. Jenkins. 1994. Time series analysis: forecasting and control (3rd Edition). Upper Saddle River, NJ: Prentice Hall.

Cameron, K., and P. Hunter. 2000. Optimization of LTM networks using GTS: statistical approaches to spatial and temporal redundancy. Brooks Air Force Base, TX: Air Force Center for Environmental Excellence.

Clemen, R. T. 1996. Making hard decisions: an introduction to decision analysis (2nd Edition). Belmont, CA: Duxbury Press.

Department of Energy (DOE). 1998. Expedited site characterization. Innovative Technology Summary Report DOE/EM-0420.

DOE. 2001. Adaptive sampling and analysis programs (ASAP). Innovative Technology Summary Report DOE/EM-0592.

DOE. 2002. Induced fluorescence sensors for direct push systems. Innovative Technology Summary Report DOE/EM-0638.

Durham, L., D. Conboy, R. Johnson, and T. Sydelko. 1999. Precise excavation—an alternative approach to soil remediation. Pp. 93–98 In: Proceedings of the National Defense Industrial Association, Denver, Colorado, March 19–April 1.

Edwards, W., and F. H. Barron. 1994. SMARTS and SMARTER: improved simple methods for multiattribute utility measurement. Organizational Behavior and Human Decision Processes 60:306–326.

Environmental Protection Agency (EPA). 1994. Assessment and remediation of contaminated sediments (ARCS) program—final summary report. EPA 905-S-94-001. Washington, DC: EPA.

EPA. 1995. Site characterization analysis penetrometer system (SCAPS). Innovative Technology Evaluation Report. EPA/540/R-95/520. Washington, DC: EPA.

EPA. 1998a. Remediation case studies: groundwater pump and treat (nonchlorinated contaminants). EPA 542-R-98-014, Vol. 10. Washington, DC: EPA Office of Solid Waste and Emergency Response.

EPA. 1998b. EPA's contaminated sediment management strategy. Washington, DC: EPA Office of Water.

EPA. 2000a. FRTR cost and performance remediation case studies and related information. EPA 542-C-00-001. Washington, DC: EPA Office of Solid Waste and Emergency Response.

EPA. 2000b. Contaminated sediment news. EPA-823-N-00-002. Washington, DC: EPA.

EPA. 2000c. Test methods for evaluating solid waste, physical/chemical methods. SW-846 Draft Update IVB. Washington DC: EPA Office of Solid Waste and Emergency Response.

EPA. 2001. Using the Triad approach to improve the cost-effectiveness of hazardous waste site cleanups. EPA-542-R-01-016. Washington, DC: EPA.

Farber, S., and B. Griner. 2000. Using conjoint analysis to value ecosystem change. Environ. Sci. Technol. 34(8):1407–1412.

Freeze, R. A., and S. M. Gorelick. 1999. Convergence of stochastic optimization and decision analysis in the engineering design of aquifer remediation. Ground Water 37(6):934–54.

Gibbons, R. D., and D. E. Coleman. 2001. Statistical methods for detection and quantification of environmental contamination. New York: Wiley.

Gilbert, R. O. 1987. Statistical methods for environmental pollution monitoring. New York: Van Nostrand Reinhold.

HydroGeoLogic, Inc. 2000. Final base-wide groundwater sampling and analysis plan, NAS Fort Worth JRB, Texas. March 2000.

James, B. R., and S. M. Gorelick. 1994. When enough is enough: the worth of monitoring data in aquifer remediation design. Water Resources Research 30(12):3499–3513.

Jennings, A. A., N. Mehta, and S. Mohan. 1994. Superfund decision analysis in presence of uncertainty. J. Environ. Engr. 120(5):1132–1150.

Johnson, V. M., R. C. Tuckfield, M. N. Ridley, and R. A. Anderson. 1996. Reducing the sampling frequency of groundwater monitoring wells. Environ. Sci. Technol. 30(1):355–358.

Johnson, R., J. Quinn, L. Durham, G. Williams, and A. Robbat, Jr. 1997. Adaptive sampling and analysis programs for contaminated soils. Remediation, Summer:81–96.

Kampbell, D. H., P. E. Haas, R. N. Miler, J. E. Hansen, and F. H. Chapelle. 1998. Technical protocol for evaluating natural attenuation of chlorinated solvents in ground water. Washington, DC: EPA.

Keeney, R. 1980. Siting energy facilities. New York: Academic Press.

Keeney, R., and H. Raiffa. 1976. Decisions with multiple objectives. New York: Wiley.

Kirtay, V. J., J. H. Kellum, and S. E. Apitz. 1998. Field-portable x-ray fluorescence spectrometry for metals in marine sediments: results from multiple sites. Water Sci. Technol. 37(6-7):141–148.

Lettenmaier, D. P. 1976. Detection of trends in water quality data from records with dependent observations. Water Resources Research 12:1037–1046.

Lettenmaier, D. P. 1977. Detection of trends in stream quality: monitoring network design and data analysis. Technical Report No. 51, NTIS PB-285 960. Seattle, WA: C. W. Harris Hydraulics Laboratory, Department of Civil Engineering, University of Washington.

Loáiciga, H. A., R. J. Charbeneau, L. G. Everett, G. E. Fogg, B. F. Hobbs, and S. Rouhani. 1992. Review of ground-water quality monitoring network design. Journal of Hydraulic Engineering 118(1):11–37.

Lorah, M. M., and L. D. Olsen. 1999a. Degradation of 1,1,2,2-tetrachloroethane in a freshwater tidal wetland: field and laboratory evidence. Environ. Sci. Technol. 33(2):227–234.

Lorah, M. M., and L. D. Olsen. 1999b. Natural attenuation of chlorinated volatile organic compounds in a freshwater tidal wetland: field evidence of anaerobic biodegradation. Water Resources Research 35(12):3811–3827.

Lorah, M. M., L. D. Olsen, B. L. Smith, M. A. Johnson, and W. B. Fleck. 1997. Natural attenuation of chlorinated volatile organic compounds in a freshwater tidal wetland, Aberdeen Proving Ground, Maryland. U.S. Geological Survey Water-Resources Investigations Report 97-4171.

Maxwell, R. M., S. D. Pelmulder, A. F. B. Tompson, and W. E. Kastenberg. 1998. On the development of a new methodology for groundwater-driven health risk assessment. Water Resources Research 34(4):833–847.

Maxwell, R., S. F. Carle, and A. F. B. Tompson. 2000. Risk-based management of contaminated groundwater: the role of geologic heterogeneity, exposure and cancer risk in determining the performance of aquifer remediation. Proceedings of the 2000 Joint Conference on Water Resources Engineering and Water Resources Planning & Management, ASCE, July 30–Aug. 2, Minneapolis.

Merkofer, M. W., and R. L. Keeney. 1987. A multiattribute utility analysis of alternative sites for the disposal of nuclear waste. Risk Analysis 7:173–194.

Meyer, P. D., A. J. Valocchi, and J. W. Eheart. 1994. Monitoring network design to provide initial detection of groundwater contamination. Water Resources Research 30(9):2647–2659.

Minsker, B. S., and J. Bryan Smalley. 1999. Cost-effective risk-based in situ bioremediation design. Pp. 349–354 In: Proceedings of the 5th International In Situ and On-site Bioremediation Symposium, April 19–22, 1999, San Diego, CA. Bruce C. Alleman and Andrea Leeson (eds.).

National Research Council (NRC). 1999. Environmental cleanup at Navy facilities: risk-based methods. Washington, DC: National Academy Press.

NRC. 2000. Natural attenuation for groundwater remediation. Washington, DC: National Academy Press.

NRC. 2001. A risk management strategy for PCB-contaminated sediments. Washington, DC: National Academy Press.

NRC. 2003. Bioavailability of contaminants in soils and sediments: processes, tools, and applications. Washington, DC: National Academy Press.

NAVFAC. 2000. Guide to optimal groundwater monitoring. Prepared for the Naval Facilities Engineering Research Center by Radian International.

NAVFAC. 2001. Guidance for optimizing remedial action operation (RAO). Special Report SR-2101-ENV. Prepared for the Naval Facilities Engineering Service Center. Research Triangle Park, NC: Radian International.

Phelan, D. J., M. P. Senus, and L. D. Olsen. 2001. Lithologic and groundwater-quality data collected using Hoverprobe drilling techniques at the West Branch Canal Creek wetland, Aberdeen Proving Ground, Maryland, April–May 2000. U.S. Geological Survey Open-File Report 00-446.

Reed, P., B. Minsker, and A. J. Valocchi. 2000. Cost-effective long-term ground-water monitoring design using a genetic algorithm and global mass interpolation. Water Resources Research 36(12):3731–3741.

Reible, D. D., K. T. Valsaraj and L. J. Thibodeaux. 1991. Chemodynamic models for transport of contaminants from sediment beds. Pp. 187–228 In: Handbook of environmental chemistry. O. Hutzinger (ed.). Heidelberg, Germany: Springer-Verlag.

Ridley, M., and D. MacQueen. 1995. Cost-effective sampling of groundwater monitoring wells: a data review and well frequency evaluation. Pp. 14–21 In: Proceedings of the Hazardous Materials Management Conference and Exhibition, April 4–6, 1995, San Jose, California.

Rizzo, D. M., D. E. Dougherty, and M. Yu. 2000. An adaptive long-term monitoring and operations system (aLTMOs™) for optimization in environmental management. Proceedings of 2000 ASCE Joint Conference on Water Resources Engineering and Water Resources Planning and Management, Minneapolis, MN. ISBN 0-7844-0517-4. R. H. Hotchkiss and M. Glade (eds.). Reston, VA: American Society of Civil Engineers.

Russell, K. T., and A. J. Rabideau. 2000. Decision analysis for pump-and-treat design. Ground Water Monitoring and Remediation, Summer:159–168.

Sohn, M. D., M. J. Small, and M. Pantazidou. 2000. Reducing uncertainty in site characterization using Bayes Monte Carlo methods. J. Environ. Engr. ASCE. 126(10):893–902.

Stallard, M. O., S. E. Apitz, and C. A. Dooley. 1995. X-ray fluorescence spectrometry for field analysis of metals in marine sediments. Mar. Pollut. Bull. 31:297–305.

Stansbury, J., I. Bogardi, and E. Z. Stakhiv. 1999. Risk-cost optimization under uncertainty for dredged material disposal. J. Water Resour. Plan. Manag. ASCE 125(5):342–351.

Starks, T. H., and G. T. Flatman. 1991. RCRA ground-water monitoring decision procedures viewed as quality control schemes. Environmental Monitoring and Assessment 16:19–37.

Stiber, N. A., M. Pantazidou, and M. J. Small. 1999. Expert system methodology for evaluating reductive dechlorination at TCE sites. Environ. Sci. Technol. 33(17):3012–3020.

Thoms, S. R., G. Matisoff, P. L. McCall, and X. Wang. 1995. Models for alteration of sediments by benthic organisms. Project 92-NPS-2. Alexandria, VA: Water Environment Research Foundation.

U.S. Army Environmental Center (USAEC). 2000. The Tri-service site characterization and analysis penetrometer system-SCAPS: innovative environmental technology from concept to commercialization. Report Number SFIM-AEC-ET-TR-99073.

U.S. Geological Survey. 2001. User's guide for polyethylene-based passive diffusion bag samplers to obtain volatile organic compound concentrations in wells. Part 1-4. Water-Resources Investigations Report 01-4060.

U. S. Army Corps of Engineers (USACE). 2000. Expedited characterization and soil remediation at the test plot area, Wenatchee Tree Fruit Research Center, Wenatchee, Washington. Cost and Performance Report. U.S. Army Corps of Engineers Hazardous, Toxic, Radioactive Waste Center of Expertise.

Wagner, B. J. 1999. Evaluating data worth for ground-water management under uncertainty. J. Water Resour. Plan. Manag. ASCE 125(5):281–288.

Wiedemeier, T. H., H. S. Rifai, C. J. Newell, and J. T. Wilson. 1999. Natural attenuation of fuel and chlorinated solvents in the subsurface. New York: John Wiley and Sons.

Wisconsin Department of Natural Resources. 2001. Remedial investigation and feasibility study for the Lower Fox River, October.

4
Evaluation and Experimentation

INTRODUCTION

An essential feature of adaptive site management (ASM) is that it allows for a change in remedy—where the chosen approach is falling short of cleanup goals—that takes into account information gleaned on other potentially more effective remedies. One or more factors generally prompt reconsideration of the remedy. As discussed in Chapter 2, the remedy may prove to be ineffective in reaching cleanup goals, which has occurred in thousands of cases. NRC (1994) found that only eight of 77 pump-and-treat systems for groundwater remediation had achieved regulatory standards. In such cases, it makes sense to look for alternatives or at least to adjust or optimize the existing remedy. Even if remediation appears headed toward long-term goals, it may take longer than desired or expected. This can be an acute problem where remediation activities are delaying property reuse, preventing beneficial use of groundwater resources, or depressing property values and discouraging economic activity in the surrounding area. At the former Moffett Naval Air Station, for example, the slow rates of removal of contaminants in groundwater are discouraging NASA, the new property owner, from considering residential construction above the plume.

Sometimes costs escalate as projects encounter unknown obstacles, labor rates rise, or other inputs become more expensive. Or the responsible party's cumulative cleanup expense may outstretch available budgets, forcing cutbacks even at sites where the original financial projections turn out to be accurate. The rising cost per unit mass removal of contamination characteristic of some remedies can inflate overall project costs enormously. When these technologies are unable to meet remedial goals in a reasonable period of time, responsible parties usually seek a change in the hope that a new innovative treatment technology has been developed that is both more economical and effective than the initially chosen technology.

The discovery of new contaminants, higher concentrations of known contaminants, and wider contamination footprints at a site merit a review of the remedy and perhaps the remedial goal. In many cases, the solution may be a simple adjustment of the remedy, like changing the location of extraction wells. In other cases, however, the new discoveries should trigger a rethinking of the entire approach. For example, at Mather Air Force Base, it was discovered after signing a Record of Decision (ROD) covering three oil/water separator sites that petroleum hydrocarbon contamination extended beyond the areas originally identified. An Explanation of Significant Difference—similar to a ROD amendment—was developed proposing to supplement the original excavation remedy with soil vapor extraction and bioventing.

Remedies for soil contamination are more often than not based upon current or reasonably anticipated future land use. When land uses change, such remedies should be reconsidered. This is especially true when a more "intense" land use is proposed that would potentially create additional exposure pathways to human or ecological receptors. Finally, regulatory milestones, whether built into the law—such as the Superfund five-year review—or established through negotiated agreement, call for the periodic review of remedies. At that time, any of the above factors may come to the fore, triggering either an optimization effort or a thorough review of the site remedy. Alternatively, there are situations where a fundamental change in cleanup policy occurs. Potentially responsible parties (PRPs) may seek to conform the remedy at a site to any new cleanup policy, particularly if the remedy has not yet been implemented.

If it is decided that the remedy or remedial goal should be revisited, several courses of action at MDP3 are possible (see Figure 2-7). Deciding on the best course is aided by the parallel track of evaluation and experimentation called for in ASM. The cleanup process at most Navy sites involves a great deal of uncertainty because of an incomplete understanding of contaminant sources, pathways, and receptors, because of the variable performance of technological solutions, and because of the limited ability to establish and maintain proper institutional controls (NRC, 1999a). Obtaining new knowledge on these issues via evaluation and experimentation can reduce the uncertainty inherent in the original remedy selection and improve the cleanup process. For example, if a remedy does not perform as intended, it is often unclear whether the problem is inherent to the remediation approach or is due to inadequate accounting of site conditions in the design of the remediation system. Devoting time and resources to better understanding the site and the remedy can help clarify the situation and suggest ways to either improve the performance of the implemented remedy or provide a rationale for introducing alternative remedies. The quantitative and empirical information generated through evaluation and experimentation is crucial to support any changes of or modifications to existing remedies.

Even in cases where no change to the remedy is anticipated, knowledge gained through activities occurring concurrently with remedy implementation can better define the risks of the remaining contamination—an issue of great importance to stakeholders.

Evaluation and experimentation refer to a broad range of activities that include literature and data interpretation, demonstration studies, and research. Ideally, this should happen on the scale of an individual site, but it can also occur at a much larger, program-level scale. At the level of an individual site, evaluation and experimentation are actions designed to verify the existing hypotheses about the site, to explore the effectiveness of other more risky remediation technologies, or to discover something that can otherwise reduce uncertainty during the cleanup process. Original research may be undertaken to formulate new hypotheses about the site that could then be tested through experimentation ranging in scale from serum bottles to bench-scale columns, pilot-scale columns, and finally field-scale tests with implemented remedies. In addition to interpreting field monitoring data collected as part of routine remedy operation, evaluation and experimentation can also involve synthesis of relevant literature, analysis of operating experiences from other sites, or seeking advice from stakeholders. For these reasons, the success of evaluation and experimentation is linked to the continued development, testing, and demonstration of innovative technologies that has been ongoing at many federal facilities.

Although the main role of evaluation and experimentation at a specific site is to support changes or modifications to remedies to increase overall effectiveness, these activities can also help to lower the costs of remediation, especially at complex sites. Incorporating evaluation and experimentation into the Navy's entire cleanup program could spur development of better technologies to allow cleanup to be accomplished at a lower cost or to a higher state than is presently possible, thereby making sites available for less restrictive uses. This chapter describes the value of evaluation and experimentation to the ASM process, existing research programs that provide information to the Navy on performance and cost of remedial technologies, and suggestions for what the Navy should do to make research part of its cleanup paradigm.

WHY EVALUATION AND EXPERIMENTATION ARE NEEDED

Evaluation and experimentation are necessitated by the ineffectiveness of many selected remedies to meet remedial goals. As documented in Chapter 2 and in previous NRC studies (1994, 1997, 1999a,b), cleanup of contaminated sites is inherently complex because of physical heterogeneity in the subsurface, the presence of nonaqueous phase liquids (NAPLs) and contaminant mixtures (e.g., organics and inorganics), limited accessibility of the contaminants, and difficulties in characterizing the subsurface, among other things. When the performance of a remedy reaches an asymptote before meeting its cleanup objective, it is not solely a function of site complexity and technical limitations, but can also result from insufficient or inaccurate characterization of the site, leading to a flawed design of the remediation system. Assuming that the chosen remedy will not succeed is reasonable for many typical remedies, particularly institutional controls (NRC, 1999a), although the factors contributing to problems at each particular site are likely to be different. The evaluation and experimentation track of ASM specifically accommodates potential problems with remedy effectiveness by improving the understanding of the site (site conceptual model) and suggesting ways to enhance the performance of the existing remedy or to guide the selection of an alternative remedy. The track is a deliberate effort to learn and produce benefits from adversity. Evaluation and experimentation can open up new opportunities to remediate and manage sites more effectively even when problems are not imminent. Examples of where such improved understanding is particularly critical are given in the section below.

If done concurrently with implementation of the remedy, evaluation and experimentation will prevent activities from "stalling" once problems arise and will allow the site managers to make forward progress. Box 4-1 gives a specific example of how a study concurrent with implementation of an initial remedy led to the use of phytoremediation to replace a pump-and-treat system that would otherwise have been operated for the foreseeable future. A more external benefit of the evaluation and experimentation efforts within ASM is that it can create an expanded database on the performance of a remedial technology that will improve user confidence that the technology can provide the desired result under a specific set of conditions. For a responsible party like the Navy that has a large number of hazardous waste sites, the external benefits of investing in learning (i.e., using what is learned in one place at other sites and in future decisions) can be substantial.

BOX 4-1
Evaluation and Experimentation in Site Management at Argonne National Laboratory

The 317 area at Argonne National Laboratory-East has volatile organic compounds (VOCs) in vadose zone soils and underlying groundwater arising from past disposal practices. In 1997 a pump-and-treat system consisting of 13 extraction wells was installed along the site boundary to prevent offsite migration of contaminated groundwater. Although this system was successful in controlling the movement of contaminated groundwater, it provided no benefit from a source remediation perspective, and it likely would have had to operate indefinitely. In 1999, the U.S. Department of Energy, through the Accelerated Site Technology Deployment (ASTD) program, funded the deployment of a phytoremediation system for the 317 area as a potential replacement for the pump-and-treat system. The phytoremediation system consists of approximately 800 hybrid willows and deep-rooted poplars spread over a two-hectare area that included the presumed VOC source zone. The purpose of the phytoremediation system is twofold: (1) to develop hydraulic control over contaminated groundwater movement and so allow the termination of the pump-and-treat system using poplars and (2) to encourage bioremediation of the VOC source area with willows. Numerical flow modeling, which is updated regularly with site-specific data, suggests that by the year 2003, the plantation would successfully control the movement of groundwater, even in the winter when the trees are dormant (Quinn et al., 2001).

VOC Source Area and Willow Plantation

Hybrid Poplars

Critical Scenarios for Evaluation and Experimentation

Certain remedial approaches involve greater uncertainty than others and necessitate evaluation and experimentation to improve understanding of the mechanisms responsible for performance of the technology and to reduce failure rates. For example, technologies that involve *in situ* reactive treatment, like *in situ* bioremediation, require information on fine-scale subsurface properties, on the presence of indigenous microorganisms and their biodegradation potential, and on the distribution of growth factors (e.g., nutrients, electron acceptors, pH, temperature, and moisture). If supplemental nutrients and electron acceptors must be delivered to the zones of contamination to support bioremediation, then the proper way to manipulate the flow field and achieve mixing must be understood. Intensive monitoring of these parameters during remedy operation is not common and should occur as part of the evaluation and experimentation track if the remedy is to be optimized and reliably used at other sites. The case study in Box 4-2 illustrates how extensive field-scale studies

---

BOX 4-2
Experimentation and Evaluation in Site Management at
Aberdeen Proving Ground

Large plumes of chlorinated VOCs in the shallow subsurface where interactions occur with tidal freshwater wetlands, creeks, and estuaries have presented a challenging environmental problem at Aberdeen Proving Ground (APG), Maryland. Because of the sensitive nature of these wetland ecosystems and the ubiquitous possibility of encountering unexploded ordnance at APG, engineered remediation methods that would require excavation or digging are prohibitively expensive, unsafe, and potentially harmful to these ecosystems. Pump-and-treat was being considered as the primary treatment/containment method for these plumes in the Canal Creek area during an early investigation that characterized the extent of groundwater contamination (Lorah and Clark, 1996). The U.S. Geological Survey (USGS) recognized the potential problems associated with pump-and-treat in this and similar areas and proposed an intensive study of natural attenuation processes in a small wetland area along West Branch Canal Creek where groundwater discharge of chlorinated VOCs was believed to be occurring. It was hypothesized that as aerobic contaminated groundwater discharged into anaerobic wetland sediments, biodegradation of the chlorinated VOCs, in addition to other natural attenuation processes such as sorption and dilution, would attenuate the contaminants before land surface in the wetland or creek channel was reached. An intensive characterization of processes occurring in the wetland porewater and sediment in one area was proposed to provide information that also could be applied to other wetland sites at APG and elsewhere. On this basis, the site manager at APG approved the study, which began in 1992 and resulted in several publications (e.g., Lorah et al., 1997, 1999a,b, 2001).

Reconnaissance-phase installation of drive-point piezometers showed no evidence of VOC contamination in the wetland porewater at some sites but the

Continued

BOX 4-2 Continued

presence of daughter compounds at other sites. It was unclear whether contamination simply was not discharging upward at all locations or if degradation and other attenuation processes had completely removed the VOCs. Closely spaced vertical porewater sampling was necessary in the wetland to adequately characterize the occurrence of biodegradation. For the final monitoring network, additional nested piezometers, screened at discrete intervals in the wetland sediment and to the bottom of the aquifer, were installed along two transects that parallel the general groundwater flow direction in the aquifer. In addition to nested drive-point piezometers, porous membrane diffusion samplers, commonly called "peepers," were used to obtain vertical profiles of redox-sensitive constituents and VOCs in the wetland porewater (Figure 4-1A,B). The peepers made for this study were based on an original design by Hesslein (1976) for investigations of redox processes in lake sediment. The USGS wetland study was the first reported use of peepers for chlorinated VOCs. The profiles from the peepers documented the production and subsequent removal of daughter products from anaerobic biodegradation along upward flowpaths in the wetland porewater (Figure 4-1B). Laboratory microcosms confirmed that vinyl chloride and 1,2-dichloroethylene were the major persistent daughter products from anaerobic degradation of 1,1,2,2-tetrachloroethane, as observed in the peepers. Degradation of trichloroethylene also produced these daughter products. Both the field and laboratory data, however, showed complete degradation of the VOCs under methanogenic conditions. Degradation rates of trichloroethylene and 1,1,2,2-tetrachloroethane measured in the laboratory experiments were extremely rapid, ranging between 0.10 and 0.31 per day (half-lives of about two to seven days) with more rapid degradation occurring under methanogenic conditions. Laboratory experiments also showed the potential for rapid aerobic degradation of 1,2-dichloroethylene and vinyl chloride by methanotrophs in aerobic microzones around plant roots and near land surface.

   At the start of this wetland study, limited environmental fate data were available in the literature for one of the major contaminants, 1,1,2,2-tetrachloroethane, and few previous laboratory or field studies had been conducted in wetland environments for any of the chlorinated VOCs. Experimentation, therefore, was especially crucial at this site. Although the results of this investigation have led to recognition by APG site managers and regulatory agencies of the feasibility of monitored natural attenuation as a remediation method for the West Branch wetland plume, a ROD has not yet been obtained, partly because of concerns about the extent of the plumes outside the initial study area. Recent Hoverprobe drilling (see Box 3-6) has allowed further characterization of the plume boundaries to resolve this issue, and APG site management is working toward a ROD that includes monitored natural attenuation as a primary treatment. In addition, early promising results of the West Branch wetland study led to an investigation of natural attenuation of chlorinated solvents in wetlands in the J-Field area of APG, and monitored natural attenuation for the surficial aquifer has been incorporated in a signed ROD for this site. The West Branch wetland study led to collaboration between the USGS and the Air Force Research Laboratory on a project (funded by ESTCP) to demonstrate monitored natural attenuation of chlorinated solvents at other wetland sites and to evaluate methodologies for wetland investigations (Lorah et al., 2002; Dyer et al., 2002). The protocol and results of this ESTCP study should assist other site managers and investigators.

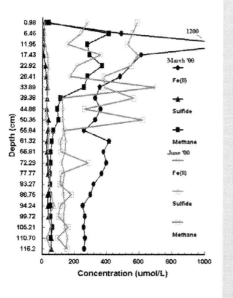 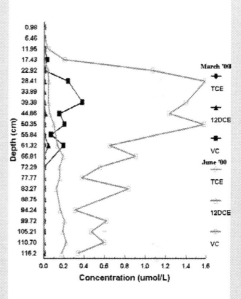

FIGURE 4-1 Concentrations of various organic and inorganic compounds with depth for two time points in 2000. Peeper profiles of (A) redox constituents and (B) VOCs at the West Branch Canal Creek wetland study area. SOURCE: Reprinted, with permission, from Lorah, et al. (2002). © (2002) Battelle Press.

established the biodegradation of chlorinated solvents at an Aberdeen Proving Ground site and led to a shift in the primary site management strategy from pump-and-treat to natural attenuation. It is an example of an approach that embraced experimentation and evaluation and in doing so revealed an opportunity, now being pursued, that may lead to more effective remediation at lower cost than was possible with the original remedy. Furthermore, the protocols for reducing uncertainty in microbial reaction processes developed during the study at Aberdeen Proving Ground should assist other site managers and investigators.

Other examples of remedial approaches with high uncertainty are the use of containment, solidification and stabilization techniques, and institutional controls to reduce the risks at a site through the elimination of one or more exposure pathways. Given the limited experience with enforcing these remedies and their legal complexities, there is a low probability of success over the long term (NRC, 1999a), such that these approaches merit additional evaluation and experimentation. For example, landfills cap designs have evolved from compacted soil to compacted clay to geomembranes and now to alternative designs such as vegetative covers. Because caps have a limited lifetime, monitoring cap performance at select sites is critical to understanding the mechanisms of failure and determining the most effective type of cap for certain environmental conditions. Despite this need, infiltration through caps has been measured at only a few landfills (Khire et al., 1997; Melchior, 1997; Licht et al., 2001). Future decisions on how to repair, replace, and select caps depend on more comprehensive information about landfill cap stability.

For other remedies like dig-and-haul that remove soil contamination, or at sites for which success is more certain (such as those with relatively simple hydrogeology and contaminant chemistry), the need for additional information through the evaluation and experimentation track is not as great.

At some sites, such as contaminated sediments and groundwater systems with extensive fractures (e.g., fractured bedrock or karst) and distributed DNAPL, contaminant inaccessibility and resistance to chemical and biological transformation greatly limit the remediation technologies that can be considered. Under these conditions, risk minimization is the short-term goal because restoration is difficult or impossible with current technologies. Evaluation and experimentation play a role at these sites to understand the current risk and reduce liability and to strive for long-term solutions in addition to optimizing current remedies. Data and information are needed to confirm the exposure pathways and other assumptions used in the risk assessment, to provide greater certainty in the risk calculation, and to determine the degree of remediation needed to reduce risk to acceptable levels. This may, for example, involve site-specific studies of contaminant bioavailability, which is a measure of the potential of a contaminant to be released and reach an ecological or human receptor (for an extensive review of such studies, see NRC, 2003). To ensure greater long-term protectiveness, studies might be undertaken to examine ways to enhance contaminant binding in sediments to prevent leaching and reduce environmental risk. If contaminated sediments are to be displaced by dredging or resuspension, then evaluation and experimentation are needed to identify what fraction, if any, of the contaminants will be released to the water column.

In all of these scenarios, the monitoring data generated during remedy implementation are critical to determining why the technology was unsuccessful and what to change. Evaluation and experimentation do not have to focus exclusively on finding a more effective remedy but can also encompass cost and time issues. Uncertainties about costs add to the reluctance to make modifications to existing remedies or implement alternative remediation systems. Studies could be designed to provide more detailed cost data based on actual site conditions.

The Role of Public Participation

Because evaluation and experimentation are important activities in establishing performance of an implemented remedy and in aiding decisions to change or modify remedies, the public should have input on what studies are conducted at a site. An engaged and informed public is better prepared to participate in the review of technology options and to understand the technical limitations. The affected community may have historical knowledge of the site that can serve as valuable input in planning evaluation and experimentation efforts. Public involvement at this stage can also facilitate studies directed toward the public's concerns, which helps to build trust. Early input from the public on research studies at the site can help to expand the number of acceptable remediation technologies that are considered.

The Role of Expert Panels

There are many situations where an asymptote in cleanup effectiveness has been reached prior to meeting cleanup goals, but there has been no ongoing evaluation and experimentation of the factors contributing to the behavior or of alternative remedial strategies that could improve the situation to help inform decision making. In such cases, expert panels can be used to conduct a short-term analysis of the available options. Although not a substitute for hiring and retaining technically trained staff, the Navy could form multidisciplinary expert panels to provide guidance on the next course of action when the selected remedy is not achieving the desired cleanup goal. Inclusion of experts outside of the Navy in addition to top Navy technical staff would enhance public confidence in the panels as providers of unbiased advice. Panels could include experts from other federal and state agencies, academia, and private consulting firms. Expert panels could be consulted throughout implementation of the remedy and could answer questions concerning the feasibility of achieving technical goals and appropriate modifications to improve performance. Panels could also provide advice on changes that are needed for sites where cleanup is underway and significant difficulties have been encountered. They can help initiate and oversee evaluation and experimentation efforts. For example, the panel might confirm the validity of a proposed site conceptual model, help with knowledge on the contaminant chemistry and behavior, help establish the effectiveness of an alternative remedial technology, and help determine the effectiveness of the monitoring plan. The panel might also be used as a resource for overcoming disagreements between responsible parties, local citizens, and regulators on technical issues. This concept is already being piloted by a joint effort between the Air Force Base Conversion Agency and Interstate Technology Regulatory Council to track remedy performance at six Air Force bases in California (M. Ierardi, Air Force Base Conversion Agency, personal communication, 2002).

MAKING RESEARCH PART OF THE CLEANUP PARADIGM

In order for the evaluation and experimentation track of ASM to be useful, site managers will have to adopt a new, prospective mindset after implementation of the remedy. It will require thinking about what remedies will be available in five years that would allow changing a remedy that is likely to not meet cleanup goals. In cases where the chosen remedy has a better track record, evaluation and experimentation will entail conducting site-specific studies (by remedial project managers or RPMs, contractors, consultants, university researchers, etc.) that occur concurrently with implementation of the remedy and will allow for future optimization of the remedy if cleanup goals are not met. In order to foster research on appropriate response strategies for those contaminants and sites posing the highest risks, the Navy should consolidate its contaminant information, determine relative risk for all of its sites, and establish priorities for site cleanup, ideally with a single database. This will help identify appropriate response strategies for those contaminants and sites posing the highest risks. Improved accessibility of site-specific data, such as in electronic format on the Internet, will help guide research that can most benefit the Navy's remedial program.

The following section describes programs that currently provide some research, development, and field-scale evaluation of remediation technologies and thus may serve an important role in the evaluation and experimentation component of ASM. Participation in these research efforts should lead to improved understanding that will help the environmental restoration program in 30 years as well as right now. The goal of this section is not to design a research agenda for the Navy, as this is generally addressed in other NRC reports (1994, 1997, 1999b, 2000). Rather, the discussion illustrates how each research program can help provide information to ASM and MDP3.

Current Programs

Many programs currently provide research and development, information transfer, and independent review functions that may serve an important role in the evaluation and experimentation track of ASM. Support for innovative technologies covers a broad range of activities. Federal agencies, either acting alone or through federal/private sector partnerships, have taken the lead on research and development of innovative remediation technologies, with over 600 innovative technologies currently under evaluation (EPA, 2000a). In the past decade, the Navy and other military services have supported field demonstration projects using innovative technologies (EPA, 2000a). Several projects at Navy facilities are shown in Table 4-1.

Research and Demonstration

There are two important types of research and development programs—one for basic and applied research to develop new technologies or provide the necessary basic understanding of processes to lead to

TABLE 4-1 Navy Field Demonstration Projects Using Innovative Technologies

Date[a]	Site	Technology	Contaminants
Soil, Sludge, and Sediment			
1993 (report)	NAS Seal Beach, CA	Ex situ bioremediation	BTEX
1999	NAS Yorktown, VA	Ex situ enhanced bioremediation (land farming)	Organic explosives, chlorinated solvents
1994	NAS Yorktown fuel farm	Bioslurping	TPH
1992	NAS Yorktown airfield	Bioventing	Hydrocarbons
open	Small arms firing range, NAS Adak, AK	Phytoremediation and soil washing	Heavy metals
1997	Pearl Harbor, HI	Ex situ extraction from porous surfaces	PCBs, metals
1999	Naval facility, Pearl Harbor, HI	Electrokinetics and electrokinetic heating	Heavy metals
1998 (report)	Hunter's Point Shipyard, San Francisco, CA	Ex situ physical separation/chemical leaching/soil washing/fluidized bed classifier	Cu, Cr, Pb, Zn
1992	Port Hueneme, CA	Solidification/stabilization	Pb, Cu
Open	NFESC, Port Hueneme, CA	Solvated electron technology	Pesticides
1995	Advanced fuel hydrocarbon national test site, Port Hueneme, CA	Thermally enhanced vapor extraction	TPH
1997	Mare Island Naval Shipyard, Vallejo, CA	Thermal desorption (both thermal blankets and wells)	PCBs
1998	NAS North Island, San Diego, CA	Photolytic destruction	Chlorinated solvents

Groundwater

1997	Port Hueneme National Test Site, San Diego, CA	Air sparging	Gasoline
open	NAS Fallon, NV	Enhanced bioremediation	Chlorinated solvents
open	NWS, Seal Beach, CA	Enhanced bioremediation	Gasoline, BTEX
1999 (report)	UST Site 23 NAS Point Mugu, CA	Enhanced bioremediation	TCE, VC, cis-1,2-DCE
1995 (report)	NAS North Island, San Diego, CA	Pervaporation	Solvents, degreasers
1993 (report)	Bangor SUBASE	Advanced oxidation process	TNT, RDX
open	Port Hueneme, CA, and other sites	Air sparging	Chlorinated compounds, petroleum
1995	Port Hueneme, CA, Naval Exchange Site	Circulation wells	BTEX
open	NAS Alameda, CA	PRB (Iron and ORC)	cis-DCE, VC, TCE, BTEX
1997	NAS Moffett Field, CA	PRB	TCE, PCE, DCE
1999	Naval facility, Pearl Harbor, HI	Surfactant-enhanced aquifer remediation (SEAR)	Fuel oil
1991	NAS Seal Beach, CA	Vapor extraction	VOCs, volatile fuel

[a]Date (year) of project; usually the start date, but in some cases, the date a report was issued. "Open" means the project is ongoing.

SOURCE: EPA (2000a).

technology development, and one for demonstration and validation of technologies that are past the initial pilot-testing stage. Both types of programs are needed to bring innovative technologies to full-scale application and widespread understanding of their utility. The Navy should support both types of programs to ensure meeting long-term cleanup challenges.

SERDP (http://www.serdp.org). Although there are a relatively large number of opportunities to obtain support for demonstrating and evaluating remedial technologies that have passed the research and development or pilot-testing stage, there are relatively few programs or agencies that support basic research through competitive grants either awarded to external parties or through internal funding. The largest programs are under a corporate Department of Defense (DoD) program, a relatively new Department of Energy (DOE) program, and under the U.S. Environmental Protection Agency (EPA). The DoD's environmental basic research and development program—the Strategic Environmental Research and Development Program (SERDP)—was established in 1990 and is conducted in partnership with DOE and EPA. SERDP operates in concert with DoD's Environmental Security Technology Certification Program (ESTCP, see below), which supports field demonstration and validation of technologies past the basic research and development stage. Total funding for SERDP/ESTCP was about $84 million in FY02. Both government and private sector parties may compete for SERDP funds, and calls for proposals are given annually to address the program's statement of need in the thrust areas of environmental compliance, cleanup, pollution reduction, and conservation. SERDP tends to favor funding of large, multiagency proposals and might only fund one or two new projects annually in each statement of need. The statements of need are selected each year after input from panels of experts (gathered from government, academia, and the private sector) that are convened within the thrust areas. In addition to this core SERDP solicitation that funds multi-year projects, annual solicitations also are released under the SERDP Exploratory Development, or "SEED," program for one-year projects with a maximum funding of $100,000. SEED is designed to provide initial funding for high-risk but potentially high-payoff projects.

ESTCP (http://www.estcp.org). Competitive research grants are provided for field demonstrations of promising innovative technologies through DoD's ESTCP. ESTCP issues two calls for proposals annually—one for DoD agencies and one for other federal agencies and the private sector—for demonstration of cleanup technologies that address their statement of need. The needs change annually and typically address specific *in situ* treatment or containment technologies such as bioremediation or phytoremediation, rapid onsite characterization technologies, and unexploded ordinance detection and remediation. ESTCP's goal is to select lab-proven technologies with broad DoD and market application and provide funding for field demonstrations at DoD facilities. The DoD site or sites used for the demonstration do not need to be selected before a grant is awarded, although it is beneficial to have a potential site identified and a promise of additional support to supplement ESTCP's award. All projects must document the cost and performance of the field trials in reports that have standardized formats.

NETTS (http://www.serdp.org/netts). The SERDP-funded National Environmental Technology Test Sites (NETTS) program also supports demonstration projects by providing three well-characterized DoD sites (Dover Air Force Base, McClellan Air Force Base, and the Navy's Port Hueneme) for applied research and demonstration projects of innovative cleanup, site-characterization, and monitoring technologies. The NETTS program provides site support, such as initial site characterization, demonstration oversight, permitting assistance, and technology assistance, and it also provides infrastructure support, such as access roads, test pads, offices, laboratories, analytical equipment, drill rigs, field vehicles, utilities, and security.

SITE (http://www.epa.gov/ord/site). EPA's Superfund Innovative Technology Evaluation (SITE) program also supports field demonstrations of technologies. There is an annual solicitation for host sites for the demonstration or evaluation of innovative technologies for hazardous waste cleanup in groundwater, soil, or sediment, and an annual solicitation for remedial technologies that can be demonstrated at previously selected sites. Although EPA does not provide funds to the host site, the SITE program assigns an EPA employee to manage each site and covers the cost associated with work plan preparation, field sampling, analysis, and reports. SITE hosts provide infrastructure support to the project and residual waste disposal. Technology vendors provide their own resources for equipment, operation, and maintenance for the demonstration or form a financial agreement with the host site. At the conclusion of a SITE demonstration, a report is prepared that evaluates all available information on the technology, analyzes its applicability to different site and waste characteristics, and presents performance and cost data.

ASTD (http://63.161.144.52/). In 1998 DOE began the Accelerated Site Technology Deployment program (ASTD) to provide additional funding to projects that use innovative technologies with the goal of providing incentives to promote multisite deployment of new technologies. This program differs from SERDP/ESTCP because it is not intended to support demonstrations; rather, it is supporting technologies that have been demonstrated elsewhere and for which evidence of their effectiveness has already been gathered. ASTD is meant to facilitate widespread deployment of these proven technologies. Technologies currently being deployed under the ASTD that are pertinent to Navy sites include permeable reactive barriers, enhanced bioremediation, alternative landfill covers, and thermally enhanced soil vapor extraction. ASTD also differs from SERDP/ESTCP and SITE in that projects are proposed only by RPMs for use specific to a site that they manage. On a smaller scale, the Navy and Air Force accomplish a similar objective of matching innovative technologies to direct use at an RPM's site through Broad Agency Announcements (BAAs) to receive proposals for demonstrations of technologies. Funding is provided for proposals of interest if they can be matched to the needs of an RPM for a site.

STAR (http://es.epa.gov/ncer/grants). EPA competitively funds basic and applied remediation research by external parties through the National Center for Environmental Science to Achieve Results (STAR) grants. Under the STAR grants, researchers are addressing numerous issues relating to remediation, including pesticide remediation, socioeconomic aspects of remediation, bioremediation, phytoremediation, soil and sediment remediation, and remediation of specific classes or constituents such as polyaromatic hydrocarbons (PAHs), polychlorinated biphenyls (PCBs), oxygenates (like MTBE), gasoline, and metals. Through this competitive grant selection process, EPA's Office of Solid Waste and Emergency Response and Office of Research and Development also fund university-based Hazardous Substance Research Centers (HSRCs), which address different theme areas related to environmental research and provide a technology transfer and community outreach function. In addition to EPA, the HSRCs can receive funding from DOE, DoD, academia, and other state and federal government agencies. The five new HSRCs established in 2001 address research on detecting, assessing, and managing hazardous substances in urban environments, on low-cost remediation technologies to remove contaminants from the environment, on developing *in situ* processes for VOC remediation in groundwater and soils, on managing contaminated sediments, and on developing new or improved methods to remediate mine waste sites.

SBRP (http://www-apps.niehs.nih.gov/sbrp/index.cfm). The Superfund Basic Research Program (SBRP) within the National Institute of Environmental Health Sciences also awards grants competitively, although proposals may be submitted only by U.S. universities. The Superfund Amendments and Reauthorization Act (SARA) of 1986 established SBRP to develop methods and technologies for detecting hazardous substances in the environment, to advance techniques for the assessment and evaluation of the effects and risks of hazardous substances on human health, and to assess basic biological, chemical, and physical methods of reducing the amount and toxicity of hazardous substances. In the area of remediation, example projects include biodegradation of PAHs in soil, abiotic dechlorination of chlorinated ethenes, and remediation of gas-phase chlorinated solvents in unsaturated sediments.

ETV (http://www.epa.gov/etv/index.htm). EPA's Environmental Technology Verification (ETV) program was instituted in 1995 to verify the performance of innovative technologies and substantially accelerate their entrance into the domestic and international marketplace (EPA, 2002). The program operates through public/private testing partnerships, and any technology vendor within technology categories selected by stakeholders for verification may participate. Test and quality assurance plans and protocols are developed with the participation of technical experts, stakeholders, and vendors. They are then made available prior to testing, peer reviewed by other experts, and then updated after testing. This program does not fund research, but it can be used by RPMs as a source of reliable information on new technology.

Other programs. The Office of Naval Research (ONR) also provides grants through BAAs for research on characterization of environmental processes and their application to remediation technologies, especially in marine/estuarine sediments. The U.S. Geological Survey provides internal research funds for studying contaminant fate and behavior in hydrologic environments under the Toxic Substances Hydrology Program. This program, which was initiated in 1982, has provided a unique niche in being able to fund long-term process-oriented field research at selected sites. One example is the ongoing project begun in 1983 at a crude-oil spill site in Bemidji, Minnesota, which has provided fundamental knowledge and methods that are widely used to characterize natural attenuation of BTEX at other sites (Cozzarelli et al., 1999). The toxics program is coordinated with EPA, DoD, DOE, U.S. Department of Agriculture, and other Department of Interior (DOI) agencies to select research priorities. Scientists from academia, other federal agencies, and industry commonly are active members of research teams for a site, although much of their funding is provided from sources outside the toxics program. The investigations cover a wide range of contaminants, including chlorinated solvents, BTEX, MTBE, metals, PAHs, and pesticides. A recent addition to the toxics program is a TCE-contaminated fractured rock site in New Jersey, providing the Navy with an opportunity to conduct long-term studies at one of its sites.

***

For ASM to be successfully implemented, data and information from the above research and demonstration programs, particularly results that are relevant to the contaminants and problems at their facilities, must be made available to RPMs. This information provides options in case an existing remedy approaches an asymptote prior to reaching cleanup goals. Although support of research and demonstrations at non-Navy sites also is critical in building a database on performance and cost effectiveness for innovative technologies, the Navy should consider emulating DOE's ASTD program. This might be done by expanding the Navy's BAA program to facilitate a direct linkage between the RPMs' need for experimentation at a particular site and available technologies and expert assistance. The model of the national test centers (such as at Port Hueneme), which have hosted technology demonstrations at Navy facilities, should be expanded to include additional facilities.

In addition, because implementing innovative technologies is different from performing the fundamental research necessary to develop innovative technologies in the first place, these test centers should also be considered as candidates for conducting basic research. The above review of the existing major research and demonstration programs shows that there are fewer programs supporting initial basic research and technology development than there are programs for supporting demonstrations or deployment of technologies that are already proven to some extent. Because no single remediation technology has been found that can take care of all or even most of the Navy's complex problems (see Chapter 5), basic research into entirely new technologies will be necessary in order to eventually attain long-term cleanup goals.

Information Transfer

Since 1990, EPA's Technology Innovation Office (TIO) has sought to increase the applications of innovative technologies to the characterization and treatment of contaminated waste sites, soils, and groundwater by acting as a leader in technology transfer. TIO gathers and assesses research ventures of other offices within EPA, of other federal agencies, and of the private sector. TIO spreads information on technologies both through traditional paper publications and extensive web-based information networks. Within the next few years, EPA, the U.S. Army Corps of Engineers, and their contractors expect to gather and evaluate baseline data on all Superfund pump-and-treat systems and optimize the operation of up to 16 systems.

The Federal Remediation Technologies Roundtable (FRTR), which is a partnership formed to exchange information on the development and demonstration of innovative technologies, includes the Army, the Navy, the Air Force, the Corps of Engineers, DOE, DOI, and EPA as members. A focus of the FRTR has been to provide a more comprehensive record of remedial cost and performance at demonstration or test sites. The FRTR also has published review and guidance documents.

A consortium called the Interstate Regulatory Technology Council (ITRC) includes members from over 35 state environmental regulatory agencies that work with federal agencies and other stakeholders to transfer technology information and to help build consistent regulation of new site restoration technologies and other environmental resource problems. The ITRC has technical teams that develop guidance documents on innovative technologies and provides classroom and Internet training on these technologies. For example, ITRC technical teams have produced guidance documents on *in situ* bioremediation, *in situ* chemical oxidation, and permeable reactive barriers. In addition, ITRC has a State Engagement Program that works to obtain multistate concurrence on the guidance documents that are produced, expediting the regulatory acceptance of new and emerging technologies.

In 1992, TIO and the EPA's Office of Research and Development established the Remediation Technologies Development Forum (RTDF), a consortium of industry, government, and academia, to stimulate collaboration between the federal government and private sector in developing innovative solutions to mutual hazardous waste problems. The partners voluntarily share knowledge, experience, equipment, and facilities while jointly participating in research and demonstration efforts with a goal of developing more effective, less costly hazardous waste characterization and treatment technologies. For example, the Bioremediation Consortium of RTDF has conducted several studies at chlorinated solvent-contaminated sites at Dover AFB, Delaware, including a cometabolic bioventing study, a natural attenuation study, and pilot tests of accelerated anaerobic biodegradation that initially used injection of substrates and nutrients and later used bioaugmentation (Grosso et al., 1999; Ellis et al., 2000). Other current RTDF teams focus on phytoremediation, permeable reactive barriers, and diffusion samplers. RTDF teams have provided training courses and manuals on technology procedures.

The individual DoD agencies also have their own divisions that provide an information transfer role to RPMs. The Air Force has supported evaluation of remediation technologies through the Air Force Center for Environmental Excellence (AFCEE) and through the Air Force Research Laboratory (USAFRL), although the USAFRL is in the process of closing down its environmental work. An example of AFCEE involvement in information transfer and implementation of innovative technologies is the development of protocols, which were later reviewed and published as EPA documents, for evaluating monitored natural attenuation at petroleum hydrocarbon and chlorinated solvent sites. AFCEE also completed an evaluation of the performance and cost of implementing natural attenuation as a remedy for fuel and chlorinated solvent contamination at multiple Air Force sites. For the Navy, the Naval Facilities Engineering Service Center (NFESC) provides training and other information transfer activities for RPMs through a variety of programs and initiatives. For example, web-based multimedia tools have been developed that are easily accessible to RPMs, that are updated and revised quickly, that provide a link to technical experts, and that track feedback from users. Furthermore, NFESC organizes the Remediation Innovative Technology Seminar, which provides training to RPMs, regulators, and Navy contractors on new and innovative technologies.

Adopting ASM will require that the Navy continue to participate in RTDF consortia and FRTR activities to remain abreast of available technologies and their applicability to different sites and media. Participation in development and dissemination of interagency guidance documents on promising technologies also would assist in providing reliable information to RPMs and increasing understanding and acceptance of innovative technologies.

Independent Review Panels

The benefits of independent review panels for facilitating decision making during ASM have been discussed previously. In 2000, EPA established an independent review program, called the Optimization of Fund-lead Ground Water Pump and Treat Systems, with the goal of assisting EPA Regions in optimizing existing pump-and-treat groundwater remedies that have been constructed and are being operated by EPA or the states with Superfund money (EPA, 2000b). Individual DoD agencies have all supported independent review panels to examine their environmental cleanup. For example, the Army established the Groundwater Extraction and Treatment Effectiveness Review program to form teams to go into the field to assess existing treatment systems at Army sites and redesign these systems to run more proficiently at lower operational costs. For the Navy, NFESC has successfully deployed "Tiger Teams" to review, evaluate, and optimize environmental restoration efforts at numerous Navy installations. Tiger Teams are panels of internal and external technical specialists that can provide guidance on the most effective strategies to achieve site closure, potentially providing solutions that may not have been conceptualized by installation staff or its contractors.

Obstacles to Research

There are significant obstacles to conducting research in the current environmental restoration program that may inhibit adoption of ASM. These obstacles, and suggestions for how to create incentives to overcome them, can be broadly grouped into the following areas: resource obstacles, regulatory obstacles, timing issues, and socioeconomic barriers.

Resource Issues

Perhaps the most important issue is how to fund evaluation and experimentation activities at an individual site that will require additional resources beyond those needed to implement the chosen remedy. Past experience indicates that military and government officials may be reluctant to provide such additional funds. Historically, there has been a clear line drawn between enforcement and cleanup expenditures and research expenditures, within both EPA and the military. Different EPA offices handle cleanup and research and, as a practical matter, research budgets are separate from cleanup and enforcement budgets within many agencies. For example, although some of EPA's research budget does come from Superfund, the agency research budget is primarily derived from general funds and is allocated annually based on broad research goals, not individual site-specific considerations. Similar trends are apparent in the military, such that it is difficult to get funding under the environmental restoration program for anything labeled as "research." At times, even the use of cleanup funds to supplement a study primarily funded through a program such as ESTCP has been stated to be inappropriate use of these funds. Thus, activities such as conducting treatability tests to later optimize a remedial action at a specific site may be allowed and funded, but it is much more difficult to fund research on a new remedy that is not part of the ongoing site-specific activity. In general, DoD discourages the linkage of actual installation restoration activities with research and development, particularly if the results are primarily useful at sites other than the site where the research is being conducted. Where restoration and research funds are legally different, these distinctions must be observed. This strongly suggests that the Navy (and all federal departments more generally) and EPA should assess the statutory, regulatory, and institutional barriers that prevent cleanup funds from being utilized for research and that prevent research projects from being located at restoration sites.

In addition to these direct funding issues, there is a human resource issue. It is natural to expect resistance to a process that expects cleanup staff to distill new information germane to an already complicated set of operational tasks. Furthermore, hosting a demonstration study at a site inevitably requires assistance from the RPM and others knowledgeable on site specifics in infrastructure, permitting, and regulatory acceptance issues. RPMs often have too large of a workload and little incentive to provide this type of support for a demonstration that may not provide a remedial solution for their sites. In fact, there can be a perception that allowing this demonstration may uncover additional problems at the site, resulting in additional work for the RPM.

Despite these drawbacks, experience at some facilities illustrates the value of combining research and development activities with cleanup. For example, the Navy initially installed the permeable reactive barrier (PRB) at Moffett Field, California, as a pilot-scale installation restoration activity. ESTCP subsequently sponsored the NFESC to validate the performance and cost effectiveness of the PRB technology at Moffett Field for eventual application at other DoD sites, and later SERDP added funds as well. Much of the detailed knowledge of the performance of PRBs is derived from this innovative partnership (NFESC, 1998; Gavaskar et al., 2001). Now that grant funding has expired, it is hoped that the Navy is committed to continuing such low-level expenditures. Thus, there are creative funding mechanisms to enable the incorporation of evaluation and experimentation into site management under the current system, although such results suggest the need to revise policy and even statutes to encourage further linkages.

Box 4-3 discusses a new DOE program—the Accelerated Site Technology Development (ASTD) program —that serves as a useful model for how to pay for evaluation and experimentation activities that focus on the development of innovative technologies. In this case, DOE will pay for a portion of cleanup at certain experimental sites where an innovative but proven technology is proposed for use. The program targets those innovative technologies for which considerable evidence of effectiveness has already been gathered but for which widespread deployment has not yet occurred.

Timing Issues

There are potential timing issues that will arise regarding the evaluation and experimentation track of ASM. For example, will it be possible to obtain research results from site-specific studies during the timeframe of remediation? If not, then the practicality of that research for informing decision making is limited. Second, site managers may perceive evaluation and experimentation as somehow delaying completion of the project because time and resources must be spent on multiple activities. However, this assumes that the evaluation and experimentation activities will not prove useful in optimizing the existing remedy or helping to better understand a technology that will replace the existing remedy. As discussed above, for cases where the technology has limited potential to succeed (as with institutional controls or at sites with DNAPLs and heterogeneous hydrogeology), concurrent study can prevent the cleanup from stalling by providing alternatives when contaminant concentrations level off above the remedial goal.

Regulatory Issues

Regulatory barriers to implementing ASM are discussed in detail in

BOX 4-3
DOE's Accelerated Site Technology Deployment Program

Several DoD programs already discussed including SERDP, ESTCP, and NETTS encourage the development and demonstration of innovative technologies for hazardous waste site remediation. Although these three programs foster technological innovation from basic research and development through demonstration and validation for the DoD complex, the final hurdle for innovative technologies is widespread deployment. Within DOE, the final deployment hurdle is addressed by the Accelerated Site Technology Deployment (ASTD) program. The ASTD program recognizes that obstacles such as regulatory and stakeholder approval, site acceptance, perceived business and technology risks, and simple inertia can prevent the application of new technologies that have the potential for saving money and/or reducing cleanup schedules. The purpose of the ASTD program is to facilitate the use of proven, innovative technologies across the DOE complex.

The ASTD program provides site managers with supplementary funding for projects if innovative but proven technologies are used. For a project to qualify for ASTD funding, the following requirements must be met:

• The site manager proposes an innovative but proven technology that has demonstrated an improvement over the existing site baseline plans.
• The site manager has made a budgetary commitment to use the innovative technology that covers at least 50 percent of the deployment costs.
• A cost/benefit analysis demonstrates the potential for significant life-cycle cost savings over baseline approaches if the innovative technology is used.
• The site manager has identified other sites willing to deploy the technology if initial deployments are successful.
• The site can provide evidence that the necessary regulatory permits will be obtained.

For individual sites, the attraction of a funded ASTD activity is the ability to obtain additional funding above and beyond baseline dollars for completing site environmental restoration obligations.

Sixty ASTD projects were initiated between FY98 and FY00 at 22 DOE sites at a cost of $255.8 million. Over one third of the funding for these projects has been through the ASTD program, with the balance provided by leveraged site restoration funds. The projected life-cycle cost savings from these 60 projects is more than $1 billion (DOE, 2001).

Chapter 6. However, it should be noted here that EPA has acknowledged that historically, many of its cleanup and regulatory schemes inhibited the use of innovative technologies (EPA, 1994a, 1997, 2000c), which is clearly instrumental to the success of ASM. To address this problem, EPA has issued a policy to "routinely consider innovative treatment technologies where treatment is appropriate" and to not eliminate "promising new technologies from consideration solely because of uncertainties in their performance and cost" (EPA, 2001). To promote the use of innovative technologies, EPA has even agreed to reimburse up to 50 percent of the cost of implementing an innovative remedy at select Superfund cleanup sites (EPA, 1996, 2001), although few private parties have offered to participate. In addition, EPA's policy is to promote the use of federal facilities as demonstration and testing centers for innovative environmental technologies (EPA, 1994b, 1998). In light of the importance of such centers to the adoption of ASM, this policy should be embraced wherever possible.

Socioeconomic Issues

Social and economic incentives to not invest in and utilize innovative technologies also present barriers to the evaluation and experimentation track of ASM. First, the market value of innovative remediation technology companies since 1990 has been poor. For example, stocks of most of the environmental technology companies that dropped in value in the mid-1990s (NRC, 1997) remain low, or the companies have gone out of business. Because most innovation in the private sector stems from research performed by small, innovative technology companies that are funded by private capital, investors will abandon a sector that consistently underperforms (in terms of profit). Second, the market is inherently fragmented in terms of the types of wastes, the size of the sites, the many different contaminated media involved, and the differences between federally owned sites, private sector sites, and sites cleaned up pursuant to state programs. The number of private sector companies involved in innovative remediation technology research is relatively small compared to the number of companies that have been named as potentially responsible parties across the country. This fragmentation means the inherent reward of investing in technology is smaller than if the market segments were broader. Third, the method by which future costs are calculated provides an incentive to clean up a site until it is health protective, but not to clean it up to unrestricted use (NRC, 1997). As EPA notes, there are "numerous financial incentives to delay remediation and few incentives to carry out remediation in a timely manner" (EPA, 2000c). Because of the private sector's limited investment in innovative technology development, only by increasing the level of federal research can there be any hope that new technologies capable of attaining cleanup goals will be developed.

The environmental arena has also begun, for many reasons, to accept more remedies where contamination is left in place, which could discourage evaluation and experimentation efforts. NRC (1997) concluded that "in the absence of assessing liability for cleaning up contaminated sites and posting this liability on the corporate balance sheets, there is no economic driver for improved remediation." As noted in Chapter 1, government regulatory agencies increasingly have accepted containment for at least part of the site. Without a clear legal mandate requiring cleanup of soil and groundwater to unrestricted use levels, there is less economic incentive for potentially responsible parties or private sector companies to invest in the development of new remediation technologies (NRC, 1997).

Many authors (including EPA) have reported a cultural bias against innovative approaches, not just within EPA, but also within the companies liable for the cleanup (EPA 1996, 2000c; NRC, 1997; Presidential/Congressional Commission on Risk Assessment and Risk Management, 1997). And historically, neither the public nor PRPs typically favor research (EPA, 2000c). For the public, a primary concern is that research will delay the onset of remediation—a concern that is addressed in ASM by having evaluation and experimentation occur in parallel with remedial activities. Many private sector companies and government PRPs prefer certainty. By definition, an innovative technology is less certain to achieve site cleanup goals. However, a directed study with the potential to increase overall effectiveness and reduce unit cost may be perceived differently.

Clearly, a bias against the evaluation and experimentation track is that the research may not necessarily be applicable to the site of interest. Public-private partnerships may aid in overcoming this obstacle. For example, at the Army's Fort McCoy, researchers from the University of Wisconsin Geology and Geophysics Department are helping to build a database on petroleum cleanup. They have conducted field workshops at one of Fort McCoy's remediation sites. An Army spokesman said, "Although study results may not aid Fort McCoy directly, the results are of value to the scientific community and do help build and improve the overall database on removing contaminants. The information can be used to help clean up other contaminated sites, which might include other Department of Defense sites." This partnership is succeeding because the Army provides the site, the infrastructure, and the cleanup activities to study, but it does not financially support the researchers' efforts.

MAJOR CONCLUSIONS AND RECOMMENDATIONS

Much of the short-term increased costs associated with implementing ASM in anticipated to be associated with evaluation and experimentation. However, if ASM is targeted to those high-risk, complex sites where large costs are at stake (as suggested in Chapter 2), the costs associated with ASM are likely to be balanced or exceeded by the savings that result from switching to a more efficient and effective technology or by overall life-cycle savings. An example is provided by the National Zinc NPL Site in Bartlesville, Oklahoma. After setting initial cleanup goals for heavy metals, several site-specific studies of lead, arsenic, and cadmium bioavailability were conducted. These included rat feeding studies using local contaminated soil as well as mineralogical and chemical extraction methods (NRC, 2002). After review by the lead state agency, a community advisory group, and an independent expert in the field, results from the study led to revised cleanup goals based on the measured limited bioavailability of the metals to humans. These revised values greatly reduced the aerial extent of soils requiring remediation and reduced the remediation costs by approximately \$40 million, with the bioavailability studies themselves costing less than one hundredth of this cost savings. Although in this case the action involved revising a cleanup goal rather than discontinuing and replacing an ineffective remedy, it nonetheless illustrates the benefit of investing in learning as part of the cleanup process.

It is important to distinguish between ASM's evaluation and experimentation track and treatability studies under the CERCLA process. Treatability studies are generally conducted during the RI/FS or the RD/RA phases, and they provide a starting point for ensuring that a certain treatment approach or specific remedy design will be effective at the site of interest (EPA, 1992). Indeed, they can be critical to evaluating a potential remedy prior to its full-scale implementation. Although treatability studies may involve the type of experimental studies discussed as part of evaluation and experimentation, they generally occur earlier in the CERCLA process (before or during MDP1) and they do not necessarily involve experimentation on less certain technologies that could be turned to in the event of failure of the initial remedy (although they could certainly be designed to do so). Thus, as narrowly defined above, treatability studies are an important component of ASM, but they are not a substitute for evaluation and experimentation, which facilitates more informed decision making during MDP3. Because ASM involves feedback loops, treatability studies may occur multiple times during the lifetime of a hazardous waste site as different technologies are proposed and implemented.

Other than suggesting that evaluation and experimentation will be most cost-effective at complex, high risk sites, it is inappropriate for this report to make specific recommendations on cost criteria for deciding whether or not to conduct evaluation and experimentation, although the Navy and other federal agencies that adopt ASM may decide to do so. For example, agencies may prefer to allot some percentage of annual costs (capital, operation and maintenance, or combined) to enabling evaluation and experimentation and building the data infrastructure necessary to support innovative research. Or such decisions might be made on a site-specific basis to take into account the certainty of initial remedy effectiveness. Other resource allocation issues will need to be addressed in order to overcome the aforementioned barriers to research. These include the creation of incentives for site managers to conduct evaluation and experimentation and of flexibility so that site managers can respond to what may be surprising results from evaluation and experimentation efforts. These issues and others should appropriately be examined by pilot testing ASM at a few select sites, as recommended in Chapter 2.

Evaluation and experimentation are integral to adaptive site management and should occur concurrently with remedy implementation. Improved understanding of a site through evaluation and experimentation can reduce the amount of uncertainty associated with the risk estimate at a site and suggest ways to enhance the performance of the existing remedy. Evaluation and experimentation of new, innovative technologies can also help guide the selection of an alternative in case the remedy is ineffective in meeting cleanup goals. The need for making adjustments in remedial technology over time should be considered the norm, and designs should be conceptualized and implemented accordingly. Employing evaluation and experimentation is most important for remedies likely to reach an asymptote prior to meeting the remedial goal, for sites with intractable contamination such as DNAPLs and metals, and for sites where containment or institutional controls are used.

Stakeholders should help define the research objectives for developing innovative technologies that can respond to difficult site-specific cleanup challenges. Public involvement during evaluation and experimentation efforts can help to expand the number of acceptable remediation technologies that are considered, build trust, and reduce uncertainty in the cleanup process. An engaged and informed public is better prepared to participate in the review of technology options and to understand the technical limitations.

DoD should better promote testing of innovative or new experimental technologies at selected sites both for site-specific application and if the results are likely to improve cleanup activities at other sites. Long-term cost and performance data are unavailable for most innovative technologies, making it impossible to fully evaluate their success or efficacy. Consequently, quantitative comparison of these technologies to more traditional remedies also is difficult, especially in terms of reduction in risk or exposure versus cost or time.

DoD should expand its programs that focus on developing and testing innovative remedial technologies and monitoring techniques. It appears certain that a number of DoD and other sites will require costly, substantive management for decades or longer. Therefore, in the absence of enabling legislative or regulatory changes, the lack of such research will result in DoD and others not having the new tools that can improve remedial programs and reduce long-term fiscal liabilities. Responsible federal agencies should collaborate closely with researchers in the public and private sectors to ensure that RPMs are trained and knowledgeable on new and innovative technologies that might be used to replace existing ineffective remedies.

Congress should make sure there are funds available to support the evaluation and experimentation track of adaptive site management. Although significant research efforts have been underway, unless the federal government provides new resources, only slow progress will be made toward finding cost-effective methods of reducing contaminant levels and meeting cleanup goals. Federal support is needed to fill the gap left as a result of lacking market incentives for the development of innovative hazardous waste cleanup technologies.

Resource, timing, regulatory, and socioeconomic obstacles need to be overcome in order to fully adopt evaluation and experimenta-

tion as a component of ASM. Combining research and development activities in conjunction with cleanup has value, but additional resources beyond those needed to implement the chosen remedy are generally not available with current cleanup programs. Site managers often perceive the results from research as yielding answers over time scales that are too long to prove useful in optimizing existing remedies or in making informed decisions about when to replace a remedy. A final obstacle to evaluation and experimentation is that social and economic incentives for investing in and utilizing innovative technologies are limited. The increasing use of containment and institutional controls has discouraged additional investment in the development of new remediation technologies.

The Navy and, more generally, DoD should make site-specific operations data for a select number of complex sites more easily accessible to the research community. Making such data available would facilitate the development of new monitoring techniques, remediation technologies, and predictive modeling for hazardous waste sites. In addition, if DoD and EPA managed site-specific data in a uniform manner and made these data easily accessible to researchers, other RPMs, and the public, it would be easier to identify what technical barriers are preventing attainment of cleanup goals at sites.

REFERENCES

Cozzarelli, I. M., M. J. Baedecker, R. P. Eganhouse, M. E. Tuccillo, B. A. Bekins, G. R. Aiken, and J. B. Jaeschke. 1999. Long-term geochemical evolution of a crude-oil plume at Bemidji, Minnesota. Pp. 123–132 In: Proceedings of the Technical Meeting, Charleston, South Carolina, March 8–12, 1999—Subsurface Contamination from Point Sources. U.S. Geological Survey Water Resources Investigations Report 99-4018C. D. W. Morganwalp and H. T. Buxton (eds.).

Department of Energy (DOE). 2001. Accelerated site technology deployment (ASTD): a vehicle to expedite cleanup through the use of innovative technology. Analysis, lessons learned, and recommendations. DOE/EM-0574. Washington, DC: DOE Office of Environmental Management, Office of Science and Technology.

Dyer, L. J., M. M. Lorah, and D. R. Burris. 2002. Effect of sampling method on measured porewater concentrations in a wetland contaminated by chlorinated solvents. In: Proceedings of the 2nd International Wetlands and Remediation Conference, September 5–6, 2001, Burlington, Vermont. Columbus, OH: Battelle Press.

Ellis, D. E., E. J. Lutz, J. M. Odom, R. J. Buchanan, Jr., C. L. Bartlett, M. D. Lee, M. R. Harkness, and K. A. Deweerd. 2000. Bioaugmentation for accelerated in situ anaerobic bioremediation. Environ. Sci. Technol. 34(11):2254–2260.

Environmental Protection Agency (EPA). 1992. Guidance for conducting treatability studies under CERCLA. EPA/540/R-92/071a. Washington, DC: EPA Office of Research and Development.

EPA. 1994a. Report to Congress on the effect of environmental regulation on hazardous waste. Washington, DC: EPA.

EPA. 1994b. Memorandum from Carol Browner, Administrator, Re: Policy for innovative environmental technologies at federal facilities (August 1994).

EPA. 1996. Memorandum from Elliott P. Laws, Assistant Administrator of the Office of Solid Waste and Emergency Response, to Superfund, RCRA, UST, and CEPP National Policy Managers Federal Facilities Leadership Council and Brownfields Coordinators, Re: Initiatives to promote innovative technology in waste management programs (April 29, 1996).

EPA. 1997. Environmental Technology Verification Program verification strategy. EPA/600/K-96/003. Washington, DC: EPA Office of Research and Development.

EPA. 1998. Promotion of innovative technologies in waste management programs. OSWER Policy Directive 9380.0-25. Washington, DC: EPA OSWER.

EPA. 2000a. Innovative remediation technologies: field-scale demonstration projects in North America (2nd edition). EPA 542-F-00-001. Washington, DC: EPA Technology Innovation Office.

EPA. 2000b. Memorandum from E. Davies, Acting Director, Office of Emergency and Remedial Response, and Walter Kovalick, Director, Technology Innovation Office, to Superfund National Policy Managers, Regions 1–10, Re: Superfund reform strategy, implementation memorandum: optimization of fund-lead ground water pump and treat systems (October 31, 2000).

EPA. 2000c. Analysis of barriers to innovative treatment technologies: summary of existing studies and current initiatives. EPA 542-B-00-003. Washington, DC: EPA Office of Solid Waste and Emergency Response.

EPA. 2001. Letter from Timothy Fields, Assistant Administrator for Solid Waste and Emergency Response to EPA's Science Advisory Board, Re: SAB's September 26, 2000 recommendations for overcoming barriers to waste utilization.

EPA. 2002. EPA's Environmental Technology Verification web page. Available at: www.epa.gov/etv.

Gavaskar, A., B. Sass, N. Gupta, E. Drescher, W.-S. Yoon, J. Sminchak, and J. Hicks. 2001. Evaluating the longevity and hydraulic performance of permeable reactive barriers at Department of Defense sites. Draft final report for NFESCE.

Grosso, N. R., L. P. Leitzinger, and C. Bartlett. 1999. Site characterization of

Area 6, Dover Air Force Base, in support of natural attenuation and enhanced anaerobic bioremediation projects. NTIS Number: PB99-166456/XAB. EPA 600-R-99-044.

Hesslein, R. H. 1976. An in situ sampler for close interval pore water studies. Limnology and Oceanography 21:912–914.

Khire, M. V., C. H. Benson, and P. J. Bosscher. 1997. Water balance modeling of earthen covers. Journal Geotechnical and Geoenvironmental Engineering ASCE 123(8):744–754.

Licht, L., E. Aitchison, W. Schnabel, M. English, and M. Kaempf. 2001. Landfill capping with woodland ecosystems. Practice Periodical of Hazardous, Toxic, and Radioactive Waste Management ASCE 4(5):175–184.

Lorah, M. M., and J. S. Clark. 1996. Contamination of ground water, surface water, and soil, and evaluation of selected ground-water pumping alternatives in the Canal Creek area of Aberdeen Proving Ground, Maryland. U.S. Geological Survey Open-File Report 95-282.

Lorah, M. M., L. D. Olsen, B. L. Smith, M. A. Johnson, and W. B. Fleck. 1997. Natural attenuation of chlorinated volatile organic compounds in a freshwater tidal wetland, Aberdeen Proving Ground, Maryland. U.S. Geological Survey Water Resources Investigations Report 97-4171.

Lorah, M. M., and L. D. Olsen. 1999a. Degradation of 1,1,2,2-tetrachloroethane in a freshwater tidal wetland: field and laboratory evidence. Environ. Sci. Technol. 33:227–234.

Lorah, M. M., and L. D. Olsen. 1999b. Natural attenuation of chlorinated volatile organic compounds in a freshwater tidal wetland: field evidence of anaerobic biodegradation. Water Resources Research 35(12):3811–3827.

Lorah, M. M., L. D. Olsen, D. G. Capone, and J. E. Baker. 2001. Biodegradation of trichloroethylene and its anaerobic daughter products in freshwater wetland sediments. Bioremediation Journal 5(2):101–118.

Lorah, M. M., D. R. Burris, and L. J. Dyer. 2002. Efficiency of natural attenuation of chlorinated solvents in two freshwater wetlands. In: Proceedings of the 2nd International Wetlands and Remediation Conference, September 5–6, 2001, Burlington, Vermont. Columbus, OH: Battelle Press.

Melchior, S. 1997. In situ studies on the performance of landfill caps. Pp. 365–373 In: Proceedings of the International Containment Technology Conference, St. Petersburg, Florida.

Naval Facilities Engineering Service Center (NFESC). 1998. Permeable Reactive Wall. TechData Sheet TDS-2047-ENV (Revised). Port Hueneme, CA: NFESC.

National Research Council (NRC). 1994. Alternatives for ground water cleanup. Washington, DC: National Academy Press.

NRC. 1997. Innovations in ground water and soil cleanup: from concept to commercialization. Washington, DC: National Academy Press.

NRC. 1999a. Environmental cleanup at Navy facilities: risk-based methods. Washington, DC: National Academy Press.

NRC. 1999b. Groundwater and soil cleanup: improving management of persistent contaminants, Washington, DC: National Academy Press.

NRC. 2000. Natural attenuation for groundwater remediation. Washington, DC: National Academy Press.

NRC. 2003. Bioavailability of contaminants in soils and sediments: processes, tools, and applications. Washington, DC: National Academy Press.

Presidential/Congressional Commission on Risk Assessment and Risk Management. 1997. Framework for environmental health risk management. Final Report, Volume 2: Risk assessment and risk management in regulatory decision-making. Washington, DC: U.S. Government Printing Office.

Quinn, J. J., M. C. Negri, R. R. Hinchman, L. M. Moos, J. B. Wozniak, and E. G. Gatliff. 2001. Predicting the effect of deep-rooted hybrid poplars on the groundwater flow system at a large-scale phytoremediation site. International Journal of Phytoremediation 3(1):41–60.

5
Technology Overview

The Navy requested an update of previous reviews of innovative technologies for cleanup of groundwater, soils, and sediment (NRC 1994, 1997a, 1999a, 2000). This chapter discusses a variety of innovative technologies the Navy might consider during adaptive site management (ASM), for example, for initial remedy selection, as replacements for existing remedies that have proved to be unsuccessful, or as additions to current remedies to better achieve cleanup goals or reduce cleanup time. Because the Navy defined sediment contamination and solvents and metals in soil and groundwater as its most pressing current problems, the focus is on these types of contamination and on applicable remedial technologies, including the concept of treatment trains designed to meet multiple goals for multiple contaminants. The emphasis is on those technologies showing the greatest promise, particularly those technologies being developed and evaluated by the Department of Defense (DoD) and by the U.S. Environmental Protection Agency (EPA) and its Technology Innovation Office in association with the Federal Remediation Technologies Roundtable (FRTR). Although petroleum hydrocarbon sites remain a significant problem because of their sheer number (as discussed in Chapter 1), they are not the focus of this chapter at the request of the Navy and because their cleanup is generally considered to be well understood.

Both *in situ* and *ex situ* technologies can be identified according to applicable contaminant groups. Using the FRTR grouping of contaminants (see Table 1-1), eight contaminants groups—halogenated and non-halogenated volatile organic compounds (VOCs), halogenated and non-halogenated semivolatile organic compounds (SVOCs), fuels, inorganics, radionuclides, and explosives—can be defined and linked to the treatment technologies listed in Table 5-1 in terms of both *in situ* and *ex situ* procedures. Contaminants and technologies germane to soils, sediments,

198

TABLE 5-1 Primary Treatment Technologies

In Situ Soil and Sediment	Ex Situ Soil and Sediment	In Situ Groundwater	Ex Situ Groundwater
Biosparging	Bioremediation— Composting	Aeration	Free Product Recovery
Bioventing		Air Sparging	
Bioremediation	Bioremediation— Land Treatment	Bioremediation	Pump and Treat with:
Capping	Bioremediation— Slurry Phase	Bioslurping	Air Stripping Bioreactors
Chemical Reduction/Oxidation		Chemical Reduction/Oxidation	Carbon Adsorption
	Chemical Reduction/Oxidation		
Dual-Phase Extraction		Circulating Wells	Chemical Reduction/Oxidation
Dynamic Underground Stripping	Contained Recovery of Oily Waste	Cosolvent Flushing	
		Dual-Phase Extraction	Chemical Treatment
Electrokinetics	Critical Fluid Extraction	Dynamic Underground Stripping	Distillation
Hot Air Injection	Cyanide Oxidation		Electrochemical Treatment
Heating		Electrokinetics	
Phytoremediation	Dehalogenation	Hot Water/Steam Flushing/Stripping	Filtration
Soil Flushing (in situ)	Hydraulic Dredging		Precipitation
Soil Vapor Extraction	Incineration (offsite)	Monitored Natural Attenuation	Reverse Osmosis
	Incineration (onsite)		Solar Detoxification
Solidification/ Stabilization	Landfill Disposal	Permeable Reactive Barrier	Solvent Extraction
Steam Extraction	Mechanical Dredging	Phytoremediation	Supercritical Water Oxidation
Thermally Enhanced Recovery (e.g., EM, in situ RF, ISTD)	Physical Separation	Surfactants/Surfactant Flushing	UV/Oxidation
	Plasma High-temperature Metals Recovery	Vertical Barrier Wall	
Vitrification			
	Pyrolysis		
	Solar Detoxification		
	Soil Washing		
	Solidification/ Stabilization		
	Solvent Extraction		
	Thermal Desorption		
	Vitrification		

SOURCE: Adapted from EPA (1997a).

and groundwater can be further categorized according to the purpose of the technology and its relative maturity. Accordingly, as indicated in Table 5-2, screening of potential technologies can be facilitated to assist remedial project managers (RPMs) in selecting a remedial alternative. Each technology is defined in a glossary at the end of this chapter.

Key reference information useful for identifying and selecting technologies and combinations of technologies responsive to cleanup needs has been consolidated into a matrix published elsewhere (EPA, 1997a; http://www.frtr.gov). Other sources of information include technology-specific fact sheets produced by a joint effort between the Department of Energy (DOE) and the Air Force Base Conversion Agency (as well as those from other federal agencies). The objective of these fact sheets is to provide RPMs with information on optimizing cleanup technologies, on presenting multiple lines of evidence about remedy performance, on preparing five-year reviews, on operating remedy demonstrations, and on communicating progress to the public. The FRTR website maintains a database of many remediation technologies, their applications, conditions of use, performance data, and cost (although it is not comprehensive). This database would be even more useful if universities, states, and the private sector were encouraged to submit additional information where appropriate. The lack of a central, comprehensive database is likely to hamper the data analysis exercises (see Chapter 3) that characterize full-scale ASM. In addition, federal facility database systems are aligned to measure progress of the cleanup *process* (see Figures 1-1 and 1-2) versus measuring cleanup *performance*—an approach to data collection and analysis that will need to shift in order for ASM to be successfully implemented.

Although site conditions and contaminant sources limit the selection of applicable treatment technologies, most sites can be remediated by three primary strategies—destruction or alteration of contaminants, extraction or separation of contaminants from environmental media, and immobilization of contaminants. Currently, destruction technologies include both *in situ* and *ex situ* thermal, biological, and chemical methods. Extraction and separation technologies include thermal desorption, soil washing, solvent and vapor extraction for soils and sediments, and phase separation, adsorption, stripping, and/or ion exchange for groundwater. Immobilization technologies include stabilization, solidification, and containment. Generally, no single technology can remediate an entire site, and the use of treatment trains, sometimes combining *in situ* and *ex situ* techniques, is common, as discussed subsequently.

The main advantage of *in situ* treatment is that it allows remediation to occur without costly removal of the contaminant source. However, *in situ* treatment generally requires more time, and there is less certainty about attaining cleanup goals in terms of the extent and uniformity of treatment because of the usual heterogeneity of the source location and problems with treatment verification. In contrast to *in situ* treatment, the main advantage of *ex situ* treatment is that it generally requires shorter time periods to complete, and there is more certainty about the extent and uniformity of treatment. However, *ex situ* treatment incurs costly source excavation/removal and possible permitting and exposure implications. The control and proper disposition of emissions and residuals from *ex situ* treatment are important considerations that require compliance with permit conditions and the application of best management practices associated with each technology or combination of technologies. It should be noted that disposal actions may also be necessary for such *in situ* technologies as permeable reactive barriers and phytoremediation. Further discussion of this issue for individual technologies can be found in the references provided in Table 5-2.

Beyond considering the potential advantages and disadvantages of *in situ* and *ex situ* technologies, an important consideration in the evaluation of a remedy is the physical/chemical properties and the behavior of the contaminant and its source. For instance, subsurface contamination by nonhalogenated or halogenated VOCs potentially exists in four phases: (1) as vapors in the unsaturated zone, (2) as compounds sorbed on soil particles in both saturated and unsaturated zones, (3) as contaminants dissolved into pore water according to their solubility in both saturated and unsaturated zones, and (4) as a nonaqueous phase liquid (NAPL). The preferred remediation may involve a treatment train approach (e.g., air sparging/soil vapor extraction, liquid-phase carbon adsorption, and catalytic oxidation for nonhalogenated VOCs, or groundwater pumping, activated carbon adsorption with adsorbate reinjection, and offsite disposal of spent activated carbon for halogenated VOCs). In the case of soils or sediments, vapor extraction, thermal desorption, and incineration exemplify a corresponding treatment train.

A similar scenario could be developed for nonhalogenated or halogenated SVOCs. They can occur in the subsurface as vapors in the saturated zone, as contaminants sorbed or partitioned onto the soil or aquifer material in both the saturated and unsaturated zones or on sediments, as contaminants dissolved into pore water in both saturated and unsaturated zones, and as NAPLs. Common *ex situ* treatment technologies for SVOCs in groundwater include carbon adsorption and UV oxidation. In

TABLE 5-2 Candidate Technologies for Soil, Sediment, and Groundwater Remediation

Technology[a]	Purpose[b]						Target Contaminants[c]					
	a	b	c	d	e	f	a	b	c	d	e	f
In situ Soil and Sediment Remediation												
Bioventing	X						X				X	
Capping			X	X					X	X	X	X
Chemical oxidation/reduction	X			X					X	X		X
In situ heating	X			X	X				X	X	X	
Phytoremediation	X			X			X	X	X	X		X
Soil flushing	X			X		X	X	X				X
Soil vapor extraction	X			X		X	X	X			X	
Vitrification			X		X		X	X	X	X		X
Ex Situ Soil and Sediment Remediation												
Composting	X						X	X			X	
Confined aquatic disposal			X		X				X	X	X	X
Hydraulic dredging	X								X	X	X	X
Incineration	X						X	X	X	X		
Landfills	X		X		X		X	X	X	X		X
Land treatment	X		X				X	X			X	
Mechanical dredging	X								X	X	X	X
Slurry-phase bioremediation	X	X		X		X	X	X			X	
Soil washing	X			X		X			X	X	X	X
Solidification/stabilization			X		X							X
Thermal desorption	X			X		X	X	X	X	X	X	
Groundwater Remediation												
Air sparging	X	X		X		X	X	X			X	
Bioremediation	X	X		X		X	X	X		X		X
Bioslurping	X			X		X			X	X	X	
Circulating wells		X		X		X	X	X			X	
Cosolvents and surfactants	X			X		X			X	X	X	
Dual-phase extraction	X	X		X		X	X	X			X	
Dynamic underground stripping	X	X		X		X			X	X	X	
Chemical oxidation/reduction		X		X			X	X	X	X	X	X
Natural attenuation		X		X			X		X		X	
Permeable reactive barriers		X		X		X	X	X	X	X		X
Phytoremediation	X	X	X				X	X	X	X		X
Pump-and-treat	X	X					X	X	X	X	X	
Steam flushing	X			X		X	X	X	X	X		X
Vertical barrier walls			X		X		X	X	X	X	X	X

[a]See Glossary at end of this chapter
[b](a) Source conversion/removal, (b) plume remediation, (c) containment,
(d) remediation enhancement, (e) isolation, (f) pretreatment
[c](a) Nonhalogenated VOCs, (b) halogenated VOCs, (c) nonhalogenated SVOCs,
(d) halogenated SVOCs, (e) fuels, (f) inorganics, (g) radionuclides, (h) explosives
[d](a) Emerging, (b) innovative, (c) established/conventional
SOURCES: Adapted from FRTR (1997, 1998).

g	h	a	b	c	Relevant References
		Maturity[d]			
				X	AAEE, 1995, 1997; FRTR, 1998
X				X	EPA, 1994; Evanko and Dzombak, 1997; NRC, 1997b, 1999a; EPRI, 1999; McLellan and Hopman, 2000
		X			NRC, 1997a; EPA, 1998a
			X		Fountain, 1998; FRTR, 1998
	X		X		Schnoor, 1998; Fiorenze et al., 2000
			X		NRC, 1999a
				X	AAEE, 1997; FRTR, 1998; NRC, 1999a
X				X	AAEE, 1997; Evanko and Dzombak, 1997; NRC, 1999a
	X			X	AAEE, 1995, 1997
X			X		EPA, 1994; NRC, 1997b; EPRI, 1999; McLellan and Hopman, 2000
X				X	EPA, 1994; NRC, 1997b; EPRI, 1999; McLellan and Hopman, 2000
	X			X	AAEE, 1994, 1997; FRTR, 1998
X				X	AAEE, 1994, 1997; FRTR, 1998
				X	AAEE, 1995, 1997; FRTR, 1998
X				X	EPA, 1994; NRC, 1997b; EPRI, 1999; McLellan and Hopman, 2000
	X			X	AAEE, 1995, 1997
	X			X	AAEE, 1993, 1997; FRTR, 1998; NRC, 1999a
X				X	AAEE, 1994, 1997; Evanko and Dzombak, 1997
	X			X	AAEE, 1993, 1997; FRTR, 1998
				X	Miller, 1996a; NRC, 1999a
			X		AAEE, 1995, 1997; FRTR, 1998; NRC, 2000
				X	Miller, 1996b
			X		Miller and Roote, 1997
			X		Jafvert, 1996
			X		AAEE, 1997
			X		Fountain, 1998; Balshaw-Biddle et al., 2000; NRC, 1999a
X			X		EPA, 1998a; NRC, 1999a
			X		NRC, 2000
	X			X	Vidic and Pohland, 1996; EPA, 1998b
	X		X		EPA, 1999e; Schnoor, 1998; Schnoor, 2002
				X	FRTR, 1998; NRC, 1999a
				X	Fountain, 1998; NRC 1999a
				X	NRC, 1999a

soil and sediment, biodegradation, incineration, and excavation with off-site disposal are typical. Associated treatment trains may involve thermally enhanced soil vapor extraction followed by *in situ* bioremediation for nonhalogenated SVOCs, and excavation, *ex situ* dehalogenation, soil washing/dewatering and land application for halogenated SVOCs.

Inorganic contaminants such as metals may be found in the elemental form, but more often exist as salts mixed in soil or sediment. The fate of metals depends on their physical and chemical properties, the associated waste matrix, and the environmental phase within which they reside. The most common reservoirs for metals are soil and sediment, and the most common treatment technologies include solidification/stabilization, excavation and offsite disposal, and extraction. Depending upon solubility and mobilization potential, metals may also exist in groundwater, and are most frequently treated by *ex situ* precipitation, filtration, and ion exchange, although *in situ* treatment by oxidation/reduction and vitrification has occurred. A representative treatment train may be the combination of electrokinetics and phytoremediation.

OPTIMIZATION OF REMEDIES

Before discussing innovative technologies, it is worthwhile to consider the optimization of existing remedies to make them more efficient and effective. This process can utilize data and information from both routine monitoring conducted during remedy implementation as well as from evaluation and experimentation efforts to better define the site conditions. Periodically reevaluating the entire remedial design to determine whether it should be adjusted is critical because the remedial system is dynamic and will lead to changes in *in situ* conditions as the remedy is being implemented. As one would expect, optimization is more developed for technologies that have been in use for longer periods, like pump-and-treat.

Experiential Optimization

As discussed in Chapter 2, the term "optimization" is used here to mean any adjustment in a *single* remedy to make it more efficient or cost-effective to implement. To distinguish it from mathematical optimization, the report further defines "experiential optimization" to mean remedy adjustments such as eliminating redundancy, replacing overdesigned components with appropriately sized ones, or relocating or adding some components. In this approach, the technical staff responsible for operation of the remedial system evaluates all components of system design and determines, using engineering judgment, whether any components are redundant, overdesigned, or poorly located and whether additional components are needed. Table 5-3 summarizes examples of experiential optimization for a variety of remedial systems, including soil vapor extraction, air sparging, bioventing, bioslurping, *in situ* chemical oxidation, reactive permeable barriers, light nonaqueous phase liquid (LNAPL) free product recovery, dense nonaqueous phase liquid (DNAPL) removal and containment, groundwater extraction for hydraulic containment, groundwater extraction for mass removal, and groundwater monitoring. The table entries specifically address optimizing existing remedies and do not include changes to alternate remedies. Additional detail can be found in NAVFAC (2001). These examples demonstrate that a good deal of engineering judgment and expertise are required to implement the suggested schemes. Seventeen case studies mentioning the use of optimization in revising cleanup strategies can be found at the FRTR website (http://www.frtr.gov), although information is not provided on how the optimization was carried out.

Mathematical Optimization

In the peer-reviewed, archival literature, optimization of a remedial scheme is defined more restrictively to mean mathematical simulation of subsurface fluid flow and/or transport coupled with a linear, nonlinear, or dynamic programming algorithm to predict an optimal configuration or management of remedial system components. Formal mathematical optimization of any remedial system is theoretically possible but in practice has principally been applied to pump-and-treat systems.

EPA has recently begun to promote the use of formal mathematical optimization coupled with groundwater modeling for pump-and-treat applications as a potential means to save funds and energy (EPA, 1999a). EPA (1999a) presents a screening model that allows a user to make a rapid determination of whether additional expenditure on a mathematical optimization is worthwhile. In cases where many wells are pumping at a significant rate, where an optimal strategy is not obvious, or where the cost of additional wells is insignificant in comparison to the total amount currently being expended on pumping/energy costs, the screening model will usually indicate that an optimization exercise is worth pursuing. In a second volume (EPA, 1999b), EPA provides details of how mathematical optimization of a groundwater pump-and-treat situation can be accomplished. The user of the available software should have access to or should be able to construct a groundwater model of the site, and in addition be able to understand and implement the optimization algorithms suggested by EPA. The level of technical competence of the user is presumed to be relatively high.

Typically, pump-and-treat systems are designed based on experience and are adapted to site-specific conditions by carrying out field-scale pilot tests. To assist in the design process, users can use 2- or 3-dimensional numerical groundwater models (e.g., MODFLOW; McDonald and Harbaugh, 1996) to predict groundwater flow paths and hydraulic head distributions at a field site in response to imposed injection or withdrawal stresses, given that site lithology is adequately characterized in terms of spatially varying soil and rock permeabilities. This allows the user to answer questions regarding the number of wells to install and the effects of well placement and pumping rates on the movement of water through the saturated zone. It is possible to find an efficient design by simulating a number of combinations of well numbers, well placement, and injection or withdrawal rates to achieve either desired hydraulic containment or water removal.

However, the best design may not be found by such an iterative procedure. There are many possible combinations of design parameters, and identification of a best set of choices for test simulations may not be readily apparent for heterogeneous soils and complicated site boundary conditions. A more advanced level of design technology that builds on the numerical simulation approach is formal optimization of the design variables, where the best combination is found by mathematical techniques used in the field of operations research (e.g., Bradley et al., 1977; Gill et al., 1981). To optimize pump-and-treat design, mathematical programming algorithms can be coupled with a 2- or 3-dimensional groundwater flow model defining the physical system to determine the optimal set of design parameters for achieving pumping or injection objectives. This approach is the topic of EPA's recent set of reports (EPA, 1999a,b) and is also the subject of textbooks written within the last decade (e.g., Gorelick et al., 1993; Ahlfeld and Mulligan, 2000).

Optimization as a formal mathematical methodology that can be used to improve system performance has been in use for some time. Indeed, a literature review reveals that the concept of coupling simulation and optimization models dates back to 1958 (Lee and Aronofsky, 1958) and has been applied in the areas of petroleum and gas production, water supply,

TABLE 5-3 Summary of Experiential System Optimization of Certain Remedies

Technology	Component Evaluated for Optimization	Recommended Action	Justification
Soil Vapor Extraction	Characterization of subsurface heterogeneity	Check for level of detail of characterization	Improved detail will aid in better placement of extraction well screens
	3D distribution of vapor monitoring probes	Check for adequate number of vapor monitoring points	Improved placement/numbers will aid in determining adequacy of (1) volume of influence of vacuum system and (2) air flow velocities
	Flow rates at extraction wells	Determine mass removal from each well; decrease flow from unproductive wells and increase flow to more contaminated areas	Improve distribution of total energy used for vacuum application
	Continued high contaminant concentration in vapor	Check for unidentified or uncontrolled source areas	Presence of continuing source area will extend cleanup times
	Economics of aboveground vapor treatment system	Check treatment efficiency	Lower vapor concentrations may cause change in existing treatment efficiency; switching of treatment technology as vapor concentrations get lower could generate cost savings

Continued

TABLE 5-3 Continued

Technology	Component Evaluated for Optimization	Recommended Action	Justification
Soil Vapor Extraction (con't.)	Location/activity of extraction wells	Conduct equilibrium tests by shutting off all wells for 3–6 weeks	Rebounding will occur in hot spots; focus additional contaminant removal on these locations
	Vertical location of extraction intervals	Vertical profile testing to determine air flow rates and contaminant concentration with depth	Determine location of unproductive screened intervals that can be packed off; also want to avoid extracting water from wells that are too close to water table
Air Sparging	Zone of influence	Check for design zone of influence. If not being achieved, increase air flow to injection wells or install additional wells outside current zone of influence; evaluate system for short-circuiting	Design zone of influence needs to be achieved to attain cleanup goals
	Increasingly high injection pressures required to maintain flow	Check wells for plugging; redevelop or replace affected wells	Cleanup will not be achieved or will be delayed if injection wells are plugged.
	Control of sparging vapors	May need to install SVE system	Need to keep sparging vapors from migrating to undesirable areas

	Parameter	Action	Rationale
	Slope of contaminant concentration as a function of time	Check for target slopes; if slopes are too shallow, increase airflow to injection wells; install additional wells; evaluate system for short-circuiting; identify uncontrolled source area; evaluate alternative technologies	Desire to reduce cleanup times
	Asymptotic contaminant concentrations due to desorption or diffusion limitations	Pulse injection wells, install additional wells in contaminated areas, or evaluate alternative technologies	Desire to reduce cleanup times
Bioventing	Percent oxygen in soil gas	Adjust air flow and blower pressure to achieve at least 5% oxygen in soil gas	Permeability of soil dictates combination of pressure and air flow required to force air into the pore space
	Excessive air flow	Reduce air flow until oxygen is between 5% and 15%	Can achieve energy efficiencies by replacing oversized blowers with properly sized blowers
	When to stop clean up	In-situ respiration testing	Measured rate of biodegradation is indicator of low hydrocarbon supply

Continued

TABLE 5-3 Continued

Technology	Component Evaluated for Optimization	Recommended Action	Justification
Bioventing (con't.)	Radius of influence	Check to see whether radius is as designed. Increase air flow, install additional wells, evaluate for short-circuiting	Desire to achieve design radius of influence to effect desired cleanup
	High contaminant concentrations	Excavate hot spots or evaluate alternative technology	Concentrations may be too high for microbial activity to be effective
LNAPL Free Product Recovery	Recovery options	Conduct pilot baildown tests, limited pump down tests, and vacuum-enhanced recovery tests	Free product recovery is usually on the order of not more than 10%
	Declining recovery rate	Check to see if well screens are clogged	Lower recovery will extend cleanup times
	Ratio of fuel to water pumped	Check placement of pumps in wells; check to see if pumping rate is greater than necessary	If ratio is too low, recovery time will be extended
	Radius of influence or containment of free product	Increase pumping rates or install downgradient interceptor trenches	Incomplete containment of free product will increase cleanup times
DNAPL Containment	Detail of site characterization	Tightly spaced soil borings; partitioning tracer test	Guidance for locating DNAPL

Bioslurping	Declining recovery rate over time	Check for biological fouling or mineral buildup at well screen	Lower recovery rate will extend recovery time and increase costs
	Design recovery rate never achieved	Check well development and well screen locations	Inability to meet design recovery rate will extend recovery time and increase costs
	Location of suction tubes	Check to make sure suction tubes intersect free product	Tubes located above the water table will cause groundwater mounding; tubes too far below the free product will expend energy pumping excessive groundwater
	Vacuum rate	If vacuum rate is below design rate, check for short-circuiting and proper sizing of vacuum pump	Operation below design rate will reduce the radius of influence and extend cleanup times
	Migration of free product	Check on adequate location and number of recovery wells	Desire to prevent free product migration
Permeable Reactive Barriers	Location of monitoring wells	Need wells upgradient, downgradient, laterally, and within reactive barrier	Desire for accurate evaluation of system performance
	Breach of reactive barrier	Upgrade or reinstall barrier; consider alternative technology; grout any leaks between barrier and funnel walls	Desire to contain/treat contamination

Continued

TABLE 5-3 Continued

Technology	Component Evaluated for Optimization	Recommended Action	Justification
In situ Chemical Oxidation	Radius of influence	Check against design radius of influence. If radius of influence is below design radius, refine permeability characterization; reassess injection volume of reagent	Permeability may be too low for reagent to effectively reach contaminant' inadequate injection volume of reagent will result in incomplete oxidation
	Chemical concentrations remaining after treatment	If chemical removal is incomplete or rebounds, check on well locations, volume of chemical reagent, refinement of site characterization, chemistry of aquifer material	Desire to attain complete oxidation reaction by having all reactant reach contamination and by having minimal interference by reactions with aquifer material
Groundwater Extraction for Hydraulic Containment	Mapping of dissolved phase	Check for level of detail of characterization; utilize direct push probes and discrete sampling for additional detail	Improved level of detail will aid in better placement of extraction well screens
	Source controls	Possible addition of source-control well, in situ chemical destruction, or in situ barriers or treatment walls	Without removal of source, rates of mass removal will become asymptotic; with source control, volume of water pumped in downgradient areas may be able to be reduced
		Evaluate potential for natural attenuation	Other source control or mass removal may not be necessary

Location of extraction wells, total pumping rates	Mathematical optimization	Identify better combinations of pumping locations, rates, and schedules. May be able to achieve objective of plume containment with lower than maximum pumping rates
Well design	Evaluation of well design, construction techniques, well materials	Possible improvement of system efficiency; identify potential of well clogging if rates have decreased over time
Monitoring wells	Check for adequate number of monitoring wells	Evaluate whether entire plume is being contained
Groundwater Extraction for Mass Removal		
Extraction rates	Evaluate mass removal for each location	Decrease extraction rates at unproductive wells, increase extraction rates in more contaminated areas
Pumping rates	Check on whether contaminant removal is limited by chemical solubility or diffusion; possibly lower or cycle pumping rates	Pumping rates in solubility-limited and diffusion-limited systems may be too high and ineffective; cost savings can be realized by reducing pumping rates
Pumping rates	Check on whether design rates have been achieved	Failure to attain design rates may prevent plume containment
Location/activity of extraction wells	Complete equilibrium tests by shutting off wells for three months	Define hot spots where remediation efforts should be focused

Continued

TABLE 5-3 Continued

Technology	Component Evaluated for Optimization	Recommended Action	Justification
Groundwater Extraction for Mass Removal (Con't.)	Vertical location of extraction intervals	Complete vertical profile testing	Identify intervals containing greatest masses of recoverable contaminants; allow determination of unproductive intervals to pack off
	Map of dissolved phase	Check to make sure that plume is being contained while being removed; increase pumping rates as needed	Contaminant plume migration increases plume volume and possible receptor exposure.
	Location of extraction wells, total pumping rates	Mathematical optimization	Find better combinations of pumping location/rates/schedule to increase mass removal and/or decrease cleanup costs
	Above-ground treatment system	Evaluate for economic efficiency	As contaminant concentrations change, an alternate treatment system may be more cost-effective
	Above-ground treatment system	Evaluate for design treatment efficiency	Unit may not be operating properly and could be repaired
	Above-ground treatment system	Evaluate monitoring versus maintenance costs	Dollars spent monitoring maybe better suited to maintenance
	Above-ground treatment system	Evaluate pumps and blowers for overdesign	Potential cost savings as concentrations begin to decrease
	Above-ground treatment system	Evaluate cost of remote monitoring vs. onsite labor	Possible cost savings via remote monitoring

Groundwater Monitoring	Number of wells	Identify redundant wells for elimination (with regulator)	Potential cost savings
	Frequency of sampling	Evaluate appropriateness of sampling less frequently based on remediation progress	Potential cost savings
	Sampling and analytical protocols	Ensure that correct protocols are being applied to monitoring well samples	Potential cost savings if all monitoring wells are not required to undergo same protocols as point-of-compliance wells

SOURCES: Adapted from Air Force (2001) and NAVFAC (2001).p

wastewater injection, excavation dewatering, and hydraulic containment of groundwater contaminant plumes. The objective functions specified in problem formulations vary widely and have included, for example, maximizing profit, maximizing production, maximizing sum of hydraulic heads, maximizing total injection/withdrawal flow rates, minimizing costs, minimizing difference in desired versus actual production, and minimizing total injection/withdrawal flow rates. For the specified objective functions, decision variables have included flow rates at wells, head or pressure at wells, and well installation (binary or yes/no decision variables). Optimization algorithms that have either been proposed or actually used to solve these problems include linear, quadratic, nonlinear, and mixed linear-integer programs; some algorithms for solving certain optimization problems are widely available as commercial software packages (e.g., Murtaugh and Saunders, 1983, available from http://www.sbsi-sol-optimize.com/Minos.htm; Schrage, 1997, available from http://www.lindo.com).

EPA presents several case studies demonstrating that application of optimization to existing pump-and-treat well fields can save on the order of hundreds of thousands of dollars per site, depending on the objective. If, for example, the objective is plume containment, often it will be found that adequate hydraulic gradients toward the center of the plume can be maintained by reducing the pumping rates of the wells at the site, thereby reducing annual energy costs. In other cases, it can be shown that additional well placement and reevaluation of pumping and injection rates can also save additional dollars beyond the present scenario. A case study of mathematical optimization is presented in Box 5-1.

EPA notes that hydraulic modeling does not address mass removal or desired contaminant concentrations. To model such contaminant concentrations or masses, contaminant transport modeling must be coupled with optimization algorithms. This approach appeared in the literature over 15 years ago (Gorelick et al., 1984) and is now being pursued by EPA. Transport modeling is more complicated in that there are more parameters that need to be specified (dispersivities, sorption coefficients, biodegradation rates) and the process is nonlinear in contaminant concentration.

The principles discussed above can be applied to mathematical optimization of remediation of the vadose zone. An optimization handbook for soil vapor extraction is under development by EPA. A recent discussion of the mathematical approach to optimization of soil vapor extraction system design is provided by Sun et al. (1998).

One deficiency in the use of mathematical optimization not widely recognized is that rarely is the uncertainty in the predicted optimal scheme quantified. Aquifers are naturally heterogeneous such that the three-dimensional spatial variability of the rock and soil structure can never be known precisely. The uncertainty in the distribution of soil properties affects predictions of flow and transport. To address the issue of uncertainty in groundwater flow and transport modeling, statistical methods are used to generate a synthetic geologic structure between points of observed head/solute concentration, and often Monte Carlo methods are employed to evaluate equally likely realizations of geologic structure that obey the assumed underlying statistical pattern. In this way, the effect of the uncertainty of the inputs (soil/rock hydraulic conductivity distribution) on the outputs (hydraulic head and solute concentrations) is quantified. The practice of quantifying uncertainty in subsurface flow and transport modeling is virtually ignored in the literature on coupling flow and transport models with optimization algorithms for improving well placement/pumping rates. Inclusion of the consideration of uncertainty would provide a range of possible optimal scenarios instead of just one scenario.

There is no documentation indicating that the Navy has been using the mathematical optimization approach championed by EPA as a method of saving remediation costs for pump-and-treat scenarios. The Navy may wish to consider implementing mathematical optimization for improving the efficiency of pump-and-treat systems and ultimately saving hundreds of thousands of dollars in pumping costs. However, a high level of technical expertise is needed to (1) calibrate a groundwater model to existing site hydrogeology and (2) couple site-specific groundwater modeling results with the mathematical optimization tools available from EPA. This of course requires an investment in personnel resources. The Navy could consider utilizing EPA's screening methodology (EPA, 1999a) to decide whether a full-blown optimization effort would be economical to undertake. According to EPA, implementation of the screening model costs about $15,000.

***

At the current time, mathematical optimization is readily available only for pump-and-treat remediation schemes, such that experiential optimization will be needed for other remedies. Although few quantitative criteria are available for implementing experiential optimization, checklists provided by, for example, NAVFAC (2001) and the FRTR should be useful until a more complete database of experience is developed.

BOX 5-1
Mathematical Optimization of a
Groundwater Pump-and-Treat System
SOURCE: EPA (1999b).

Figure 5-1 illustrates contamination contours of a 1,2-dichloroethane (EDC) plume in a sand, clay, and gravel aquifer, beneath a site adjacent to a river in Kentucky. The saturated thickness of the aquifer varies from 100 feet at the southern border to 30–50 feet at the river. At the time of the study, a pump-and-treat system had been operating at the site since 1992. Twenty-three wells (18 from an original design plus five added subsequent to the original design, all labeled BW on Figure 5-1) had been installed principally for preventing migration of groundwater contaminants to the river, eight ("SW") wells were installed near the plume centers for the purpose of accelerating mass removal, and eight "OW" wells were installed to prevent plume migration to adjacent properties. The typical pumping rates for the three kinds of wells were 420–580 gallons per minute (gpm), 80–160 gpm, and 25–100 gpm, respectively. A range of pumping rates for each type of well reflects adjustments in the system to respond to variations in the water table elevation caused by variations in the river level.

EPA chose this site as a case study for illustrating the application of mathematical optimization because of the large number of existing wells in operation as well as the high annual expense of operation. Contaminants removed from the aquifer were being treated by steam stripping, and the treated water was discharged to the river. The cost of pumping and treatment by steam stripping was on the order of $1.8 million per year in 1999. A screening analysis by EPA (1999a) determined that it would be economically justifiable to expend funds ($40,000) to conduct groundwater modeling and optimization analysis of the current system to see if cleanup objectives could be attained at a lower cost by installing new wells and/or utilizing different pumping rates at existing wells. The screening analysis suggested that a change in pumping rates and/or in the number of wells pumped could save millions of dollars over the planning horizon (20 years), even if new wells costing $20,000 each were added to the system.

The goals of the hydraulic optimization were to evaluate the following: (1) the potential for reducing pumping rates at the BW wells with continued prevention of plume migration to the river, (2) the tradeoff between the total number of BW wells operating and the total pumping required for containment, (3) the total pumping required for containment with BW wells pumping only, (4) the pumping required for containment as a function of variation in the hydraulic head constraint required, and (5) the tradeoff between adding SW wells and reducing pumping rates at BW wells.

The code used to conduct the optimization was "MODMAN," consisting of the U.S. Geological Survey groundwater flow code MODFLOW (McDonald and Harbaugh, 1996) coupled with a linear programming algorithm LINDO (Schrage, 1997), to find the optimal set of pumping rates given the physical constraints of the system (EPA, 1999b). The mathematical objective function specified was minimization of the total sum of the pumping rates at the site, which is a surrogate for minimizing costs, since electricity usage is proportional to pumping rate. The annual steam stripping costs were equivalent to about $2000/gpm of water pumped. The physical characteristics of the hydrogeology are captured by first calibrating the groundwater flow code to the site, and subsequently determining site-specific aquifer responses to unit pumping rates at various locations, which are then built into the coefficients of the specified objective function. (This method of including the physical system characteristics as coefficients in the objective function is termed the matrix-response method, see Gorelick et al., 1993). Physical constraints that were mathematically defined included (1) hydraulic head at locations where hydraulic containment was desired, and (2) maximum desirable pumping rates at each well. In the case of the BW wells protecting the river, a hydraulic head constraint along a line between the river and the BW wells was specified, as shown in Figure 5-2 by the cross marks. The numerical value specified was 0.01 ft lower than the head of the river, in order to guarantee a solution that would contain a hydraulic gradient pointing toward the plume and away from the river at the desired locations.

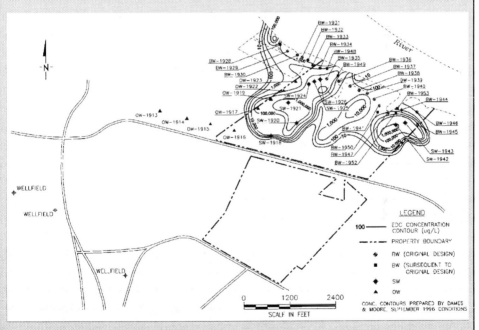

FIGURE 5-1 1,2-dichloroethane concentrations in September 1996 at a facility in Kentucky. SOURCE: EPA (1999b).

Continued

BOX 5-1 Continued

The results of the first two goals of the optimization runs are discussed here. A first set of runs examined whether the pumping rates of the original 18 BW wells could be reduced, holding the pumping rates of the SW and OW wells constant at their original design rates, to achieve the specified hydraulic constraint of 0.01 ft at the noted locations. The optimization algorithm found that only 17 of 18 BW wells were needed, and that the total pumping rate required at these wells to achieve hydraulic containment was 273 gpm instead of the original design total pumping rate of 549 gpm. This scenario resulted in a savings of $552,000 per year in operating costs. Further runs limiting the total number of wells allowed to operate (runs each with a maximum of 10–16 wells specified) indicated that as few as 14 wells could be pumped (275 gpm or a cost savings of $548,000 per year), with a more modest incremental savings as the number of wells was further limited to be as few as 10 (see Figure 5-3). Only when the number of wells was limited to nine was the solution found to be infeasible, that is, the constraints could not be met. If the optimization algorithm had been used in the design

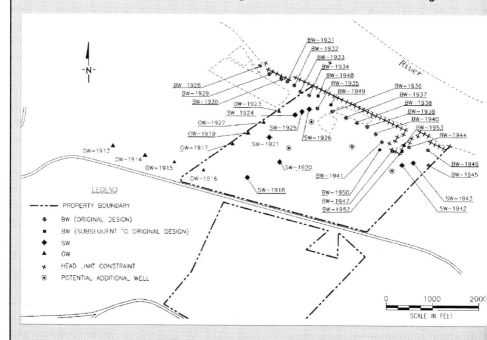

FIGURE 5-2 Hydraulic head constraint locations and potential additional well locations specified in the hydraulic optimization modeling. SOURCE: EPA (1999b).

mode before the wells had been installed, the designers would have found it optimal to install 14 instead of 18 wells to achieve the containment objective, thereby also saving an additional $80,000 in well installation costs for the original design ($180,000 total including the later modification that added five BW wells). This indicates the power of using optimization algorithms to infer information about the physical system that may not otherwise be obvious. Based on these illustrative cost savings, in the summer of 2000 EPA issued two directives requiring that all Superfund sites at which pump-and-treat remediation was being conducted be evaluated using optimization to assess potential cost savings (EPA, 2000a,b), although the emphasis of the guidance is on experiential optimization rather than modeling.

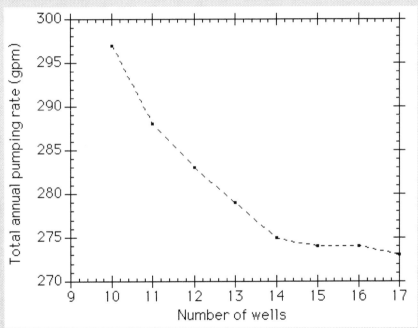

FIGURE 5-3 Total pumping rate versus maximum number of wells allowed to pump for the containment problem in the Kentucky case study. SOURCE: EPA (1999b).

PROMISING TECHNOLOGIES FOR THE NAVY'S PRESSING CONTAMINATION PROBLEMS

The following sections describe specific innovative technologies applicable to the types of contamination problems encountered at Navy and other federal facility sites. The discussion includes several technologies because there are more than just two or three that would suffice to cover all of the Navy's critical problem sites. The innovative technologies for treating solvents in soil and groundwater were chosen because they have recently garnered intense interest from potentially responsible parties (PRPs), including the Navy, and they have proven promising based upon previous applications. Thus, pump-and-treat and other conventional technologies are not included. On the other hand, a broad overview is given of technologies for treating metals sites and contaminated sediment sites that reflects the committee's professional experience regarding their potential use and efficacy.

Cost issues are not discussed in subsequent sections, primarily because complementary cost data for remediation technologies are not readily available for every type of application. However, a recent cost compendium has been prepared to include current information about the costs of bioremediation, thermal desorption, soil vapor extraction (SVE), onsite incineration, groundwater pump-and-treat, and permeable reactive barriers (PRBs) based upon about 150 projects (EPA, 2001a). The overall findings regarding remediation costs indicated that:

- correlations between unit costs and quantity of material treated or mass removed were evident for bioventing, thermal desorption, SVE, and pump-and-treat systems,
- economies of scale were observed for bioventing, thermal desorption, and SVE in that unit costs decreased as larger quantities of soil were treated,
- costs of technology applications are site-specific and thus are affected by many factors (e.g., properties, distribution, and concentrations of the contaminant; character of treated matrix and hydrogeological setting; market forces; maturity of technology; regulatory requirements; etc.), and
- some technologies (e.g., PRBs) could not be quantified with respect to cost due to lack of information concerning the longevity of the project, the contaminant quantity treated, and the mass of contaminant removed.

It will be important to the eventual success of ASM (particularly at management decision period [MDP] 3) to have greater understanding of the labor, utility, chemical, and disposal costs of different technologies. Presently, most financial data systems do not break down cleanup costs in this way, such that new budgeting requirements and formats will be needed to produce data that can support ASM.

Technologies for Remediation of Organic Contaminants in Soil and Groundwater

Recalcitrant organic contaminants are found at over 54 percent of all Navy facilities (NRC, 1999b), and are common contaminants at federal facilities in general. As discussed in Chapter 1, they pose significant challenges to site remediation, particular when found in karst and fractured rock environments. Three of the innovative technologies discussed below (*in situ* chemical oxidation, thermal treatment, and enhanced bioremediation) are broadly classified as *source removal* technologies because their goal is to reduce substantially the source term (be it solidbound, free-phase or dissolved contamination). *In situ* oxidation and thermal treatment in particular are noteworthy for reducing contaminant mass over a short timeframe. Barrier walls, in contrast, are effective primarily for contaminant plume treatment and control. To date, they have been developed for a limited number of organic compounds and metals.

In Situ Chemical Oxidation/Reduction

In situ chemical oxidation/reduction (ISO) is a groundwater remediation technology for toxic organic chemicals that has largely been used for source removal and control. The oxidants most commonly employed include peroxide, ozone, and permanganate. Hydrogen peroxide is capable of directly oxidizing organic contaminants—by free radical formation when ferrous iron is used as a catalyst (Fenton's Reagent). Fenton's Reagent oxidation is most effective under very acidic conditions, such that the need for pH adjustment is a disadvantage during the application of the technology. The advantages of peroxide include relatively low regulatory resistance, more field experience than for either ozone or permanganate, and a sparsity of byproducts of oxidation.

Ozone gas also can oxidize contaminants directly or through free radical formation, and it is the strongest viable chemical oxidant available. Because ozone is a gas, it is most suitable for treating the vadose zone, or possibly contaminant accumulations (e.g., LNAPL) in the capillary fringe. Like peroxide, ozone reactions are most effective in systems with acidic pH, where they proceed with extremely fast, pseudo first-order kinetics. Because of ozone's high reactivity and instability, it is produced onsite and requires closely spaced delivery points. *In situ* decomposition of the ozone can lead to beneficial oxygenation and biostimulation, and it is less costly than either peroxide or permanganate. However, because ozone must be applied as a gas, vapor recovery and possible treatment can add to the cost of the technology.

Permanganate is typically provided as a liquid or solid potassium or sodium salt that dissolves directly in the groundwater, and its reaction stoichiometry in natural systems is complex because of its multiple valence states and mineral forms. Depending on pH, the reaction can include direct oxidation or free radical enhanced oxidation. The reactions proceed at a somewhat slower rate than for peroxide or ozone according to second-order kinetics. Permanganate has a lower cost than peroxide and is effective over a broader pH range, and it is more stable than ozone. However, oxidation via permanganate also produces manganese oxide, which can precipitate and potentially cause reduced porosity. Increased dissolved manganese levels are also a regulatory concern, as is the purple color of groundwater containing unreacted permanganate.

The rate and extent of oxidation of a target contaminant are determined by the properties of the chemical itself and its susceptibility to oxidation as well as by the reaction matrix and its conditions (e.g., pH, temperature, oxidant concentration, other reduced compounds, minerals, and free radical scavengers). Generally, the technology is used on chlorinated volatile organic compounds (CVOCs) such as trichloroethylene (TCE) and on petroleum hydrocarbons. The method of oxidant delivery throughout the reaction matrix is of paramount importance; vertical and horizontal injection wells and sparge points with forced advection to rapidly move the oxidant, particularly for peroxide and ozone, into the subsurface are often deployed. Moreover, all three oxidation reactions (Box 5-2) can lead to (1) a decrease in pH if the system is not effectively buffered, (2) genesis of colloids with reduced permeability, (3) mobilization of redox-sensitive and exchangeable sorbed metals, (4) possible formation of toxic byproducts, (5) evolution of heat and gas, and (6) biological perturbations.

BOX 5-2
Simplified Stoichiometry for Oxidation of TCE by
Peroxide, Ozone, and Permanganate

Peroxide: $3H_2O_2 + C_2HCl_3 \rightarrow 2CO_2 + 2H_2O + 3HCl$

Ozone: $O_3 + C_2HCl_3 \rightarrow 2CO_2 + 3HCl$

Permanganate: $2KMnO_4 + C_2HCl_3 \rightarrow 2CO_2 + 2MnO_2 + 2KCl + HCl$

The stoichiometric relationships, like those shown in Box 5-2, can be used to estimate the amount of oxidant that would theoretically be needed to destroy the target contaminant. However, for site-specific oxidant demand estimates, bench-scale treatability tests based on soil slurry systems are often conducted to evaluate the feasibility of *in situ* oxidation and to calculate potential oxidant loading requirements. Results from slurry systems do not take into account preferential flows that are likely to occur in the subsurface, such that in reality, an excess of oxidant is often applied. Example bench-scale testing results are provided in Gates and Siegrist (1995).

Single, multiple, and continuous injections using recirculation of amended fluid have been used to apply the technology. For single or multiple injections, permanent or temporary injection points are used to deliver an aqueous solution containing the oxidant and any needed catalyst under pressure. The oxidant (and catalyst) concentration, the target pH, the injection well spacing (i.e., radius of influence), the number of injections, and the injection pressure are all important design parameters that can affect cost and performance. The use of recirculation, with injection and extraction wells, is intended to increase subsurface mixing and reaction opportunity, but the costs are likely to be higher. In addition, thin screen intervals at different depths more fully saturate the target zone and reduce the need for vertical migration of the oxidant. High injection pressures may be used to create fractures in tighter subsurface materials and thereby encourage migration and mixing of the reactants. Finally, in some cases, vapor extraction is used in conjunction with oxidation in the vadose zone to relieve off-gas pressure, to encourage oxidant migration, and/or to capture any volatile emissions (ESTCP, 1999). Despite these measures, it should be noted that *in situ* oxidation reagents, particularly Fenton's Reagent and ozone, are relatively short-lived compared to the rate of groundwater flow in most aquifers, such that oxidant contact with and treatment of contaminants is not significantly mediated by groundwater advection and oxidant dispersion. Table 5-4 provides example calculations of the minimum volume of injectant that must be delivered in an active form to achieve cleanup of different target treatment volumes. In summary, significant volumes of liquid oxidants may be required to treat relatively small areas.

Measuring Performance. Performance measurement should be based on multiple lines of evidence. Contaminant concentration changes over time and space are the most common and useful measurement. However, because contaminant concentration reductions can be caused by oxidation, simple displacement, and/or dilution effects, the measurement of geochemical indicators, tracers, and contaminant-destruction byproducts (e.g., chloride), as well as the use of control wells, should be considered. Geochemical indicators such as dissolved oxygen, redox potential, and conductivity (background chloride, etc.) provide an initial geochemical fingerprint that will change if the oxidant is delivered to a specific monitoring location. Tracer compounds that should be considered for the evaluation of oxidant distribution include both visual tracers and a semi-conservative dissolved tracer (i.e., Mn^{2+}, K^+, Na^+, etc.); bromide and iodide should be considered when applying liquid oxidants like peroxide. The release of halogenated ions, such as chloride or bromide, from target contaminants is a useful line of evidence if original contaminant concentrations are high enough to result in a significant increase in halogen ion concentrations as a result of contaminant destruction. All injection trials should include one or more control wells where water and tracer are injected into a contaminated zone in order to differentiate dissolved contaminant displacement or dilution from destruction.

TABLE 5-4 Volumes of Liquid Oxidant required to affect Target Radius of Influence

Assumed radius of influence (ft)	Target or injection well screened interval (ft)	Volume of aquifer affected (gal)	Required volume of injectant to achieve assumed radius of influence (gal)[a]	Approximate Number of injection wells/acre[b]	Approximate Total volume injectant/ acre (gal)[c]
10	10	23,500	5,900	140	826,000
10	20	47,000	11,800	140	1,652,000
20	20	188,000	47,000	35	1,645,000
50	20	1,170,000	294,000	6	1,764,000

[a]Entries equal Column 3 multiplied by an assumed porosity of 0.25.
[b]Number of wells per acre is approximated by dividing the surface area of an acre by the surface area coverage of a single well.
[c]Entries equal Column 4 multiplied by Column 5.

Although the theoretical stoichiometry of ISO is known, nontarget materials in the subsurface (e.g., natural organic acids, reduced iron and manganese, and sulfides) can all consume oxidant. Moreover, these sources can affect the potential for heat and off-gas generation or foaming and for rebound of contaminant levels caused by diffusion from untreated sources. Thus, data on the concentrations, masses, and fluxes of these materials in the treatment zone are essential to both rational design and measuring performance.

Because most subsurface environments are highly heterogeneous, the effectiveness of injection and/or reinjection needs to be evaluated both at the initial location and at possible new locations. The presence of target contaminants in lower-permeability layers that are adjacent to more permeable, preferential flow paths should be of special concern since oxidant delivery may be incomplete in lower-permeability zones. Thus, performance monitoring should be conducted in unique lower-permeability and/or high organic carbon layers. Likewise, monitoring for contaminant concentration rebound can guide the design of any subsequent remediation strategies by defining the remaining contaminant reservoir that was not treated by ISO. However, such information should be coupled with measurements of soil pore water chloride concentrations during injection to provide supporting evidence of dechlorination reactions and concomitant loss of the contaminant. For such a soil confirmation program to be useful, it needs to appropriately consider potential spatial and temporal variability of contaminant distribution, and recognize the associated mechanisms (e.g., chemical oxidation, volatilization/air stripping/gas phase partitioning, and dilution) of contaminant reduction.

Technology Evaluation. A recent status review of *in situ* oxidation (ESTCP, 1999) at 42 government (DoD and DOE) and private sites is summarized in Table 5-5. The review was conducted in two phases; phase I consisted of a survey of sites to identify where ISO had been used. The survey involved contacting ISO vendors and reviewing government (DoD, DOE, and EPA) databases and websites to determine current status of the project, scale, contaminants and media, responsible parties and regulators involved, extent of any available site data, and initial response indicating relative success or failure to satisfy facility-specific performance objectives. Accordingly, 19 sites were deemed successful, six failed, and 17 were uncertain. Of the 42 sites, 19 were partially or primarily contaminated with CVOCs, with TCE being the most prevalent

TABLE 5-5 Characteristics of *In Situ* Oxidation (Phase I) Field Sites

Characteristics	Number of Sites			
	DoD[a]	DOE	Private	Total
Contaminants				
CVOC	6	3	12	21
BTEX/TPH	6	-	13	19
Both	1	-	-	1
Unknown	1	-	-	1
Media Treated				
Soil only	0	0	0	0
Groundwater	2	0	17	19
Both	10	3	7	20
Unknown	2	-	1	3
Oxidant				
Hydrogen peroxide	12	1	24	37
Potassium permanganate	1	2	1	4
Ozone	1	0	0	1
Vendor				
Geo-Cleanse	8	1	4	13
Clean-Ox	3	0	13	16
ISTEC	1	0	7	8
Other	2	2	1	5
Scale				
Pilot/Demo Only	9	3	15	27
Full Only	1	0	4	5
Both	4	0	6	10
Outcome[b]				
Success	5	3	11	19
Failure	6	0	0	6
Uncertain	3	0	14	17
Totals	14	3	25	42

SOURCE: ESTCP (1999).
[a]DoD Breakdown: Navy (NFESC) = 5; Corps of Engineers/Air Force = 7; Army (Base Contract) = 2
[b]Outcome determinations are relative terms based on available Phase I information provided by facility representative (e.g., direct comments or pilot-scale tests that led to full-scale operations). These terms denote the ISO technology's ability or lack thereof to satisfy facility-specific program performance objectives.

contaminant of concern. Hydrogen peroxide was used at 37 sites, potassium permanganate at four sites, and ozone at only one site. Five of the 42 were Navy sites.

The results of the Phase I survey were then used to select several sites for more detailed Phase II evaluation, consisting of a review of available site characteristics, design, and performance data to more fully investigate and understand the site conditions and characteristics, the reasons why ISO was selected, the design parameters and rationale, the cost and performance of ISO under real-world conditions, the reasons for success or failure of ISO to meet the project objectives, and the specified technological concern. The Naval Submarine Base at Kings Bay, Georgia (Box 5-3), and the Naval Air Station at Pensacola, Florida, received such detailed site profiling and evaluation of results. At both of these sites, natural attenuation appeared promising after ISO treatment. The location, area and contaminant of concern, regulatory driver, oxidant, scale, remediation objectives, ability to meet objectives, and follow-up actions for these and the other Phase II sites are summarized in Table C-1 in Appendix C.

Collectively, the experiences with ISO indicate varying degrees of success, largely based on the sufficiency of site characterization and technology deployment. Various key factors have been identified and relate to site characterization needs and design and operational issues. The success of ISO is dependent upon effective contact and mixing with target contaminants, compatible subsurface geochemistry, and the maintenance of sufficient oxidation capacity to overcome oxidant losses from nonspecific oxidation reactions (e.g., reactions with the aquifer matrix and spontaneous oxidant decomposition). Major unanswered issues regarding the technology include:

- the absence of a well-defined screening procedure to evaluate site-specific geochemical parameters for compatibility with ISO techniques,
- the lack of properly designed pilot-testing procedures,
- differentiation between dissolved contaminant displacement and dilution versus treatment,
- oxidant loss due to consumptive reactions with soils and naturally occurring organic and inorganic materials, and estimations of the amount of oxidant necessary to overcome these losses so as to achieve the desired contaminant destruction,
- effectiveness of ISO for dissolved versus sorbed contaminants,
- credible analyses of contaminant rebound effects, and
- compatibility with anaerobic biodegradation processes.

BOX 5-3
In Situ Oxidation for Remediation of Chlorinated Solvents in Soil and Groundwater at Kings Bay Naval Submarine Base, Georgia

The site under consideration is a leaking former sanitary landfill, under which a perchloroethylene (PCE) plume has developed that is 120 feet long by 40 feet wide, with a 30- to 40-foot horizon below ground surface (bgs). The plume is moving toward a residential area through sandy soils that have a hydraulic conductivity in the impacted zone of 30 ft/day. The PCE concentrations detected in the landfill source area were as high as 8,500 µg/L, with breakdown products TCE, dichloroethylene (DCE), and vinyl chloride detected at concentrations of more than 9,000 µg/L in groundwater.

The remediation strategy chosen was to conduct full-scale *in situ* chemical oxidation with Fenton's Reagent (Geo-Cleanse) at 50 percent peroxide and an equivalent volume of ferrous iron catalyst delivered by injection to the subsurface. A total of 44 injection wells (23 in Phase I; 21 in Phase II) were installed at both deep (40–42 ft bgs) and shallow (32–35 ft bgs) depths. Phase I included two injections of oxidants totaling 12,045 gallons (8,257 gallons November 2–21, 1998; 3,788 gallons February 8–14, 1999). Phase II included two additional injections totaling 11,247 gallons (8,283 gallons June 3–11, 1999; 2,964 gallons July 12–15, 1999). The estimated volume of groundwater treated during Phase I was 78,989 gallons (based on treatment volume of 11,778 cubic yards and a porosity of 22 percent). During both phases, the design injection rate of oxidant was 0.2–1 gpm, while air was injected at 3 cfm to disperse the catalysts.

Following the *in situ* oxidation treatment, total VOCs in the primary treatment area were reduced from 9,074 µg/L to 90 µg/L, a 99 percent reduction. Subsequent results have shown that concentrations have remained below 100 µg/L. The natural attenuation capacity of the aquifer is expected to polish residuals outside the source area that are present in concentrations of less than 100 µg/L. Modeling exercises are predicting plume collapse in five years, barring the existence of other source areas outside the primary treatment zone. Based on the apparent success of *in situ* oxidation, the existing pump-and-treat system was discontinued.

Accordingly, uncertainties that have emerged during the demonstration and applications have indicated a need to provide comprehensive information on several factors (ESTCP, 1999). First, there must be better delineation of the contaminant's location and extent and of its sorption potential, particularly for DNAPL accumulations. The degree of soil layering versus the distribution of contaminants is an important parameter to understand because the distribution of oxidants will be limited to more permeable soil horizons unless injection/distribution approaches are tightly controlled. Mass and volume estimates of total CVOCs before and after treatment are needed to determine the efficacy of natural attenuation, and to be able to estimate the injected fluid volumes of oxidant. Another requirement is vapor monitoring, including detection of potentially explosive vapors in the subsurface, to safeguard against possible health and safety hazards during treatment. The prior consideration of these factors in formulating and optimizing a remedial action plan will enhance the potential success of ISO applications at contaminated sites (NAVFAC, 2001).

Design considerations include determining the radius of influence of injection wells to ensure adequate contact and the enhancement of mixing to promote contact between oxidant and contaminant. It may be necessary to consider multiple injections into the same or preferably new locations to accommodate matrix heterogeneities and circumvent problems with the development of preferential flow paths and short-circuiting caused by plugging of flow paths. Comparisons of the estimated *in situ* half-life of the oxidant to the groundwater flow velocity will help determine whether natural or induced groundwater flow can significantly distribute the oxidant. Finally, it will be important to incorporate ISO into an overall site management strategy, particularly at DNAPL sites, where source removal or reduction can be complemented by more cost-effective residuals treatment (e.g., natural attenuation or sparging).

Thermal Treatment

There are three general methods that can be used to inject or apply heat to the subsurface to enhance remediation: injection of hot gases such as steam or air, hot water injection, and electrical resistance heating (Davis, 1997, 1998). Steam, hot air, and hot water injection rely on contact between the injected fluid and the contaminant. Steam injection will displace mobile contaminants in front of the steam as well as vaporize volatile residual contaminants, and therefore can recover volatile and semivolatile contaminants in both the liquid and vapor phases. Hot air injection has been used to recover contaminants only in the vapor phase, and it is applicable to water-soluble volatile and semivolatile organics. Because steam has a heat capacity approximately four times that of air and a heat of evaporation of more than 2,000 kJ/kg, steam is often preferred to enhance the recovery of volatile contaminants and oils in soils and aquifers. However, for contaminants that have a high solubility in water, residual contamination remains after steam injection, unlike with hot air injection. Hot water injection is applicable for contaminants in the nonaqueous liquid phase, which tend to have low volatility and very low solubility in water, and this is most effective when the nonaqueous phase is present in quantities greater than the residual saturation.

Electrical energy has been applied to the soil in the low frequency range used for electrical power, that is, electromagnetic (EM), alternating current (AC), or resistivity heating, as well as in the radio frequency (RF) range. In each case, electrical energy heats the soil, increases the volatility of contaminants, and may induce the groundwater to boil and form steam (Fountain, 1998). The contaminants are driven out of the source zone by a combination of volatilization and thermally induced vapor-phase transport. Hence, electrical heating is usually coupled with soil vapor extraction (SVE) or steam injection to recover the volatilized contaminants. DNAPLs will be volatilized if the soil is heated to near the contaminant boiling point; the contaminant may also be mobilized by a reduction in viscosity. For semivolatile organic contaminants, higher temperatures (300°–400°C) obtained using RF energy are required to achieve greater removal and transformation efficiencies.

Electrical heating has proved to be effective in sandy media, and it also has a greater potential than steam or hot water injection in less permeable media such as clays. The higher water content generally found in clay will aid in directing the electromagnetic energy to the clay and promotes both a faster heating rate and higher temperatures. RF heating, however, is limited to the unsaturated zone, and for contaminants trapped below the water table, dewatering would have to be conducted prior to electrical heating.

Each of these thermal treatment methods is applicable only to certain types of sites and contaminants. The permeability of the media, the amount and type of heterogeneity, the amount of sorption, and the solubility of the contaminant must all be considered. For example, electrical heating may be favored in low-permeability media and when there is significant heterogeneity. Hot air or RF heating may be more applicable for highly soluble contaminants where drying of the soil may be necessary, and higher temperatures and/or longer remediation times may be necessary when adsorption is significant. Figure 5-4 can be used to determine which of the techniques is most applicable in a given situation; in some cases, more than one technique may be applicable, such that the selected technology is often the least severe in terms of temperature and pressure requirements (Davis, 1997).

A second important point is that each of these thermal treatment

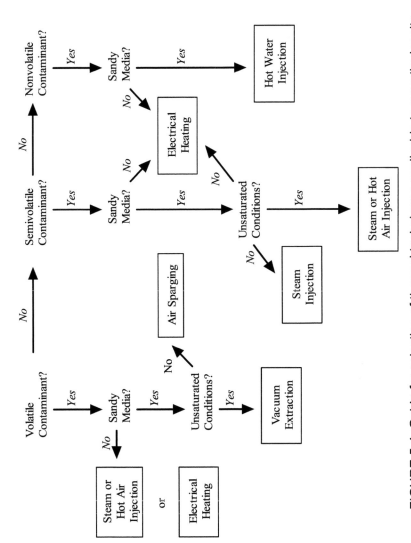

FIGURE 5-4 Guide for selection of thermal techniques applicable to a particular site.
SOURCE: Davis (1997).

methods is completely dependent on the capture effectiveness of the newly mobilized contaminant. With the possible exception of hot water flooding, all thermal remediation technologies require a highly effective soil vapor extraction system as the ultimate contaminant removal mechanism. The soil vapor capture system must be capable of overcoming condensate formation *in situ* and in above-ground equipment; it must be capable of fully capturing the "flash" volatilization of heated nonaqueous phase liquid; and, where applicable, it must be designed to effectively capture contaminants mobilized in the saturated zone. Groundwater extraction systems are often used in concert with soil vapor extraction systems at sites where contaminants are present in or adjacent to the saturated zone.

Measuring Performance. Primary remedial performance evaluation parameters for thermal treatment are media-specific contaminant concentrations before and after heating as well as mass removal versus time. The reliance on contaminant concentrations before and after treatment raises important questions regarding the defensibility of using standard groundwater and soil sampling and analysis techniques on "hot" samples. Techniques to cool or otherwise address uncontrolled contaminant loss from volatilization are being developed, but they have not been validated or widely applied. Rebound testing data are limited, but should only be considered valid if a sufficient period of time has elapsed since the return of the subsurface to ambient conditions. The elapsed time necessary for rebound effects to be exhibited is greater with (1) increased degree of soil layering, (2) greater degree of differences in hydraulic conductivities of distinct soil layers, (3) lower groundwater flow velocities, and (4) lower contaminant solubility or volatility.

Given that the success of thermal remediation technologies is directly dependent on capture system effectiveness, performance evaluation of the extraction system is critical. This evaluation should be completed using the standardized techniques for radius of influence or capture zone analysis (pressure profile for SVE—USACE, 1995; potentiometric surface for hot water). However, the evaluation of capture effectiveness of tracer compounds would provide a far more rigorous performance measure. Thus, the injection of an inert gas tracer (e.g., helium) into various areas of the extraction/heating array would provide useful data as to whether volatilized contaminants would be captured by the extraction system. Water-soluble conservative tracers (e.g., bromide and iodide) would verify that the flow path of injected water and its respective capture efficiency were acceptable. These techniques have seen limited application, and technical guidance regarding their appropriate use is sparse, if not nonexistent.

Temperature and pressure monitoring should be maximized since these monitoring networks are relatively cost-effective to install and are particularly informative and straightforward regarding the collection and analysis of these data.

Technology Evaluation. Of the thermal treatment options, steam injection, electrical resistance heating, 3- or 6-phase heating, and microwave and RF heating have been applied for *in situ* remediation of subsurface contamination. Selected steam injection projects are presented in Table C-2 in terms of target contaminants, treatment designs, and outcomes, and Table C-3 provides a compilation of full-scale and demonstration *in situ* thermal treatment projects. (Both tables are in Appendix C.) The *in situ* thermal desorption (ISTD) project at the Naval Facility, Ferndale, CA, is further described in Box 5-4. Given the newness of the

---

BOX 5-4
***In Situ* Thermal Desorption of Polychlorinated Biphenyls (PCBs) in Soil at Naval Facility, Ferndale, CA**
Source: Davis (1997)

The site of interest is on a 30-acre military base used for oceanographic research and undersea surveillance. There are approximately 1,000 cubic yards of PCB-contaminated silty and clayey colluvial soils under and adjacent to a former transformer/diesel generator building. Contamination underneath the building was 2–15 feet below ground surface (bgs), while PCBs adjacent to the building occurred 5–15 feet bgs. In this location, the depth to groundwater is greater than 60 feet. Concentrations of PCB-Aroclor 1254 were found to be 0.15–860 ppm, and PCDD/Fs was detected at levels up to 3.2 ppb 2,3,7,8-TCDD Toxicity Equivalents (TEQ).

The remediation strategy chosen at this site was based on TTEMI thermal well technology. This consists of heater-only and heater-vacuum wells installed at a depth of 15 feet in a hexagonal pattern with 6.0-foot spacing over an area 40 x 30 feet. The cleanup goal for PCB concentration was 1 ppm or lower; for dioxins and furans, the total concentration goal of 2,3,7,8-TCDD TEQ was less than 1.0 ppb. Remedial operation began on November 5, 1998, and ceased on January 15, 1999. Interim sampling was subsequently conducted, followed by a shutdown of soil heating on February 26, 1999.

Confirmation sampling to detect residual levels of contaminant was conducted in April 1999. This revealed that the target treatment area achieved remedial objectives for all samples. Additional contamination was identified outside of the thermal treatment zone because of the presence of unknown utility structures; this contamination was removed by excavation after limited thermal treatment.

technology and limited data on performance, it is too early to pronounce judgment of its overall and general efficacy.

Nonetheless, some general guidelines are suggested to improve the chances that thermal treatment is successful. First, it is important to adequately characterize the site with respect to the horizontal and vertical distribution of the contaminant, the heterogeneities of the medium, and the preferred flow paths (e.g., of the injected steam). This information is important for the design of the delivery and extraction systems and also for anticipating the monitoring and analysis requirements. The location and physical/chemical properties of the contaminant will determine the degree of solubilization, volatilization, and desorption and, hence, the removal opportunity. Unless adequately accounted for, subsurface heterogeneities may result in nonuniform heating and incomplete remediation.

With rare exception, all thermal technologies involve the production or transport of steam through the subsurface with the potential to volatilize contaminants, resulting in significant vapor production. The flow path of the steam and mobilized contaminant is determined by the relative permeability of soils. However, the relative permeability of soils and other properties (e.g., moisture and electrical resistance) can change dramatically over time *as a direct result* of thermal technologies. Condensate will be produced *in situ* at any location where steam contacts soil at a temperature that is lower that the boiling point of water or of the contaminant. The air permeability of soils is highly sensitive to degrees of water saturation. Thus, the soil vapor extraction network must be designed such that paths of vapor movement are made available even if condensate is formed *in situ* and air permeability is reduced in certain zones.

Unlike soil vapor capture efficiency, which can be negatively impacted by high water content, the performance of electrical resistance heating technologies can be negatively impacted by low water content. As soils dry out from heating and evaporation, the resistivity of the soils increases. An increase in resistance requires an increase in power input to maintain the original or desired heat input (Balshaw-Biddle et al., 2000). Thus, the application of electrical resistance heating in the vadose zone will likely require electrode irrigation systems and specialized electrode design. However, field experience to date has proved that electrode irrigation alone may not overcome soil power delivery problems because the power is consumed in boiling the irrigation water as opposed to heating of the subsurface at a distance from the electrode.

The compatibility of thermal remediation technologies with subsurface utility networks or underground wastes must be assessed in detail. Major considerations include subsurface utilities acting as conduits for mobilized contaminants, melting of polyvinyl chloride (PVC) and similar materials, vapor excursion to the surface, and spatial changes in soil conductivity (e.g., metal materials).

Barrier Walls

Barrier walls remove, transform, or otherwise prevent groundwater contaminants from migrating offsite. There are two general types of walls: nonreactive and reactive. Nonreactive barrier walls include slurry walls, sheet-pile walls, and grout walls whose primary function is to prevent offsite movement of contaminated groundwater. Although improvements in nonreactive barrier walls are being made, they are generally not considered innovative technology. Reactive barrier walls have been called *passive reactive barriers* or, more recently, *permeable reactive barriers (PRB)*. PRBs are used (1) to control migration of and to treat contaminated plumes, (2) to control migration of contaminants from source areas (followed downgradient, perhaps, by pump-and-treat or monitored natural attenuation), or (3) as a polishing step following other *in situ* technologies (e.g., flushing).

The general concept of a PRB is shown in Figure 5-5. In the most commonly applied approach, a trench is dug and backfilled with permeable, reactive material. Contaminated groundwater then naturally flows (termed *continuous wall*) or is made to flow using pumping and/or impermeable barriers that direct flow (termed *funnel and gate*) through the barrier where reactions occur. More recently, techniques have been developed that allow injection of these materials into the subsurface to create reactive barriers at depth.

Contaminant removal can take place by chemical reaction, sorption, precipitation, or biotransformation. This technology is based on reduction, in that highly oxidized contaminants are transformed to nontoxic or immobile products. The most common reactive material is zero-valent iron ($Fe(0)$), which will be the focus of this section. As the $Fe(0)$ corrodes, electrons are released that can be used to reduce highly oxidized contaminants (Box 5-5).

The range of organic and inorganic pollutants treated and reactive materials used in PRB are summarized in several recent reviews (Sacre, 1997; Gavaskar et al., 1998, 2000, 2001; EPA, 1998b; Scherer et al., 2000). These reviews also discuss advantages and limitations of the

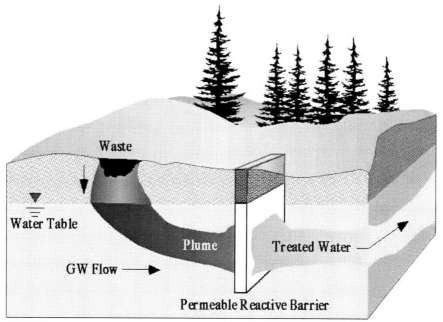

FIGURE 5-5 Permeable reactive barrier. SOURCE: EPA (1998b).

technology, design and implementation aspects, barrier emplacement methods, and principles of barrier media selection. Data from Scherer et al. (2000) summarize the general status of reactive barrier technology in terms of the contaminants treated and the materials used (see Table 5-6). Of 124 projects reviewed, 72 were laboratory studies, 26 were field demonstrations, 20 were commercial installations, and six were pilot studies (Sacre, 1997). Cr(VI) and halogenated aliphatics—primarily TCE—are the most common pollutants treated. There is documentation that reactive materials other than Fe(0) are being used in field installations. For example, sodium dithionite is being used to treat a Cr(VI) plume at Hanford, WA. Activated carbon is being used to remediate groundwater contaminated with a mixture of pesticides, xylene, and ethylbenzene. A mixture of municipal compost, leaf compost, and wood chips is being used to remove nickel, iron, and sulfate from mine-tailings contamination. These PRBs have been in operation for only a few years, and although they show promise, their long-term efficacy can not yet be ascertained. It has recently been suggested that bioaugmentation of Fe(0)-PRB may be advantageous for some contaminants (Weathers et al., 1997; Till et al., 1998; Scherer et al., 2000).

BOX 5-5
Reduction of Highly Oxidized Contaminants by Fe(0)

Removal of contaminants by Fe(0)-PRB is based on reduction. As Fe(0) corrodes, electrons are released that can be used to reduce highly oxidized contaminants such as trichloroethylene ($CHCl=CCl_2$) and hexavalent chromium (e.g., CrO_4^{2-}):

$$3Fe(0) \rightarrow 3Fe^{2+} + 6e^-$$
$$CHCl=CCl_2 + 3H^+ + 6e^- \rightarrow CH_2=CH_2 + 3Cl^-$$

$$3Fe(0) + CHCl=CCl_2 + 3H^+ \rightarrow 3Fe^{2+} + CH_2=CH_2 + 3Cl^-$$

$$1.5Fe(0) \rightarrow 1.5Fe^{2+} + 3e^-$$
$$CrO_4^{2-} + 8H^+ + 3e^- \rightarrow Cr^{3+} + 4H_2O$$

$$1.5Fe(0) + CrO_4^{2-} + 8H^+ \rightarrow 1.5Fe^{2+} + Cr^{3+} + 4H_2O$$

Trivalent Cr (Cr^{3+}) produced becomes immobilized as the solid $Cr(OH)_3$ within the barrier.

The electrons can also be used to reduce water-derived protons:

$$Fe(0) \rightarrow Fe^{2+} + 2e^-$$
$$2H_2O + 2e^- \rightarrow H_2 + 2OH^-$$

$$Fe(0) + 2H_2O \rightarrow Fe^{2+} + 2OH^- + H_2$$

Thus, treatment with Fe(0)-PRB will result in an increase in groundwater pH.

Measuring Performance. Performance of PRBs is typically assessed by measuring contaminant concentration (and potential products) upgradient and downgradient of the barrier. Other geochemical parameters of interest include pH (which increases dramatically across Fe(0)-PRB), dissolved oxygen, total dissolved solids in general and dissolved iron in particular. Measurements of hydraulic conductivity are also important because of the potential for clogging the barrier with mineral precipitates (e.g., $CaCO_3$, iron oxides and carbonates, etc.), hydrogen gas (produced from water during the corrosion of Fe(0)), and microbial growth. Ongoing monitoring should allow for determination of the adequacy of plume capture and of desired residence times (Gavaskar et al.,

TABLE 5-6 Organic and Inorganic Contaminants Treated with Permeable Reactive Barrier Technology, and Materials Used for PRBs (percentages are based on a total of 124 projects)

Organic Compounds Treated	%	Inorganics Treated	%	Materials Used for PRBs	%
TCE	26	Chromium	31	Zero-Valent Iron	45
DCE	13	Lead	11	Peat	6
PCE	12	Molybdenum	9	Zeolites	6
CCl$_4$	9	Arsenic	9	Lime	5
VC	6	Cadmium	9	Geochemical	5
TCA	5	Nitrate	7	Fixation	
Other Halogenated	4	Selenium	4	Ferric Oxyhydroxide	4
Organics		Nickel	3	Surfactant Modified	2
CHCl$_3$	4	Copper	3	Silicates	
Benzene	3	Vanadium	3	Zero-Valent Iron	2
Halogenated	2	Other	9	and Sulfur-	
Methanes				Containing	
DCA	2			Materials	
Toluene	2			Sawdust	2
Nitroaromatics	2			Microbes	2
PCBs	2			Chitosan Beads	2
Naphthalene	2			Hydrogen Sulfide	2
Nonhalogenated	2			Other	17
VOCs					
1,2,3-TCP	2				
CFC-113	2				

SOURCES: Scherer et al. (2000) and adapted from Sacre (1997).

1998, 2000, 2001). Water-level measurements, *in-situ* flow sensors, and other flow measurements should be used. PRBs are typically designed for a 20- to 30-year life.

As with most remediation technologies, a best-case scenario for process monitoring would be to complete mass and water balances. With PRBs, this is difficult at best. If the contaminant were, for example, TCE, the suite of daughter products of reduction (*cis*-DCE, vinyl chloride [VC], and ethane) could be measured. If the products are unknown (e.g., from carbon tetrachloride reduction) or are immobilized in the wall, mass balances are not possible. With Fe(0)-PRB, transformation is based on reduction. An electron balance is thus possible, in theory, by measuring Fe(II) entering and exiting the barrier. However, Fe(II) can be oxidized if O$_2$ is present, such that both Fe(II) and Fe(III) minerals are typically precipitated within the barrier (Phillips et al., 2000). Efforts are underway to better understand such phenomena (Liang et al., 2000).

Modeling the geochemical and hydraulic behavior of PRBs will be helpful in monitoring and predicting performance (Gavaskar et al., 2001; Morrison et al., 2001).

Technology Evaluation. Because field-scale application of the technology is only seven years old, design procedures and protocols are not yet well developed, although some guidance documents are available (e.g., Gavaskar et al., 1998, 2000; U.S. Air Force, 1997; NAVFAC, 2001). The current technology is limited in applicability to contaminated groundwater (i.e., it is not for soils and sediments). The most successful application is the use of Fe(0)-PRB for remediating groundwater contaminated with chlorinated ethenes. Some success has been reported for Cr(VI) reduction to Cr(III).

Several websites contain summaries of the current status of PRB installations (e.g., http://www.rtdf.org, http://clu-in.org, http://www.frtr.gov, http://www.gwrtac.org/, http://erb.nfesc.navy.mil/). EPA (2001b) indicates that PRBs are part of the remedial action at eight Superfund sites. There was also a recent updated review of some 38 full-scale Fe(0)-PRBs (Vidic, 2001) and a review of DoD installations. The Remediation Technologies Development Forum (RTDF) website and EPA (1999c) summarize a number of field-scale installations of PRBs. Chlorinated solvents, primarily the chlorinated ethenes PCE, TCE, *cis*-DCE, and VC, and 1,1,1-TCA, are being treated using Fe(0)-PRB at 12 of these installations. The oldest of these installations (Intersil Semiconductor Site, Sunnyvale, CA) has been in operation since 1995 and is briefly described in Box 5-6. Summaries of these full-scale installations indicate that most are working much as designed and are meeting treatment goals, at least for Fe(0)-PRB treating chlorinated solvents. Some success has been reported in the immobilization of Cr(VI) and U(VI) via reduction and precipitation.

From these different reviews, it is clear there are issues not yet resolved regarding long-term performance. These include but are not limited to, potential clogging due to chemical precipitation (some of which is caused by increased pH) or biological growth; competency of the confining layer beneath the PRB in preventing escape of contaminants under the PRB; deterioration of water quality downgradient of the PRB, including the release of incomplete reduction products (e.g., VC from TCE), high pH, and potentially high soluble Fe(II) levels; remobilization of chromium and uranium; the role of microbes in enhancing or reducing treatment effectiveness; longevity of the Fe(0) (i.e., when and how often it will have to be replaced); and hydraulic capture. Some of these issues

BOX 5-6
PRB Case Study: Intersil Semiconductor Site, Sunnyvale, CA

Intersil manufactured semiconductors at the site from the early 1970s until 1983 (http://www.rtdf.org/public/permbarr/; Gallinati et al., 1995; Warner et al., 1998). The primary contaminants resulting from this activity are TCE (50–200 µg/L), cis-DCE (450–1,000 µg/L), VC (100–500 µg/L), and Freon 113® (20–60 µg/L). Air stripping was used for remediation at the site until an Fe(0)-PRB was installed in 1996. The contaminated area, a semiconfined aquifer, is 2–4 feet thick. The lower aquitard is clay and silty clay. Low-permeability walls were installed to direct the flow to the PRB (Figure 5-6).

The PRB is 4 feet wide, 36 feet long, and 20 feet deep and is filled with 220 tons of granular iron to a depth of 11 feet. Installation cost $1 million. The cleanup goals are 5 µg/L for TCE, 6 µg/L for cis-DCE, 0.5 µg/L for VC, and 1,200 µg/L for Freon 113®. Since installation of the PRB, concentrations of these VOCs have been below the cleanup goals within the barrier. Some hydraulic mounding has occurred above the PRB, but it has not yet adversely affected performance. An unexpected benefit was observed as a result of placing a pea gravel zone upgradient of the PRB to aid in flow distribution. Some limited mixing of Fe(0) into this zone resulted in conditions favorable for some chemical precipitation of minerals and pretreatment of chlorinated solvents. It is possible that this will extend the life of the barrier itself.

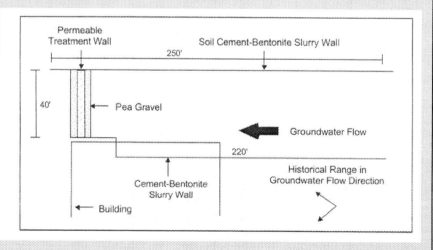

FIGURE 5-6 Fe(0)-PRB at the Intersil site. SOURCE: EPA (1998b).

were addressed in a recent report summarizing PRB performance at DoD sites, particularly Moffett Field (Box 5-7).

In summary, PRBs offer three potentially significant advantages: (1) conservation of water and energy, (2) low operation and maintenance costs, and (3) *in situ* application. However, capital costs may be high, only a few types of redox-sensitive pollutants are amenable to PRB

BOX 5-7
PRB Case Study: Moffett Field Naval Air Station, Mountain View, CA

A pilot-scale PRB facility has been operated at the Moffett Field Naval Air Station since 1996. A funnel and gate system treats a groundwater plume containing PCE, TCE, and *cis*-DCE as major contaminants. Steel sheet piles with interlocking joints make up the funnel and sides of the gate. Iron filings are the reactive media with pea gravel upgradient and downgradient from the filings. A recent report evaluated various aspects of longevity and hydraulic performance at Moffett Field, as well as at other DoD PRB installations (Gavaskar et al., 2001).

The report indicated that flow through the PRB was progressing as designed. After five years of operation, concentrations of PCE, TCE and *cis*-DCE were below their respective maximum contaminant levels. However, there has not yet been a front of clean water downgradient from the PRB, although there were signs that this would happen in the future. Several reasons were proposed, one being that the PRB was not tied into an impermeable layer and contaminated groundwater is leaking under and around the PRB. Analyses indicated that Fe(0) reactivity deteriorates with time, although it is still not possible to predict when the Fe(0) will need to be replaced. Analysis of hydraulic performance indicated an average residence time of nine days. The presence of an upgradient pea gravel zone helped to create a more uniform flow entering the PRB. Analysis for mineral precipitates indicated the presence of calcite, geothite, and some calcium-aluminum precipitates, although no discernable effect on flow velocities was reported. A cursory assessment of microbial activity indicated considerably less diversity downgradient of the PRB.

The major lesson learned from the assessment was that geochemical characterization of site groundwater is important. Because of the loss of Fe(0) reactivity, a thicker PRB is needed for groundwater with higher total dissolved solids (TDS>500–1,000 mg/L). Hydrogeologic modeling and monitoring (e.g., water level measurements) before and after installation should help assess hydraulic capture. The report indicates that when PRBs are located within plume boundaries, it is likely that some time will pass before downgradient contaminant concentrations will decrease (i.e., a "clean" front of groundwater appears). Finally, the report recommends additional research to help assess the rate at which Fe(0) loses reactivity and why this happens. Such information is required to estimate how much Fe(0) will be needed and how long it will be before Fe(0) needs to be replaced.

treatment (at least in its present form), immobilized contaminants may not be immobilized "forever," and there is a lack of long-term performance data for these systems. The technology currently is applicable only for contaminated groundwater remediation under appropriate geochemical and hydrologic conditions. The technology is most appropriately applied when plumes are less than about 1,000 feet wide and less than about 30 feet below ground surface (Gavaskar et al., 1998, 2000, 2001). The depth is limited because it is quite expensive with current technology to dig deeper than about 30 feet. However, developments in construction techniques (e.g., hydrofracturing, etc.) may overcome this limitation (Vidic, 2001). Proper design must ensure hydraulic capture of the contaminated plume (i.e., ensure that the plume does not pass over, under, or around the PRB). Thus, PRBs are most effective when they are keyed into an impermeable formation at depth (e.g., bedrock).

Enhanced Bioremediation

Several terms are currently used to describe the use of biological processes to remediate contaminated sites *in situ*. Enhanced bioremediation is taken to mean that enhancements are made to stimulate the growth of indigenous, subsurface microbes to increase the rate of contaminant removal or immobilization (e.g., Cr(VI)). The Navy currently uses enhanced bioremediation to describe the addition of oxygen and other nutrients and/or cometabolic substrates (e.g., carbon and energy source) to stimulate the growth of indigenous microbes and increase the rate of aerobic biodegradation (http://erb.nfesc.navy.mil). Enhanced bioremediation is to be distinguished from intrinsic biodegradation, where no additions are made to the contaminated site, but rather indigenous microbes are allowed to degrade and/or immobilize contaminants at the rate dictated by the *in situ* geochemical environment. As noted in a recent NRC report (NRC, 2000), the term being used today is "natural attenuation," where all naturally occurring processes that act to decrease the concentration and mass of a contaminant are included. (Natural attenuation will be discussed in a separate section.) Bioaugmentation involves adding specific organisms to the subsurface environment. Other *in situ* bioremediation processes include bioventing, biosparging, bioslurping, and air sparging.

The focus of this section is enhanced bioremediation, which typically involves adding nutrients to the subsurface environment to increase the rate at which contaminants are biodegraded by indigenous organisms.

Examples include carbon and energy sources such as sugars, lactate, volatile acids (e.g., acetic acid), complex materials such as molasses and vegetable oils, proprietary compounds that release molecular hydrogen slowly, and gaseous hydrogen; electron acceptors such as oxygen (applied as air, pure oxygen, hydrogen peroxide, or proprietary slow-oxygen-release compounds), nitrate, and sulfate; nutrients, primarily nitrogen and phosphorus; and perhaps buffers. In some cases, it may be advantageous to create anaerobic zones followed by aerobic zones to ensure more complete removal of contaminants and their daughter products. Several excellent reviews are available to describe the fundamentals of enhanced bioremediation and natural attenuation (NRC, 1993, 1997a, 2000; Rittmann and McCarty, 2001). Figure 5-7 provides a general schematic of the process.

Enhanced bioremediation has been reported to remove contaminants aerobically both as a primary substrate—for example, addition of oxygen and nutrients for degradation of petroleum hydrocarbons (Brown et al.,

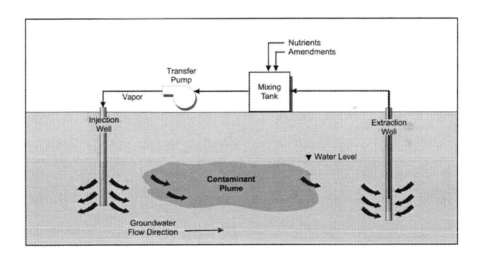

FIGURE 5-7 Schematic of enhanced bioremediation system. SOURCE: NAVFAC (2001).

1993)—and as a cometabolic substrate—for example, addition of toluene and oxygen for removal of TCE (McCarty et al., 1998). Similarly, enhanced bioremediation has been used to remove contaminants anaerobically as a primary substrate—for example, removal of PCE via dehalorespiration by addition of vegetable oil (Boulicault et al., 2000) or H_2 (Newell et al., 2000)—and as a cometabolic substrate—for example, addition of acetate for the removal of carbon tetrachloride in the presence of nitrate (Semprini et al., 1992). Although there are many reports of success with enhanced bioremediation in field demonstrations, few reports present the types of evidence described below as needed for conclusive proof that bioremediation is responsible for most of the contaminant removal. There are exceptions (e.g., Beeman et al., 1994; McCarty et al., 1998).

The Navy considers enhanced aerobic bioremediation of petroleum hydrocarbons to be "conventional" (i.e., established) treatment and enhanced aerobic cometabolism of chlorinated aliphatic hydrocarbons to be an "emerging" technology, which is in agreement with recent NRC reports (NRC, 1993, 2000). There is considerable current interest in adding electron donors to stimulate reductive dechlorination of chlorinated aliphatic hydrocarbons, primarily the chlorinated ethenes. The recent discovery of dehalorespiring bacteria that use chlorinated ethenes (e.g., PCE, TCE) as electron acceptors and H_2 as a preferred energy source in support of growth has led to intensive efforts to discover ways to deliver a slow, steady supply of H_2 to these organisms. Molecular H_2 can be delivered directly to the subsurface (Newell et al., 2000), or it can be produced via fermentation and anaerobic oxidation of a variety of substrates. Many substrates have been tried in the laboratory, and there are several reports of field tests and full-scale operations. Soluble substrates that have been tried include acetate, propionate, butyrate, lactate, benzoate, methanol, and simple sugars. Insoluble substrates, typically termed "slow-release compounds," include biomass, compost, molasses, edible oils, tetrabutyl orthosilicate, wood chips, and proprietary compounds (typically polymers that hydrolyze and dissolve slowly in water). One reason for using substrates that are fermented or oxidized to release H_2 is to minimize competition for H_2 among the dehalorespiring bacteria and methanogens and sulfate reducers. Dehalorespiring bacteria have been shown to outcompete other organisms for available H_2 when H_2 concentrations are low (Fennell and Gossett, 1998; Yang and McCarty, 1998). Soluble substrates such as benzoate, lactate, and propionate are useful for water recirculation systems (EPA, 2000c; Leigh et al., 2000; Yang and McCarty, 2000a). Insoluble (or slightly soluble) substrates are useful for more passive approaches that create "biologically active zones" (biobarriers) in the subsurface. Examples include the use of molasses (EPA, 2000c; Hansen et al., 2000), plant biomass (Haas et al., 2000), soybean oil (Boulicault et al., 2000), and proprietary compounds (Boyle et al., 2000).

It should also be noted that recent evidence indicates that source zones or near-source zones may be treated with enhanced anaerobic bioremediation (Carr et al., 2000; Yang and McCarty, 2000b). Dehalorespiring bacteria can enhance the dissolution of DNAPL.

Measuring Performance. Several recent reports (NRC, 1993, 2000; NAVFAC, 2001; Rittmann and McCarty, 2001) have outlined how performance of enhanced bioremediation should be measured. It is not a trivial undertaking for a number of reasons. For example, success is defined differently by the wide variety of parties involved. These include regulators, buyers of bioremediation, the public, researchers, and developers of bioremediation. No single measure is universally applicable to the wide variety of sites being addressed. Finally, contaminated sites are frequently heterogeneous, making it impossible to fully characterize them. Thus, it is difficult to conclusively prove the success of *in situ* bioremediation.

The key to evaluating success is to *directly* link observed loss of contamination with microbial activity. NRC (1993) recommends an approach relying on three types of evidence:

1. documented loss of contaminants from the site,
2. laboratory assays showing that microorganisms at the site have the *potential* to transform the contaminants under the expected site conditions, and
3. one or more pieces of evidence showing that the biodegradation potential is *actually realized* in the field.

It is this third type of evidence that is the most crucial and, unfortunately, the most difficult to obtain. Details describing the scientific bases for these measurements are included in NRC (1993) and updated in NRC (2000).

There are techniques that provide *principal evidence* and those that provide *confirmatory evidence* (Rittmann and McCarty, 2001). Principal evidence is that which is capable of proving success or failure—for example, stoichiometric consumption of electron acceptors, formation of inorganic-C that originated from organic-C, and/or increased degradation rates over time (quantification is very important). Confirmatory evidence is that which can support the proof of success, but its absence does not prove failure. Examples include an increase in protozoan population, the detection of intermediate metabolites, and an increase in the ratio of nondegradable to degradable components.

Technology Evaluation. Potential advantages of enhanced bioremediation are that it destroys contaminants (e.g., via mineralization or conversion to benign organics such as ethene), it can be applied *in situ*, and it is less expensive than other technologies. However, it remains a challenge to deliver the enhancements, monitor effectiveness, and demonstrate conclusively that biodegradation is responsible for contaminant removal. As currently employed for contaminated aquifers, the technology is best applied in situations where the hydraulic conductivity is sufficiently high (say greater than 10^{-3} cm/sec), residual NAPL is absent or has been removed, contaminants are not overly hydrophobic, and contamination is not too deep. At the present time, enhanced bioremediation has been shown to work best with petroleum hydrocarbons—primarily benzene, toluene, ethylbenzene, and xylene (BTEX)—in relatively simple hydrogeologic environments. Increasingly, amendments, primarily electron donors that release H_2, are being added to stimulate dehalorespiring bacteria. The Navy is particularly interested in the potential of slow-release compounds for this purpose.

Many slow-release compounds are low-cost; for example, molasses and edible oils may be quite inexpensive compared to the alternatives (Harkness, 2000). However, there are several concerns regarding the use of so-called slow-release compounds. Typically, less than about 20 percent (perhaps less than 5 percent) of the substrate may be used for dehalogenation (Yang and McCarty, 2000a), resulting in an excess of organic carbon (e.g., organic acids such as acetate, propionate, etc.) being available for transport with the groundwater or for degradation by other organisms. During fermentation, and depending on the geochemistry, alkalinity may be consumed, and the pH may decrease significantly. This may trigger other undesirable water quality changes (e.g., metal dissolution). Subsequent degradation of the organic acids will increase downgradient alkalinity and pH. Production of undesirable hydrogen sulfide from sulfate reduction may occur. There is the potential for the production of methane gas, which could decrease hydraulic conductivity, as could excess microbial growth. For some slow-release compounds (e.g., edible oils), some removal from the aqueous phase will result simply from partitioning of the chlorinated organics into the nonaqueous phase rather than from degradation. Although this should not affect long-term performance, it must be considered in the design, operation, and monitoring of such a system. Addition of slow-release compounds may simply displace the polluted groundwater, giving the illusion of success. It is very important to characterize the hydrogeology of the site to determine if the slow-release compounds can be distributed effectively (Hansen et al., 2000). If the site is aerobic, this technology may not be appropriate. Hansen et al. (2000) also point out the importance of increased biosurfactant production as a result of stimulating microbial growth in the subsurface. Such surfactants can solubilize contaminants and temporarily increase aqueous concentrations.

A case study of enhanced bioremediation is given in Box 5-8. It provides "proof-of-concept" evidence for enhanced anaerobic bioremediation. It also indicates that limited data are available on the effectiveness of this technology, and although results are promising, it is too soon to assess long-term cost and performance.

Several issues need to be resolved in order to declare enhanced anaerobic bioremediation a proven technology. First, it is clear that its major application is for bioremediation of PCE and TCE. Success will be determined by the complete conversion of these compounds into ethene; conversion to vinyl chloride is an unacceptable endpoint. Although the addition of slow-release electron donors can stimulate anaerobic cometabolism of chlorinated ethanes and chlorinated methanes, complete anaerobic conversion of these compounds to ethane and methane, respectively, has not been demonstrated. Using techniques described above, it must be demonstrated conclusively that biodegradation, not physical displacement or dilution, is responsible for decreases in contaminant concentration. To date, there has been no effort to assess the effect of the degradation of slow-release compounds on downgradient water quality (possible changes include decreased pH, increased dissolution of metals, and increased biological oxygen demand from volatile acids). Finally, the technology needs to be proven cost-effective.

Technologies for Remediation of
Inorganics in Soil and Groundwater

The most frequently occurring metal contaminants at Navy sites are lead, zinc, copper, nickel, barium, cadmium, vanadium, aluminum, and beryllium. These heavy metals and other inorganic contaminants (e.g., arsenic, cyanides, perchlorate, and radionuclides) pose a great challenge

BOX 5-8
Enhanced Bioremediation Case Study: Emeryville, CA

Enhanced anaerobic bioremediation is being used to remediate an abandoned manufacturing facility in Emeryville, CA (EPA, 2000c). Metal-plating operations resulted in contamination of the groundwater with degreasing solvents and metals. The primary pollutants of concern are TCE and Cr(VI). Monitoring in 1995 indicated PCE concentrations in the tens of $\mu g/L$, TCE concentrations of up to 17,000 $\mu g/L$, *cis*-DCE concentrations as high as 900 $\mu g/L$, and VC concentrations generally lower than 20 $\mu g/L$. Most of the data focus on the chlorinated ethenes, although Cr(VI) concentrations in excess of 100,000 $\mu g/L$ were also found at the site. The soil type is interbedded sand and clay units, the depth to groundwater is between 3.5 and 8 feet, and the groundwater velocity was 60 ft/yr. Monitoring data indicated that limited reductive dechlorination was occurring, and the rate was limited by lack of organic carbon (electron donor) and/or the environment was not sufficiently reducing. A pilot study was undertaken to establish a reactive anaerobic zone by adding a mixture of molasses, anaerobic digester supernatant, and tap water. The supernatant was added because preliminary analyses indicated low bacterial counts in the subsurface. The pilot study lasted approximately six months and indicated that an anaerobic zone could be created that would support reductive dechlorination of the organic compounds and the reduction of Cr(VI) to immobile Cr(III).

The full-scale system was installed and has been operating since April 1997. It consists of 91 injection points installed to a depth of 24 feet below ground surface using a Geoprobe™. During the first injection event, a mixture of 25 gallons of molasses, 1 gallon of supernatant, and 125 gallons of water was made. October 1998 data indicated that concentrations of PCE, TCE, *cis*-DCE, and VC near the source area fell below 5 $\mu g/L$. It was reported that Cr(VI) concentrations were reduced by approximately 99 percent, and that in some areas where historic concentrations were above 100,000 $\mu g/L$, concentrations are now below 5 $\mu g/L$. Data concerning organic carbon levels (fate of added molasses), pH, and other changes in geochemistry were not available. Thus, although it appears that molasses addition is stimulating conversion of TCE to ethene, long-term monitoring is needed to confirm lasting effectiveness.

for remedial efforts. Unlike many organics, chemical and biological transformations of inorganics can change the form of the contaminant, but cannot destroy it (Evanko and Dzombak, 1997; EPA, 1997b). Furthermore, the chemical form of the inorganic contaminant influences its solubility, mobility, and toxicity in the subsurface (EPA, 1997c). The form, or speciation, of inorganic contaminants depends on the source of the waste and the geochemistry of the subsurface at the site. For example, zinc usually occurs in the +2 oxidation state and tends to form soluble compounds at neutral and acid pH values (Evanko and Dzombak,

1997). Under these conditions, zinc can become a mobile metal in groundwater. At higher pHs, zinc can form carbonate and hydroxide complexes, which reduces its solubility. Furthermore, zinc readily precipitates under reducing conditions—for example, ZnS(s). The inability of transformation reactions to destroy inorganic contaminants and the influence of geochemistry on inorganic contaminant mobility present major challenges for remediation of sites containing inorganic contaminants.

Because inorganic contaminants cannot be destroyed, site management strategies focus either on containing the contaminants by decreasing compound mobility and toxicity, separating/extracting the contaminants from the subsurface, or using institutional controls to restrict human exposure to the contamination (NRC, 1997a; Evanko and Dzombak, 1997; EPA, 1997b). This discussion of remedial technologies for inorganics is confined to the former two approaches. Also, because the top nine inorganic contaminants at Navy sites are metals, this discussion will exclude remediation of nonmetal contaminants such as perchlorate, cyanides, and nitrate. Management of radioactive wastes is also excluded.

The listing of candidate technologies for soil, sediment, and groundwater remediation in Table 5-2 identifies 16 technologies for the remediation of sites contaminated with inorganics. Descriptions of these technologies appear on several web sites (e.g., http://clu-in.org, http://www.epareachit.org, http://www.frtr.gov, http://www.gwrtac.org, http://www.rtdf.org, and http://erb.nfesc.navy.mil/). Based on the strategy used to control the contamination, technologies for remediation of inorganics in the subsurface can be grouped into the following five categories:

1. **Excavation.** Contaminated materials are removed by digging and are transported to an offsite disposal facility.

2. **Containment.** Containment technologies attempt to prevent the transport of contaminants by isolating or solidifying them within a designated area. Examples are capping, subsurface barriers, solidification/stabilization, and vitrification.

3. **Toxicity and/or mobility reduction.** Chemical or biological reactions are used to alter the form (speciation) of metal contaminants in order to decrease their toxicity and/or mobility. Examples are chemical treatment, permeable reactive barriers, biological treatment, and phytoremediation.

4. **Physical separation.** *Ex situ* processes are used to selectively remove contaminated material from the soil usually for the purpose of reducing the amount of material requiring subsequent treatment. Examples are screening, classification, gravity concentration, froth flotation, and magnetic separation.

5. **Extraction.** The metal is removed from the rest of the soil or groundwater by *ex situ* or *in situ* techniques such as soil washing, soil flushing, and electrokinetic treatment.

Measuring Performance

As in all remediation technologies, documenting the success of technologies for control of inorganic contaminants requires evidence that the technology reduces risk by decreasing the mass, concentration, mobility, and/or toxicity of the contaminants and requires identification of the operative mechanism(s). The latter is needed to ensure that there is a cause-and-effect relationship between the implementation of the technology and the observed reduction in risk.

A complete mass balance both before and after remediation provides the most confidence in the assessed performance of a technology. Concentration data with monitoring well and soil core samples are typically used to determine contaminant mass. For stabilization and containment technologies, the most important mass balance information is to demonstrate immobilization of the contaminants. Furthermore, the integrity of the stabilized material must be determined for the site-specific groundwater flow and chemical conditions. For technologies that transform the inorganics to less harmful or less mobile species, monitoring must prove that the reaction processes are taking place. Here, the mass balance analysis must confirm the stoichiometry between the reactants and products. Finally, for technologies that rely on *in situ* or *ex situ* extraction, a mass balance analysis must be conducted to determine the contaminant extraction efficiency and to confirm that the mass extracted in the outflow stream is correlated with the mass removed from the subsurface.

Technology Evaluation

In the past, the typical remedy for inorganics-contaminated sites has been excavation, transport offsite, and burial at an approved disposal facility (dig-and-haul). Although newer *in situ* treatment technologies may soon replace excavation, there are nonetheless recent advances that have improved excavation and all *ex situ* remedies that depend on it—mainly by focusing exclusively on soils (or sediments) that do not meet cleanup standards. Traditional excavation activities for soils rely on work plans that specify the required excavation footprint based on existing characterization data. When excavation is complete, the final dig face is sampled to ensure that remaining soils attain cleanup goals. The characterization data used for excavation design are typically the product of remedial investigation/feasibility study (RI/FS) sampling work, and such data are often inadequate for accurately delineating contamination footprints. Evidence suggests that excavation activities at hazardous waste sites have resulted in significant overexcavation (Durham et al., 1999).

Precision excavation techniques, an alternative to traditional excavation approaches, attempt to limit excavation and subsequent remediation to only those soils that fail to attain cleanup standards. Precision excavation differs from traditional approaches in three key ways: (1) it makes broader use of data collection during the excavation process to provide *in situ* segregation of soils, (2) its work plans do not specify excavation footprints, but rather identify the decision-making process that will be used to screen soils as excavation work moves forward, and (3) its excavation work is designed in lifts or phases that allow for dig face screening before work proceeds. The feedback and self-adjustment mechanisms built into precision excavation programs are consistent with the adaptive management concepts described in this report.

The viability of a precise excavation approach for soil- or sediment-contaminated sites depends on the availability of rapid field analytical techniques appropriate for the contaminants of concern and their action levels. Chapter 3 discusses advances in field data collection technologies and analytical technologies in more detail and their pertinence to ASM. Within the RCRA program, EPA's SW-846 contains guidance on acceptable analytical techniques. The latest draft (EPA, 2000d) includes several additions pertinent to precise excavation and Navy contaminants of concern. These include portable x-ray fluorescence spectrometry systems for metals, and portable GC, calorimetric, and immunoassay technologies for addressing explosives, PCB, and PAH contamination in soils and sediments. Differentially corrected global positioning systems are capable of providing relatively accurate locational control for data collection efforts in real time.

Beyond excavation, there are several promising onsite and *in situ* technologies being developed for addressing contamination by inorganics. Many of these alternatives to dig-and-haul have been successful when tested in beakers at the laboratory scale or in small-scale pilot tests. Limited documentation exists on the performance of treatments for inorganics contamination at the field scale or at Navy sites. For example, only seven Navy sites with contamination by inorganics in soil, sludge, and sediment are included in the compilation by EPA (EPA, 2000e) of over 600 innovative remediation technology demonstration projects in North America. No Navy case studies were listed for groundwater contaminated by inorganics. The technologies employed include phytoremediation, soil washing, *ex situ* extraction, electrokinetics, solidification and stabilization, and *ex situ* physical separation/chemical treatment. A review of containment technologies, biological treatment technologies, and physical/chemical treatment technologies approved for use by the Navy (from http://erb.nfesc.navy.mil/restoration) identified case studies employing capping, biotransformation, constructed wetland, phytoremediation, electrokinetic extraction, and solidification/stabilization for control of contamination by inorganics. Several of these case studies involved small-scale experiments. Additional information on four of the most frequently cited technologies for inorganics contamination is presented below and more detail is found in NRC (1999a). Solidification/stabilization is applicable to a wide range of inorganic contaminants and site conditions. The latter three technologies—electrokinetics, phytoremediation, and chemical treatment—are much more restrictive for certain contaminants and conditions.

Solidification/Stabilization.   Solidification/stabilization refers to processes that encapsulate a waste to form a solid material or that involve chemical reactions that reduce the leachability of a waste. Examples include chemical additives (e.g., cements and polymers) and thermal fusing/glassification.   From FY1982 through FY1998, solidification/stabilization projects were the second most common type of source control treatment technology implemented at Superfund sites, representing 24 percent of all source control projects (EPA, 2000f). Solidification/stabilization projects were mainly implemented for metal contaminants.   Fifty-six percent of the applications were used to treat metals only, whereas 90 percent of the applications were used to treat metals alone or in combination with organics or radioactive metals. The major limitations of solidification/stabilization are uncertainty in long-term effectiveness, the need for long-term monitoring because untreated contaminants remain on the site, and questions about future site use with containment technologies in place (EPA, 1999d).

Electrokinetics. Electrokinetic remediation describes technologies that separate and extract heavy metals, radionuclides, and organic contaminants from saturated and unsaturated soils, sludges, and sediments (EPA, 1995; NRC, 1999a) as well as groundwater (GWRTAC, 1997). The strategy uses an electrical field imposed by electrodes implanted in the vicinity of the contaminant source. It has been proved successful in the laboratory and in small-scale pilot tests, particularly for removal of metals in low-permeability soils that are difficult to flush. Many issues remain to be resolved prior to full-scale commercialization of electrokinetic remediation. Testing at the Naval Air Weapons Station, Point Mugu, California, indicated that small-scale experiments give a false indication of the applicability for remediation of chromium and cadmium in the field (EPA, 2000c). Additional research is needed to determine the effect of naturally occurring ions on mobilization and removal of the target metals, to identify the site-specific factors that control the performance, to determine the relationship between electrode design and electric field shape, and to determine the optimum configuration of the electrodes in the field.

Phytoremediation. Phytoremediation employs metal-accumulating plants to either remove inorganic contaminants from the shallow subsurface or to withdraw soil moisture through evapotranspiration, which can provide hydraulic containment of contaminants (EPA, 1999e; Lasat, 2002). Phytoremediation is in the early stage of commercialization and is best suited for sites with (1) widely dispersed contamination (large land area), (2) low contaminant concentrations to prevent plant toxicity, and (3) and contaminant depths not exceeding the root zone. The effectiveness of phytoremediation depends upon the interaction among contaminants, soil, plants, and microbes. A variety of factors affect this complex interaction, such as climatic conditions, site hydrogeology, and agronomic practices. The case study described in Box 5-9 demonstrates some reduction in total soil lead with phytoremediation, but several growing cycles/seasons will be required to address all of the hot spots and realize near complete lead removal. Much greater knowledge of plant/soil/contaminant interactions is needed in order to optimize phytoremediation performance at a given site.

BOX 5-9
Case Study: Phytoremediation at Simsbury, Connecticut

Phytoremediation of lead in soil was tested in 1998 at the Ensign-Bickford Company site, Simsbury, Connecticut (available at: http://bigisland.ttclients. com/frtr/00000164.html). Near surface soils at the site were contaminated with lead from open burn/open detonation activities. The test area of 2.35 acres initially contained an average total lead concentration of 635 mg/kg; concentrations were higher than 1,000 mg/kg in many areas of the site, with some areas exceeding 4,000 mg/kg. Treatment crops of Indian mustard (*Brassica juncea*) and sunflower (*Helianthus annus*) were cultivated over a six-month period. Lime and fertilizers (nitrogen, phosphorus, and potassium) were tilled into the soil to a depth of 15–20 cm. An overhead irrigation system was used to provide moisture. Supplemental foliar fertilizers were added through the irrigation system.
Plant growth for each of the treatment crops was generally good. However, certain soils within the treatment area remained saturated, which caused less than optimal plant growth and required replanting. The six-month test was considered a success because average lead uptake measured in the sunflower and Indian mustard plant materials from all crops was approximately 1,000 mg/kg dry weight. Total lead concentrations in the surface soils decreased from an average of 635 mg/kg to 478 mg/kg. After phytoremediation, no collected soil samples exceeded 4,000 mg/kg. Initially, 7 percent of the treatment area had had soil lead concentrations in excess of 2,000 mg/kg, and after six months of phytoremediation, only 2 percent still exceeded 2,000 mg/kg. Further treatment cycles are planned for the site.

Chemical Treatment. Two case studies are available that demonstrate the use of chemical addition to achieve *in situ* reduction of Cr(VI) to Cr(III) (EPA, 2000c). At White Sands Missile Range, New Mexico, H_2S gas was injected into the subsurface in an attempt to reduce the hexavalent chromium to the less mobile trivalent chromium. Test results indicated that channeling of the H_2S occurred through strata having higher relative permeability. Furthermore, observed consumption of H_2S was higher than predicted from small-scale laboratory column tests. At the DOE's Hanford Site, a dithionite solution was injected to react with natural iron in the subsurface and form reduced iron (Fe(II)). The Fe(II) reacted with Cr(VI) and reduced it to Cr(III). Concentrations of chromium in groundwater were decreased to less than 8 μg/L in one month. Two years after treatment was complete, the treatment zone remained anoxic and Cr(VI) remained below detection limits. The anoxic zone is estimated to have a life of 7–13 years without further addition of dithionite. A major uncertainty is the long-term ability of these technologies to prevent remobilization of the chromium.

Summary

The Navy has inorganic contaminants at many of its sites, which poses a great challenge. Because the behavior and speciation of inorganics is strongly coupled to site-specific conditions, it is not possible to make generic statements about the application of a given technology. The Navy's problems with inorganic contaminants are not unique, but the level of priority the Navy has given to remediation of metal contaminants does not appear to be commensurate with the high frequency of occurrence of these contaminants at their sites. The Navy should devote more effort to developing strategies for managing inorganic contaminants. Implementation of the remediation technology needs to be coupled to a specific performance goal or objective. Because inorganic contaminants cannot be destroyed, efforts need to continue to focus on containment and on performing a complete mass balance.

Technologies for Remediation and Management of Contaminated Sediment

Throughout its history, the Navy has focused significant activity in coastal ports. Contaminated sediments have resulted from handling and disposal of fuels, bilge water, antifoulants, and other compounds, and from handling of wastewater on shoreline facilities. As discussed in Chapter 1, as many as 110 Navy facilities have identified sediment contamination, with most cleanup efforts still in the RI/FS stage.

For a variety of reasons, management of contaminated sediments poses one of the most difficult site remediation issues faced today. The technologies applied to contaminated sediments, for example containment by clean capping layers or removal by dredging and disposal, are often conventional. The safe and effective application of these conventional technologies in a dynamic subaqueous environment, however, requires innovation in design and care in implementation. Contaminated sediments typically reside in spatially variable and dynamic systems subject to seasonal flow variations and episodic storm events. The volume of sediments that must be managed often exceeds a million cubic yards, dwarfing many contaminated soil sites. These sediments are also associated with equally daunting volumes of water, and efforts to remove the contamination typically entrain even more water. All management options leave a residual risk that must be evaluated and managed. In some cases, the sources that led to the contamination may not be completely controlled, leading to the potential for recontamination of the sediments. In addition, long periods are usually required to observe the resources at risk and demonstrate recovery, making the assessment of success difficult.

A range of options is applicable to the management and remediation of contaminated sediments. Among the regulatory and nonregulatory approaches to reducing and managing risks of contaminated sediments are:

- control of environmental releases contributing to sediment contamination,
- socioeconomic options including reduction of exposure through fish advisories, institution of catch and release fisheries, relocation of exposed communities, and the introduction of economic or other acceptable offsets[1],
- natural attenuation including intrinsic biodegradation and natural capping by deposition of clean sediment,
- other *in situ* management via containment or treatment, including capping, and
- removal and *ex situ* management, which requires application of dredging technologies, pretreatment technologies, *ex situ* treatment and disposal technologies, and technologies for the management of residual contaminants, including contaminated gaseous and liquid effluents or the residual contaminants in the treated dredged material.

Control of the environmental releases leading to sediment contamination is a critical first step in managing contaminated sediments. Unlike most sources of soil contamination, the sources of sediment contamination may not have been fully characterized; even if they have been fully characterized, they may be difficult or impossible to adequately control. Thus, the degree to which these sources continue to contribute to sediment contamination must be assessed and incorporated into a conceptual model of the system. This should include a clear understanding of the source of sediment contamination, the vertical and areal distribution of contaminated sediments, key fate and transport processes, and how these

---

[1] An example of such an offset would be the industrial development of contaminated sediment areas following an approach similar to that of land-based Brownfields.

relate to the risks to the resources of concern. After assessment or control of releases leading to sediment contamination, the identification and selection of management options should be based on a wide range of considerations, including effectiveness, permanence, implementability, risks associated with implementation, cost, and state and community acceptance. All of these factors must be evaluated relative to the identified goals and based upon site-specific conditions (NRC, 2000).

Natural Attenuation

Natural attenuation occurs via any of a number of processes that can contain, destroy or dilute contaminants in the environment. In contaminated sediments, contaminants are often relatively immobile and refractory, and the most important natural attenuation process is often the stable burial of contaminants by sedimentation or deposition of clean sediments. Although the concentration or mass of the contaminant may be unchanged, significant reduction in exposure and risk may occur via this process. Some natural attenuation processes, such as dispersion, dilution, and volatilization, may transfer the risk from one location to another, which may or may not reduce overall risk. Other processes, such as biotransformation, sorption, and containment by burial with clean sediments, may directly reduce risk. Biotransformation processes that may significantly reduce risk include biodegradation of organic compounds, reductive dechlorination of halogenated compounds, and binding of metals into insoluble or non-bioavailable sulfides. Sorption of contaminants into a soil fraction that limits the rate or extent of desorption may also reduce bioavailability and subsequent exposure and risk.

Even in situations where natural attenuation is not the primary management approach, it still serves to manage the residual contamination not addressed by other approaches, which includes marginally contaminated areas outside of the zone being actively remediated or the residual contamination remaining within the remediated zone. It is anticipated that if natural attenuation is coupled with other more active remedial approaches, the duration of the monitoring may be shorter, although monitoring intensity should be unchanged. Because natural attenuation must be relied upon to some extent at all contaminated sites, an evaluation of the change in risk with time posed by natural attenuation processes should be a component of all remediation proposals.

Other In Situ Treatment

In situ treatment options are often designed to enhance natural attenuation processes. Biological degradation or transformation processes could be encouraged *in situ*, but most sediment contaminants degrade only slowly or to a limited extent in sediments, and an adequate delivery system for nutrients and other required reagents has not been identified. An effective delivery system would likely involve mixing of the sediment, which would encourage resuspension and loss of both sediments and contaminants. The lack of an effective delivery and homogenization system has also hindered the application of *in situ* stabilization systems. A demonstration in Manitowoc Harbor, Wisconsin, revealed difficulties in the management of pore water released by the solidification process (Fitzpatrick, 1994, as referenced in EPA, 1994).

In situ treatment options that do not involve delivery of chemicals to the sediments have also been proposed. *In situ* vitrification employs electricity to raise sediments to sufficiently high temperatures to produce a glasslike product. The energy costs of heating high-moisture-content sediments to glass formation temperatures are formidable, and the technology has not been used except for small volumes of highly contaminated sediments. Electrochemical geooxidation employs electricity to encourage redox reactions in sediments. This technology is under development and has not been demonstrated on a large scale for sediment remediation. In general, *in situ* treatment and stabilization technologies that are effective and commercially available have not been demonstrated (PIANC, 2000).

The remainder of this section is devoted to capping, which is the process of placing clean sediment or sand on top of the contamination, much like as occurs with natural deposition. *In situ* sediment capping is primarily designed to stabilize or contain contaminated sediments, isolate contaminants from benthic organisms, and slow contaminant migration out of the underlying sediments. Guidance exists for the design, placement, and monitoring of a cap as a sediment management option (Palermo et al., 1998). This guidance includes quantitative information on design of armoring layers, design for contaminant containment, and stability analysis during cap placement.

After placement, contaminants will migrate by diffusion in the pore space or by advection due to consolidation or groundwater seepage through the cap. After an initial transient movement, quasi-steady release rates are realized that are typically much lower than release rates prior to capping. The length of the transient period is longer for caps that contain sorbing material or for thick caps. The quasi-steady release rate depends primarily on the thickness of the cap and the extent to which groundwater seepage drives contaminant transport. The amount of contaminant that accumulates in the upper layers of the cap depends upon the sorptive characteristics of the cap. Thus, a weakly sorbing cap such as sand will tend to reach quasi-steady release rates relatively rapidly but not accumulate significant contamination in the upper layers of the cap. A strongly sorbing cap, however, will release essentially no contaminant for a long time but will ultimately accumulate contaminants in the upper layers of the cap, although typically at concentrations much below those originally found in the contaminated sediments.

Thin layer (5 to 15 cm) capping to enhance natural attenuation is the process most closely related to natural deposition processes. By placing a thin layer of clean sediment over the contaminated sediment, the process is potentially less disruptive to the benthic community. Because the sediment-water interface tends to approach an equilibrium state, the small modification provided by a thin layer cap is potentially more stable than thicker caps without additional armoring. A layer of only 5–15 cm will generally isolate the bulk of the contaminants from the benthic community and the overlying water. Isolated penetrations of a thin layer cap can still occur, but they are unlikely to lead to aquatic organism exposure to significant contaminant mass. Primary concerns associated with thin layer capping are the long-term stability of the capping layer without armoring and the ability to accurately place a thin layer of sediment.

Thick layer capping is the conventional approach to containment of contaminated sediments. Cap thickness is normally 20 cm to as much as 1 m. The greater thickness helps ensure that an isolating cap layer remains even if there is significant heterogeneity in placement thickness or small amounts of post-placement erosion. The larger depth of capping material, however, may result in load bearing problems for an underlying soft sediment or require placement in multiple layers to allow the underlying sediment to consolidate and develop sufficient strength to support the cap layer. A number of examples exist of successful cap placement over very soft sediments—that is, undrained shear strength of 0.2 kPa or less (Zeman and Paterson, 1997; Palermo et al., 1998).

Clean sediment caps can be used to contain sediments and contaminants *in situ* or *ex situ*. The *ex situ* application, referred to as confined aquatic disposal (CAD), involves dredging of the contaminated sediment and placement in a submerged pit, followed by placement of additional clean dredged material to serve as a cap. This approach is most useful where the sediments must be dredged for navigation purposes or when capping in the original location is inappropriate. Resuspension losses during removal and during placement must be assessed, but maintaining the contaminated sediments in the aquatic environment may be advantageous, especially for contaminants that are mobilized upon exposure to air. Box 5-10 illustrates the use of a CAD cell for managing contaminated sediments.

Because capping does not normally encourage degradation or transformation of the contaminants, long-term monitoring to ensure cap stability and contaminant containment would normally be required. These requirements may be more extensive for capping than for removal

---

BOX 5-10
Management of Contaminated Sediments
with a CAD cell at Puget Sound Naval Shipyard

The Puget Sound Naval Shipyard, Bremerton, Washington, has been in operation for more than 100 years, and over that time a variety of contaminants have been introduced to the near-shore sediments. The Shipyard was placed on the National Priorities List in 1994. Sediment sampling in 1994 and 1995 identified a number of locations with contaminant concentrations in excess of State of Washington sediment quality standards. Cleanup of these sediments was complicated by the need for navigational dredging to allow large aircraft carriers to dock at the port. Navigational needs may place additional constraints upon sediment remedial approaches and may encourage removal of material beyond that required or desirable based upon environmental concerns. In this case, dredging was required even though the most desirable disposal method was deemed to be subaqueous disposal, resulting in some contaminant release and exposure during removal. Cost for the combined navigational and cleanup dredging and upland disposal was estimated to be $44 million. Ultimately, the chosen remedy was dredging followed by placement in a submerged pit and containment by a clean sediment cap (confined aquatic disposal, CAD) at an estimated cost of $14 million. This return of the contaminated sediments to a waterbody also resulted in measurable exposures.

The combined cleanup and navigational dredging produced >390,000 yd^3 of material for placement in the CAD cell. The CAD cell was 36 ft deep and 400 x 415 ft in a region with 30-ft water depth. The dredging and capping project was complicated by daily tidal exchange, a tight schedule, and the need to strictly control dredging to ensure adequate CAD cell capacity. After dredged-material placement, a 1-ft interim cap was placed, followed by a second 3-ft cap. The project was completed in fall 2001, and monitoring plans evaluating effectiveness and long-term containment are being developed. To date, the dredging and sediment and cap placement appear to be successful, and the result is a remedial project that satisfies the needs of both navigation and remediation.

Additional information is available at http://yosemite.epa.gov.region10.

options because more of the contamination would be expected to remain in place after capping. It should be noted, however, that because of residual contamination, long-term monitoring is a requirement of any remedial approach.

Removal Technologies—Dredging

Options that involve removal of contaminated sediments from a waterbody are significantly more complicated than *in situ* approaches because of the train of treatment that is triggered. Removal options generally require a pretreatment step for dewatering of the dredged materials, treatment or disposal of the removed materials, and treatment and disposal of any residuals left in the waterway or produced during treatment or disposal. Often, the feasibility, cost, and potential risks are determined by these "downstream" technologies in the treatment train rather than by the dredging activity itself. The costs associated with dredging are typically less than $20 per cubic yard while treatment and disposal of the dredged material and associated water may cost $100–$1,000 per cubic yard or more.

Dredges for removal of contaminated sediment fall into one of two basic categories: (1) hydraulic dredges that primarily use suction and hydraulic action to remove sediments and (2) mechanical dredges that remove sediments by direct, mechanical action. Hydraulic dredges are generally capable of high production rates, and they minimize sediment resuspension. Mechanical dredges are generally preferred for high solids content, low water production, greater accuracy, and improved performance in the presence of debris and obstructions. Hybrid dredges have also been used that are predominantly mechanical in action but also withdraw water to control migration of a resuspension plume. For small areas or areas where water can be removed or diverted, dry excavation of the contaminated sediments may be an option. Although this simplifies the excavation process, the potentially greater mobility of contaminants after exposure to air by vaporization or oxidation should be assessed before dry excavation is undertaken. The selection of a particular dredging technology and the risks associated with dredging relative to other remedial alternatives are dependent upon site-specific factors, and only limited general guidance can be provided. Some of the site-specific factors include sediment grain size and cohesiveness, the presence of debris, access to the site, and the conditions controlling the relationship between contaminant release and the exposure and risks faced by sensitive organisms.

Dredging is likely to be most effective when the contaminants are present in well-defined areas of relatively homogeneous, debris-free sediments. Debris, bedrock, and large areas of diffuse contamination all work to make dredging less effective as a sediment management option. Even when dredging may be the preferred option, resuspension of contaminated sediments, residual contamination, and dredged material pre-treatment, treatment, or disposal requirements may limit the feasibility of the approach, as discussed below.

Resuspension. A significant factor in the selection of dredges for removal of contaminated sediments is the resuspension potential. Sediment contaminants are largely associated with the solid phase, and therefore resuspension of particles leads to resuspension of contaminants. Hydraulic dredges that are operated slowly and with care generally give rise to less resuspension than mechanical dredges or dredges operated to maximize production rate (McCellan et al., 1989). Contaminant losses from resuspension have been estimated to be as low as 0.1 percent to 0.3 percent (Kauss and Nettleton, 1999; Hayes et al., 2000). This loss can be significantly higher with dredging of fine-grained sediments, dredging at high rates without maintaining close operational control, and dredging in the presence of debris. Debris may prevent a bucket from closing during dredging by mechanical means or cause shutdown with hydraulic dredging. Both may lead to significant short-term releases of resuspended sediment. The importance of resuspension losses depends upon a variety of site-specific factors including dredging rate, water flow, and the distribution of contaminants in the sediment column.

In general, improvements (i.e., reductions) in sediment resuspension and contaminant release come at the expense of volumetric efficiency and production rates. The low resuspension rates of the enclosed cable arm dredge noted by Kauss and Nettleton (1999) were aided by continuous monitoring and in-water cycle times of 2–6 minutes during normal operation, much slower than would be expected during navigational dredging. As another example, during hydraulic hot-spot dredging in New Bedford Harbor in 1994 and 1995, efforts to control resuspension led to the capture of 160 million gallons of water, which had to be decanted and treated while targeting the dredging of only 10,000 yd^3 of sediments (an average solids concentration based upon targeted sediments of little more than 1 percent) (Foster-Wheeler, 1999). Thus, the cost and time of dredging projects can be strongly influenced by the effectiveness and rate of subsequent wastewater treatment.

Residuals. Potentially more important than resuspension is the residual sediment contamination left on the surface after dredging. Because of the mixing that occurs with dredging operations, it is difficult to reduce the residual sediment column concentration below the depth-averaged initial concentration without significant overdredging to ensure clean underlying sediment. Under certain conditions, this layer may be very thin and of little consequence, but data are insufficient for predicting when significant residual contamination will occur. When overdredging is limited by "hardpan" or bedrock, the overall reduction in surficial sediment concentrations that can be obtained could be limited. The residual contamination may require further management by either monitored natural attenuation or more active remedial efforts.

Pretreatment and Water Treatment. Dredged material requires subsequent treatment or disposal, the first step of which is pretreatment to remove and treat excess water and reduce volume (except in the case of CAD). Although mechanical dredging does not add as much water to sediment as hydraulic dredging, some dewatering from *in situ* densities is generally required. The produced water content adds to the cost and complexities of dredged material disposal and may pose significant concerns for water quality upon return to the waterbody. Dredging is also normally subject to significant variations in production rate. Subsequent treatment or disposal steps often cannot maintain effectiveness if the feed rate is widely variable, and so a temporary storage system is normally required to serve as a basin for watering and volume equalization. The variations in sediment conditions and production rate also make it difficult to provide more aggressive pretreatment of dredged material—for example, by the addition of nutrients or dewatering agents in a dredged material pipeline during hydraulic removal operations.

Water separation and treatment is generally accomplished in a primary settling basin. Potential contaminant concerns in such systems are evaporation of contaminants from the exposed sediment and overlying water (Valsaraj et al., 1995) and carryover of dissolved and suspended contaminants with the effluent water (Myers et al., 1996). Secondary treatment can be accomplished via a variety of conventional approaches, but such treatment is complicated by the variations in sediment quality and by the variety of contaminants that may be present. Conventional pollutants such as oxygen-demanding organic matter and ammonia may pose more serious problems in the water than the toxic contaminants that are the primary focus of the sediment remediation—issues that should be taken into consideration and perhaps will require additional monitoring.

Dredged Material Disposal and Treatment. In most sediment remediation activities that have been completed or are underway, the dewatered dredged material is either left in a confined disposal facility (CDF) or transported to a landfill. If disposed of in an upland landfill, dredged material is not subjected to further treatment, other than perhaps further dewatering. For a CDF, the ultimate disposal is typically in the same facility in which primary dewatering has taken place and some treatment may occur. Particular treatment technologies that have been considered in a CDF include biodegradation, phytoremediation, and solidification/stabilization. Problems include the heterogeneity of the dredged material and the difficulty of applying biodegradation and phytoremediation to the entire column of dredged material, which may be tens of feet thick. A completed CDF may be capped for control of leachate production and vaporization, and to provide a physical barrier to direct contact by terrestrial animals. Additional development and field testing are required before the approach will receive widespread acceptance.

Although in principle the public is more supportive of treatment technologies that destroy contaminants, the costs of sediment treatment alternatives have generally not been competitive with landfill or disposal facility placement. A recent review of eight technologies (PIANC, 2000) suggested that contracts of ten or more years involving the treatment of a million or more cubic yards of dredged material per year were required for sufficient economies of scale to make the technologies commercially viable. These volumes are available only in large harbors subject to navigational dredging of sediments that cannot be disposed of in open water (e.g., New York/New Jersey Harbor) or in a few large contaminated sediment sites. It may be possible to build centralized facilities capable of processing the contaminated dredged materials from multiple sites, although significant public acceptance and regulatory barriers would need to be overcome. A second consideration is that treatment technologies require development of a market for the effluents from their processes and regulatory standards that define the acceptability of these effluents for certain uses. Finally, treatment technologies, even those that destroy contaminants, may generate residuals that are released to the environment or need to be disposed of in landfills.

Measuring Performance

Regardless of the remedial option chosen, the most direct indicator of remedial performance is monitoring of the resource at risk from sediment contamination, which is often the body burden of contaminants in edible fish. Monitoring concentrations in fish over time can be used to indicate remedial performance and to compare remedial alternatives as shown in Figure 3-4. A relative comparison of the potential risks, including the risk of no remedial efforts, can be effective in identifying absolute risks. A convenient comparison for any active remedial approach is the effectiveness or potential risks of the approach relative to those expected with natural attenuation processes. Prognostic models are required to predict contaminant processes far into the future to effectively compare outcomes from different remedial alternatives. The ability to accurately predict future conditions in sediments, including identification of key sources of uncertainty, is reviewed in Reible et al. (2002).

Exposure and risk in sediments are largely limited to the near-surface sediments. The layer subject to erosion, even during large storm events, rarely exceeds a few centimeters. As indicated previously, bioturbation is typically limited to the upper 10–15 cm of sediment. In addition, sorption-retarded advection and diffusion in sediments are generally so slow that only the upper few centimeters contribute significantly to flux. Thus, exposure and risk from sediment contaminants is largely related to surficial sediment concentrations. As an indicator of average surficial sediment concentration, the surface area weighted average concentration (SWAC) has been used at some sites (e.g., in the Sheboygan ROD, the Fox River RI/FS, and the Shiawassee ROD) as a surrogate measure of exposure and risk. For those sites where a large fraction of the contaminant mass is deeply buried, the use of SWAC metrics for remediation will lead to significantly different remedial designs than total concentrations. These metrics are most useful and accurate where exposure is controlled by widespread contamination over large surface areas and not where erosion or contaminant release from small hot spots control risk.

The risks associated with different sediment remediation approaches may vary with time. Dredging may increase short-term risk through resuspension of the contaminants, incomplete water treatment, or elevated residual surface concentrations. *In situ* approaches such as capping, however, may give rise to an elevated risk far into the future. In such case, time-integrated measures of performance may prove useful. For example, the time integral of predicted fish body burdens may represent an average indication of exposure. Similarly, the integral of the SWAC may indicate average exposure for risks controlled by average surficial sediment concentrations.

TREATMENT TRAINS

Much of the literature and guidance that describe innovative technologies address one particular technique and its application for a specific class of contaminants. For example, the Naval Facilities Engineering Service Center (NFESC) website provides the user a wealth of information on individual technologies that are grouped by biological, physical/chemical, and containment and removal technologies. However, most hazardous waste sites, including Navy sites, are rarely contaminated by a single chemical group. Rather, mixtures of contaminants with varying physical and chemical properties like chlorinated solvents (VOCs), petroleum hydrocarbons (fuels, oils, and polycyclic aromatic hydrocarbons), PCBs, and one or more heavy metals are much more likely to be present. Furthermore, the contaminants typically reside in different media, such as surface soils, sediments, aquifer solids, groundwater, and surface water. A single remedial technology is normally effective at treating only a subset of the contaminants in a waste mixture or treating one type of media. Two or more remedial technologies applied in combination or sequentially are likely to be necessary to attain cleanup goals for a waste mixture and multimedia contamination. From this perspective, each remedial technology should be viewed as a unit operation that can be linked as part of a treatment train. Creating treatment trains should be considered part of optimizing or adding to existing remedies— a key decision-making point in ASM. The same treatment train strategy is used to treat drinking water or wastewater. In drinking water treatment, processes such as coagulation, sedimentation, filtration, and disinfection are usually combined to achieve potable water. In wastewater treatment, particle removal, biodegradation, and disinfection are often coupled to meet effluent standards.

Another factor contributing to the need for treatment trains is that each waste site is unique. The efficacy and adequacy of any remedial option depends on site-specific characteristics, such as the chemical properties, horizontal and vertical extent of contamination, and hydrogeologic setting. Often, multiple technologies need to be employed for managing the risks at waste sites because of the dynamic and complex nature of the subsurface. The behavior and response with one technology often improve our knowledge of the site, which helps guide the implementation of a second technology.

The concept of treatment trains is now commonly implemented for remediation of certain classes of contaminants. The NFESC website identifies four remedial technologies under the category of "Combined

Mechanism," including constructed wetland, Lasagne™ Process, natural attenuation, and vacuum-enhanced recovery (bioslurping). For example, bioslurping uses vacuum extraction together with *in situ* biodegradation to remove contaminants. The vacuum extraction component of bioslurping aims to remove the bulk of the volatile compounds and is capable of removing separate-phase globules of contamination. The introduction of oxygen can stimulate the biodegradation of residual contamination. The FRTR website provides reference guides for common treatment trains associated with the eight contaminant groups listed in Table 5-2. For example, free product recovery, venting/air stripping, and *in situ* biodegradation can be coupled to effectively manage gasoline spills (Lee and Raymond, 1991).

The most common treatment train that is being invoked for addressing organics-contaminated soils and plumes in groundwater is a source treatment technology in conjunction with monitored natural attenuation (MNA) (see Box 5-11 for a description of MNA). MNA can be a primary (stand-alone) technology in some cases, usually for petroleum hydrocarbons. However, remediation of petroleum hydrocarbon contamination is not the focus of this report. Consequently, the application of MNA discussed here is its use as part of a treatment train. The Navy's approach of coupling a variety of source removal/containment technologies with MNA is consistent with national trends and is described in greater detail below.

At sites with metal contaminants, two or more remedial options applied sequentially to contaminated soil often increase the effectiveness while decreasing the cost of remediation (EPA, 1997b,c). Treatment trains for metal contaminants include soil pretreatment, physical separation designed to decrease the amount of soil requiring treatment, and treatment of process residuals or off-gases. A promising treatment train for remediation of metal-contaminated soil is the combination of electrokinetics and phytoremediation. Electrokinetics is used to remove metals from deep soil and groundwater, whereas phytoremediation is effective at removing metals in surface soils.

The above examples of treatment trains pertain to contaminated soils and groundwater, but the concept is equally applicable to contaminated sediments, especially when the remedial options involve removal of contaminated sediments from a waterbody. Removal options involve not only dredging, but other technologies as well to manage the dredged material, the water produced, and any residuals left in the waterway. Combining processes can also be effective for *in situ* treatment of contaminated sediments, such as capping and natural attenuation. Cap installa-

BOX 5-11
Description and Application of Monitored Natural Attenuation
for Groundwater Contamination

Natural attenuation refers to a variety of natural processes that result in a decrease in contaminant concentration and mass. These processes can be physical (dispersion, dilution, sorption, volatilization), chemical (oxidation, reduction, immobilization by precipitation), or biological (biodegradation by indigenous microorganisms and perhaps plants). *Monitored natural attention* (MNA) refers to the use of natural attenuation as a remedial option. NRC (2000) provides an extensive review of MNA and suggests that proper application of MNA should include three basic steps:

1.   Develop a conceptual model that characterizes the site (e.g., where the contaminant is, how the groundwater is moving, etc.) and identifies what processes, if any, could potentially be responsible for decreasing the concentration and mass of contaminants.
2.   Gather sufficient site-specific data to demonstrate that contaminant loss is due to a given attenuation process (e.g., biodegradation, immobilization of heavy metals by precipitation), and determine whether natural attenuation is sustainable and will meet remediation goals.
3.   Implement a long-term monitoring program to ensure attenuation processes continue to occur and remediation goals are being met.

NRC (2000) lists characteristics required for a comprehensive protocol for MNA that cover three broad subject areas: community concerns, scientific and technical issues, and implementation issues. The NRC reviewed 14 of the available natural attenuation protocols, including the Navy guidance document for MNA (Dept. of the Navy, 1998) and found that none met all the characteristics of a comprehensive protocol. The principal findings concerning the current state of practice of MNA can be summarized as follows:

•   MNA is an established remedy for a limited number of contaminants—primarily BTEX and perhaps chlorinated ethenes under some conditions (NRC, 2000).
•   MNA should only be accepted as a remedial option when attenuation processes have proved to be working and sustainable (see items 1–3 above); it should never be considered a default or presumptive remedy or "no-action" alternative.
•   Rigorous protocols need to be developed to ensure that MNA is analyzed properly.
•   MNA cannot be achieved solely by dilution and/or dispersion.
•   The reaction processes must avoid accumulation of harmful daughter products.

- Affected communities must be involved in the process and have access to all relevant information (e.g., proof that the attenuation processes are working and sustainable).

(Involving affected communities should apply to all remediation activities.)

Measuring Performance

Measuring the efficacy of MNA is similar to measuring the efficacy of enhanced bioremediation. Effective long-term monitoring is absolutely required to ensure that attenuation processes continue to result in reduction in contaminant concentration and mass and in the protection of human and environmental health. Such monitoring may be required for many years, even decades.

Technology Evaluation

Based on the knowledge of natural processes that can affect the movement and fate of contaminants in groundwater, NRC (2000) summarized the likelihood of success of MNA for various classes of contaminants. Only for BTEX, low-molecular-weight alcohols, ketones, esters, and methylene chloride is the likelihood of success rated as "high." In some cases metals can be immobilized. MNA may be appropriate for sites with contaminant classes with a lower likelihood of success, such as most halogenated aliphatic and aromatic compounds, nitroaromatic compounds, and toxic metals, but evidence for success will usually require extensive effort in site characterization, laboratory studies, modeling, and monitoring. Because natural attenuation processes are always site-specific, information contained in NRC (2000) can only be used as a general guide for the potential of MNA to be successful. Each site must be studied individually to determine if MNA is effective for remediation and for controlling risks from contaminated groundwater.

The Science Advisory Board (SAB) of EPA recently released *Natural Attenuation in Groundwater* (EPA, 2001c), a report that builds on NRC (2000). The SAB states that MNA, when properly used, is a "knowledge-based remedy in which the engineering informs the understanding, monitoring, predicting, and documenting of the natural processes, rather than manipulating them." It also makes specific recommendations as to how the science base of MNA needs to be enhanced by EPA for better application to chlorinated solvents, underground storage tanks, inorganics, and sediments.

tion stabilizes and contains contaminants for immediate reduction in risk, and then slower natural attenuation processes (e.g., biodegradation) reduce contaminant mass over the long term.

Treatment Trains and Source Removal

A feature of contaminated sites that necessitates the use of treatment trains is that conceptually, a waste site consists of two distinct components. The first component is a contaminant source area where the bulk of the contaminant mass is usually located. The second component is the plume of dissolved contaminants that emanates from the source area. The source/plume concept for subsurface contamination has been well documented in previous reviews (NRC, 1994; Cherry et al., 1997). The approaches and prospects for cleaning up the plume of dissolved contaminants are much different than they are for cleaning up the source areas, necessitating a coupling of treatment technologies that usually involves some measure of source control along with some measure to restore the contaminated groundwater.

Contaminant source areas include near-surface sources such as surface spills, leaking drums and storage tanks, and landfills, but they also include deep subsurface pools or ganglia of NAPLs and metals that have precipitated in mineral phases having low solubility. Sorbed contaminants also constitute a long-term source of dissolved-phase contamination. Source areas are difficult to characterize and locate in their entirety because of poor knowledge of site operating history along with the complexity of the subsurface (NRC, 1994, 1999b). The source areas at waste sites persist for a very long time and are capable of contaminating groundwater over time scales of decades to centuries.

One approach to managing such sites is immobilization or containment of contamination by hydraulic and/or physical barriers, followed by restoration of the dissolved plume (NRC, 1994). The success of this two-step treatment train (source containment and aquifer restoration) relies on maintaining the integrity of the containment system; Jackson (2001) points out that ensuring containment is both technically difficult and costly to achieve at many waste sites. Another concern with such approaches is that the source remains in the subsurface, so there is a long-term threat of slow dissolution of contaminants into groundwater should the containment system fail. Consequently, source removal is perceived by many stakeholders to be a more desirable cleanup approach. Technologies appropriate for source removal are listed in Table

5-2. As reviewed in the previous sections, popular technologies being used by the Navy for source removal include *in situ* chemical oxidation, thermal treatment, enhanced bioremediation, and methods to extract inorganic contaminants.

A reduction in contaminant mass from a source zone is expected to provide several benefits including a decrease in cleanup time, a possible reduction in risk, a decrease in the extent of contaminated groundwater, and improvement in the performance of natural attenuation processes (Sale and McWhorter, 2001). However, there is no consensus among the technical community on the benefits derived from partial contaminant mass removal from source zones. The research results presented in Box 5-12 illustrate that under the assumption of aquifer homogeneity and uniform groundwater flow, even substantial amounts of mass removal may have little impact on the time for cleanup, on groundwater concentrations, and on the exposure pathways for a site. In contrast, other investigators (also Box 5-12) present modeling results for heterogeneous flow fields that demonstrate significant reductions in contaminant fluxes to groundwater (and corresponding significant reductions in risk) for modest degrees of source removal. Additional studies are certainly needed to resolve the disagreement that currently exists regarding the relationship between partial source removal and its impact on contaminant fluxes to groundwater and site risk.

Source removal technologies at contaminated sites are normally implemented without an understanding of how much mass removal is needed to be effective (e.g., in meeting water quality goals, in restoring the plume, and in reducing risk). It is recommended that the Navy perform site-specific analyses of the effectiveness of source zone mass removal to better guide and justify the selection of source removal technologies being implemented at Navy sites. This analysis will also help the Navy determine if enough of the source mass can be removed to warrant the expense of implementing the technology.

Compatibility of Technologies in Treatment Trains

In the selection of technologies that are combined or sequenced into a treatment train, the impact of one process on the performance of other processes must be considered. In some instances, the combination of technologies does not cause any compatibility issues. For example, MNA or enhanced bioremediation could follow downgradient from a PRB as long as the products from the PRB (e.g., Fe(II) species, high pH,

BOX 5-12
Impact of DNAPL Source Zone Treatment upon
Contaminant Concentration and Flux

There is disagreement among researchers on the relationship between degree of contaminant mass removal from source zones and contaminant fluxes to downgradient groundwater. Sale and McWhorter (2001) demonstrate the impact of DNAPL source zone treatment upon downgradient contaminant concentrations. They conceptualize the source zone as containing multiple subzones of DNAPL, as shown in Figure 5-8. A key assumption in this analysis is that the aquifer is homogeneous and thus groundwater flow is uniform within the DNAPL source zone.

The model developed by Sale and McWhorter computes the steady-state concentration distribution in the downgradient groundwater resulting from rate-limited dissolution of DNAPL. As clean groundwater contacts a subzone, DNAPL dissolves at a rate that is proportional to the difference between the aqueous solubility and the local solute concentration. Once dissolved, the fate of the DNAPL is controlled by advection and longitudinal and transverse dispersion. Sale and McWhorter use their model to explore how the steady-state DNAPL concentration is affected by various parameters (e.g., velocity, dispersivity) and also by the size and configuration of the source zone.

Because influent groundwater is not contaminated, this model predicts that most dissolution occurs in the upgradient region of the source zone, because that is where the "driving force" for dissolution is greatest. Moreover Sale and McWhorter find that dissolution rates are sufficiently fast so that groundwater solute concentrations increase to near the solubility limit after only a short travel distance, thus effectively shutting down dissolution from downgradient portions of the source zone. Referring to Figure 5-8, this implies that the pollutant concentration exiting the source zone will remain close to its solubility limit even if a

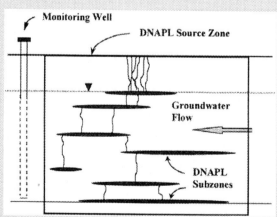

FIGURE 5-8 Schematic of a DNAPL source zone. SOURCE: Reprinted, with permission, from Sale and McWhorter (2001). © (2001) American Geophysical Union.

large portion of the DNAPL mass in the source zone is eliminated. The modeling results indicate that removal of the vast majority of the DNAPL will likely be necessary to achieve significant near-term reductions in groundwater concentrations and reductions in source longevity (Sale, 1998).

Other investigators have argued that such conclusions regarding the efficacy of source-zone mass removal are overly pessimistic because of the assumption of a uniform flow field. In particular, Rao et al. (2002) present modeling results for heterogeneous flow fields that demonstrate significant risk reduction for more modest degrees of source zone treatment. In this model, a random distribution of both hydraulic conductivity and DNAPL saturation was assumed in the source zone; the model also allows for statistical correlation between hydraulic conductivity and saturation. During a simulation, the researchers allow the DNAPL to dissolve away, and at various times they compute the downgradient mass flux and the DNAPL mass remaining. Rao et al. (2002) argues that contaminant mass flux across the downgradient boundary of the source zone is a more meaningful metric for risk than simply the groundwater concentration.

Figure 5-9 shows the results of the Rao et al. model simulations. The fractional flux reduction is plotted on the ordinate, and the fractional mass reduction is plotted on the abscissa. For a given reduction in DNAPL mass, the largest flux reduction is for the negative correlation case where the higher DNAPL saturations are associated with the lower-velocity regions (i.e., less-permeable zones). Achieving the same level of flux reduction for the positive correlation case requires a larger fractional mass reduction since the DNAPL is preferentially located in the high-velocity regions (i.e., more-permeable zones) that are making a large contribution to the flux. Because DNAPL source zone treatment technologies tend to preferentially target the more permeable portions of the aquifer, the conclusion based upon these modeling results is that contaminant mass flux could be significantly reduced even for modest reductions in the DNAPL mass.

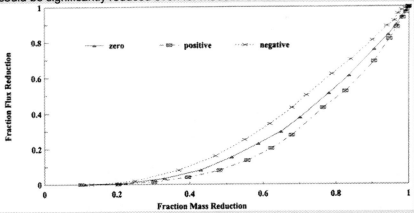

FIGURE 5-9 Mass flux reduction versus the fraction of mass removed. SOURCE: Rao et al. (2002).

and associated geochemical changes) do not inhibit native microbes. Enhanced bioremediation can be used in conjunction with surfactant flushing, with a variety of oxidation processes, and with low-temperature physical processes such as soil vapor extraction. Because the properties of inorganic contaminants differ markedly from organic contaminants, remediation technologies for inorganics are potentially compatible with a variety of other remediation techniques. Physical separation and extraction processes for inorganics could reduce the amount of material requiring subsequent treatment for organics. If containment in the source zone is effective at immobilizing both inorganics and organics, then the spread of contamination will be eliminated and the remediation of the groundwater plume will be facilitated.

Many of the approaches to managing contaminated sediments are compatible with each other when they are used on different portions of the site. For example, *in situ* capping in a portion of a site is not likely to interfere with dredging or natural attenuation in other portions of the site. Certain management approaches may also be compatible for application on the same portion of a site. For example, capping may be a useful adjunct to dredging to eliminate any negative consequences of the residual contamination. It may also be appropriate to employ capping as an interim risk reduction measure until additional remedial decisions can be made. However, if dredging is subsequently implemented, the volume, cost, and complexity of the dredging process would likely increase. At many contaminated sediment sites, natural attenuation is necessary as a complementary remedial approach to ultimately achieve risk-based goals. Indeed, it is expected that at the Lower Fox River site risk-based concentration goals will not be achieved until after decades of natural attenuation subsequent to the implementation of the initial remedy (WDNR, 2001).

In other instances, caution must be exercised in combining technologies. This is especially critical when using MNA subsequent to a source removal technology, as source treatment efforts may directly and adversely impact the microbial activity and hence the performance of MNA. One possible detrimental impact is the alteration of the electron acceptor available for microbial metabolism. For example, active source removal technologies that introduce oxygen to the subsurface are likely to shut down the anaerobic biodegradation processes necessary for natural attenuation of chlorinated solvents and that are operative for certain petroleum hydrocarbons and inorganic contaminants (see Box 5-13). Technologies that could introduce oxygen include *in situ* chemical oxidation, in-well stripping, air sparging, soil vapor extraction, cosolvent or

---

BOX 5-13
Impact of Source Removal on Natural
Attenuation of Perchloroethylene

A glacial outwash aquifer in Minnesota was contaminated with PCE from a former dry cleaner supply company (Ferrey et al., 2001). The groundwater chemistry at the site is conducive for reductive dechlorination of PCE [e.g., low dissolved oxygen (DO) concentrations, anaerobic electron acceptors, and reducing conditions]. Consequently, the initial remedial strategy for the site was to install a vacuum vaporizer well in the source area to remove high levels of PCE and to rely on MNA for plume treatment. A vacuum vaporizer well uses in-well sparging with air to strip chlorinated solvents from recirculating groundwater. Operation of the vacuum vaporizer well reduced source area groundwater PCE concentrations from 9,900 to 25 μg/L, but elevated the DO levels to between 2.9 and 3.3 mg/L. DO concentrations prior to PCE source removal were less than 0.7 mg/L. The aquifer aeration by the vacuum vaporizer well caused elevated levels of PCE transformation products in downgradient monitoring wells. At a monitoring well 360 feet downgradient of the vacuum vaporizer well, TCE concentrations increased from <10 to 35 μg/L, *cis*-DCE concentrations increased from 70 to 370 μg/L, and vinyl chloride concentrations increased from below the detection limit to 83 μg/L. The latter finding is especially problematic as vinyl chloride is more harmful than PCE, the parent compound. After operating for three years, the vacuum vaporizer well was shut down, and concentrations of TCE, *cis*-DCE, and vinyl chloride returned to pre-sparging levels. The conclusion from this study is that the benefit of a remediation system that alters the groundwater chemistry (i.e., introduction of air to create oxidative conditions) should be balanced against the potentially negative effect that the remedial technology may have on the natural attenuation mechanisms already existing.

---

surfactant flushing, and thermal treatment. A second possible negative impact is that source removal could remove a contaminant that is used to enhance the biodegradation of another contaminant. An example is the inadvertent removal of petroleum hydrocarbons, phenols, alcohols, or ketones that are serving as primary substrates for microbes involved in the intended biodegradation of chlorinated solvents in the downgradient groundwater plume, which could slow down or completely stop natural attenuation of the chlorinated solvents. Another negative impact that can arise in coupling source control with MNA is alteration of the flow field or mobility of the contaminant, which can enhance contaminant spreading, reduce time available for attenuation reactions, and sterilize the site for an indeterminate period. Potential effects of other remediation activities on MNA are tabulated in a recent NRC report (NRC, 2000, Table 3-2).

The first step in establishing the compatibility between a source removal technology and MNA is to create a conceptual model for the site. Development of a conceptual model is an iterative process that involves characterizing the groundwater flow system, delineating the contaminant source and plume, and identifying the reactions contributing to natural attenuation. Data generated by site monitoring are then coupled to the conceptual model to establish if adequate loss of the contaminants is possible and to identify the processes responsible for this loss. Techniques for this data analysis include graphical and statistical analysis of trends in concentrations of contaminants and substances, mass balances to verify reaction stoichiometry, simple modeling of solute transport, and comprehensive flow and solute transport models (NRC, 2000). A mass balance analysis can be used to determine if a given source removal technology will unfavorably alter the chemical environment for natural microbial reactions (e.g., modify electron acceptor conditions). Furthermore, data analysis is needed to establish whether MNA can achieve the desired remediation goal at the appropriate downgradient receptor(s). For example, solute transport modeling can be used to determine if the concentrations emanating from the source zone after treatment are low enough for MNA to be sufficiently protective of human health and the environment.

***

In summary, caution needs to be exercised when combining treatment technologies in order to ensure compatible performance. Incompatibility issues are especially important when source treatment is coupled with MNA as the primary site management approach. In theory, if one can completely delineate the source area and succeed in removing or destroying most of the contaminant mass, then a significant benefit can be achieved when negative effects on natural attenuation are not expected. However, source treatment can interfere with the present or future performance of natural attenuation, principally through disruption of environmental conditions required for the biodegradation reactions (e.g., availability of electron donors and acceptors) and destruction of the microbes (e.g., sterilization via chemical oxidation and thermal treatment). In the former situation, source treatment is undesirable. The latter situation requires selection of an alternate technology. If the attenuation reactions are sustainable for a long period of time, then MNA alone may serve as a long-term remedy for the site.

MAJOR CONCLUSIONS AND RECOMMENDATIONS

This chapter has reviewed several innovative technologies that are applicable to the most recalcitrant contamination scenarios found at Navy installations (and other federal facilities). The information is most relevant to those stages of ASM that involve the optimization, replacement, and addition of remedies, particularly MDP3 and MDP4. Although all the technologies have their place, there is no clearly superior single remedy that can address even a small fraction of the Navy's contamination problems. Remedy selection must remain site-specific. In general, for the innovative technologies reviewed here, there is a lack of refined evaluation procedures and peer-reviewed literature on their cost and performance—partly because their development is vendor-driven—making it impossible to fully evaluate their success or efficacy. Thus, as mentioned in Chapter 4, further testing of innovative or new experimental technologies at selected sites is needed, both for site-specific application and if the results are likely to improve cleanup activities at other sites. When evaluating remedial options and technologies, the full life cycle of the technologies and the management and disposition of all residuals that may be generated by the technologies should be considered.

A routine part of ASM is reevaluation of the current remedy design for possible optimization. Optimization can be as simple as ensuring that system components are still appropriate and are operating at design efficiency. Formal mathematical optimization can be used to evaluate well configuration and pumping rates in pump-and-treat or soil vapor extraction systems for potential cost savings. In the course of taking such action, the remedy must remain protective of human health and the environment. More detailed instruction for site managers on how to optimize various remedial systems is required, because existing information in DoD guidance manuals is presented in very general terms and can be used only by persons who are already quite technically knowledgeable in the remediation field. Recommendations below pertain to specific innovative technologies that hold promise for addressing contamination scenarios identified by the Navy as problematic.

***In situ* oxidation holds promise for removing organic compounds from the subsurface, although greater confidence in this technology awaits the creation of standardized bench-scale and field-testing protocols.** Such protocols should specify that early site screening be conducted for compatible geochemical, natural attenuation, and hydrogeologic conditions. Bench-scale tests should evaluate multiple oxidant dosage rates to include ones that can be realistically implemented at the field scale. Oxidant requirements should be calculated based on the scale of the target treatment volume. Field testing should include an experimental control to assist in the evaluation of contaminant dilution, displacement, and rebound, with rebound being utilized as the ultimate determinant of success or failure. A sufficient time period (often greater than one year) must elapse to allow rebound effects to be exhibited, particularly for sites with low relative groundwater seepage velocities (<100 feet/year) and/or with multiple soil layers across the contaminated region. The Navy should compile the lessons learned and the technical data obtained during Navy field applications of *in situ* oxidation.

Thermal treatment technologies provide aggressive and potentially successful remediation options for subsurface contaminants. Thermal technologies usually involve production or transport of steam through the subsurface, with the potential to volatilize contaminants. The flow paths of the steam and mobilized contaminants are determined by the heterogeneity and permeability of the subsurface matrix, which are also sensitive to degrees of water saturation. Thus, the application of thermal treatment technologies should be approached in a site-specific fashion, with a primary focus on site characterization and the design of effective vapor capture and dewatering strategies, particularly for sites where contaminants could exist in the saturated zone.

Permeable reactive barriers can effectively treat a limited number of groundwater pollutants under well-defined hydrogeologic conditions. These pollutants include PCE, TCE, *cis*-DCE, and perhaps Cr(VI). The technology has been applied in the field for approximately seven years, so data on long-term performance are limited. Hydraulic capture remains a key issue in determining effectiveness, and the long-term integrity of these systems is not known.

Enhanced anaerobic bioremediation has considerable potential for treating various types of organic contaminants in the subsurface, although it is not yet a proven, field-tested technology. Enhanced bioremediation can destroy contaminants via mineralization or conversion to benign organics, it can be applied *in situ*, and it is less expensive than other treatment technologies. Initial applications offer promise for *in situ* bioremediation of PCE and TCE. Significant questions remain concerning electron donor selection and delivery, long-term effectiveness, and cost.

Because metal contaminants cannot be destroyed and their behavior and speciation is strongly coupled to site-specific conditions, remediation approaches for metal contaminants remain a challenge. Given that metals are frequently reported contaminants of concern at Navy sites, the Navy should devote resources to accelerate the development of and field-scale testing of cost-effective technologies for mitigating risks from metal contaminants.

Presently, the only options that are routinely available for managing contaminated sediment include natural attenuation, capping either *in situ* or after dredged material removal, and dredging with disposal in confined disposal facilities or in upland landfills. Dredged material treatment options are under development and may be commercially available and viable in the future. Because of the large volumes of sediment dredged to maintain navigation projects in many harbors, it is likely that economies of scale will encourage substantial application of these technologies.

Treatment trains for the remediation of many contaminated sites is an important component of adaptive site management. Most sites are contaminated with multiple contaminants that may require different treatment processes. Treatment trains can often increase the effectiveness in achieving remedial goals while decreasing the cost of remediation. A common treatment train is source control in conjunction with monitored natural attenuation. This approach must be implemented with caution as source removal can disrupt microbial metabolism via redox changes, removal of primary substrates, and creation of inhibitory conditions.

Site-specific analyses of the effectiveness of source removal are needed to better guide and justify remedy selection. Additional studies including controlled field demonstrations are needed to evaluate the benefits (e.g., to groundwater quality) derived from partial contaminant mass removal from source zones and the compatibility of such treatment with natural attenuation. This analysis will also help PRPs determine if enough of the source mass can be removed to warrant the expense of implementing the technology.

GLOSSARY

Air sparging. Removal of volatile chemicals from the subsurface by injecting air beneath the water table and extracting vapors with vacuum and sometimes subjecting the vapors to subsequent treatment.

Bioremediation. Exploitation of the metabolic activities of micro-organisms to transform or destroy contaminants. *Enhanced* bioremediation refers to the addition of carbon and energy sources and/or electron acceptors to stimulate the growth of indigenous microbes and increase the rate of intrinsic biodegradation. Enhanced aerobic bioremediation of petroleum hydrocarbons is an established treatment, while enhanced aerobic cometabolism of chlorinated aliphatic hydrocarbons is an emerging technology. Enhanced anaerobic bioremediation is an innovative technology that involves adding compounds to stimulate reductive dechlorination of chlorinated aliphatic hydrocarbons.

Bioslurping. Simultaneous application of vacuum-enhanced extraction/recovery, vapor extraction, and bioventing to remove/transform contaminants, particularly LNAPLs.

Bioventing. Passing air through the soil to stimulate biodegradation of organic material with minimum volatilization.

Capping. Providing an impermeable barrier to surface water infiltration into contaminated soil to reduce further contaminant release and transport, or controlled placement of a clean, isolating material cover over contaminated sediments without relocating or causing major disruption to the original bed.

Chemical Oxidation/Reduction. Use of chemical oxidants or reductants to oxidize or reduce organic and inorganic contaminants.

Circulating wells. Creation of a groundwater circulation cell around a well through which contaminated groundwater is cycled and treated by the action of an air stripping process.

Composting. Bioremediation of contaminated soils or sediments in the presence or absence of oxygen.

Confined disposal. Placing dredged materials within diked near-shore, island, or land-based confined disposal facilities and enclosing with a cap to provide isolation.

Cosolvents and surfactants. Mobilization or solubilization of NAPLs or sorbed contaminants for facilitated removal after injection and flushing of cosolvents or surfactants into the vadose and saturated zones.

Dual-phase extraction. Use of a screened vertical well with or without a drop tube under applied vacuum to extract contaminated vapor and both aqueous and nonaqueous liquid above and below the water table, possibly augmented with air injection.

Dynamic underground stripping. Combination of steam injection and electrical heating for vacuum extraction of nonaqueous phase liquid contaminants from the subsurface.

Electrokinetics. *In situ* process that separates and extracts inorganic and organic contaminants from saturated and unsaturated soil, sediments, and groundwater under the influence of an imposed electrical field.

Incineration. *Ex situ* thermal process primarily for the destruction or removal of organic compounds from contaminated matrices.

Hydraulic dredging. Employing centrifugal pumps to draw up sediment in a liquid slurry form for transfer through a pipeline to a placement site.

***In situ* heating.** Raising the temperature of soils by electrical resistance, microwave, and/or radio frequency heating to increase volatility of contaminants and to form steam for vapor-phase transport.

Landfill disposal. Placing contaminated materials, with or without pretreatment, in or on the land with liners and covers or caps for containment.

Land treatment. Managed treatment and disposal involving tillage of contaminated materials into the surface soil to allow natural assimilation for conversion and containment.

Mechanical dredging. Using bucket-like equipment to scoop up sediment by mechanical force to minimize sediment dispersion and other effects on sediment properties prior to transfer to the placement site.

Natural attenuation. *In situ* reduction in mass or concentration of contaminants in groundwater, soil, or surface waters from naturally occurring physical, chemical, and biological processes.

Permeable reactive barriers. Emplacement of reactive materials in a subsurface structure designed to intercept a contaminant plume, provide flow through the reactive media, and transform contaminants.

Phytoremediation. Use of natural or engineered vegetation for *in situ* plant uptake and containment of contaminated soils, sediments, and water.

Pump-and-treat. Use of a series of wells to pump large amounts of contaminated groundwater to the surface for treatment before ultimate surface discharge or reinjection.

Slurry phase bioremediation. Biological treatment of contaminated solids and groundwater in suspended growth bioreactors.

Soil flushing. *In situ* soil treatment of contaminants using chemical amendments and fluid pumping to mobilize and recover contaminants.

Soil vapor extraction. Use of induced air flow through the unsaturated zone to vacuum-remove volatile compounds from soil in the vapor phase with subsequent treatment and discharge to the atmosphere.

Soil washing. *Ex situ*, water-based process employing chemical and physical extraction and separation to remove contaminants from excavated soil.

Solidification/Stabilization. Reduction of hazard by converting contaminants into less soluble, mobile, or toxic forms using chemical, physical, and/or thermal processes.

Steam flushing. Injection of steam into the saturated and unsaturated zones to mobilize and volatilize contaminants before recovery through extraction wells and *ex situ* treatment.

Thermal desorption. Direct or indirect *ex situ* use of heat to physically separate and transfer contaminants from soils and sediments before subsequent collection and treatment.

Vertical barrier wall. Isolation of contaminant source from flowing groundwater with confinement trenches, grouts, or sheet piling to reduce risk and enhance opportunities for remediation.

Vitrification. Application of electrical heating to elevate temperature sufficiently to melt the soil and form a glass upon cooling for extraction/destruction and containment of contaminants.

REFERENCES

Ahlfeld, D. P., and A. E. Mulligan. 2000. Optimal management of flow in groundwater systems. Academic Press.

Air Force. 1997. Design guidance for application of permeable reactive barriers to remediate dissolved chlorinated solvents. Publication No. DG 1110-345-117. Prepared by Battelle for the U. S. Air Force.

Air Force. 2001. Final remedial process optimization handbook. Prepared for the Air Force Center for Environmental Excellence, Technology Transfer Division, Brooks Air Force Base, San Antonio, Texas, and Defense Logistics Agency, Environmental Safety Office, Fort Belvoir, VA.

American Academy of Environmental Engineers (AAEE), Innovative Site Remediation Technology, W. C. Anderson (ed.).

Vol. 1 Bioremediation, 1995, ISBN 1-883767-01-6
Vol. 2 Chemical Treatment, 1994, ISBN 1-883767-02-4
Vol. 3 Soil Washing/Soil Flushing, 1993, ISBN 1-883767-03-2
Vol. 4 Stabilization/Solidification, 1994, ISBN 1-883767-04-0
Vol. 5 Solvent/Chemical Extraction, 1994, ISBN 1-883767-05-9
Vol. 6 Thermal Desorption, 1993, ISBN 1-883767-06-7
Vol. 7 Thermal Destruction, 1994, ISBN 1-883767-07-5
Vol. 8 Vacuum Vapor Extraction, 1994, ISBN 1-883767-08-3

American Academy of Environmental Engineers (AAEE), Innovative Site Remediation Technology (Design & Application), W. C. Anderson (ed.).

Vol. 1 Bioremediation, 1997, ISBN 1-883767-17-2
Vol. 2 Chemical Treatment, 1997, ISBN 1-883767-18-0
Vol. 3 Liquid Extraction Technologies; Soil Washing, Soil Flushing, Solvent/Chemical, 1997, ISBN 1-883767-19-9
Vol. 4 Stabilization/Solidification, 1997, ISBN 1-883767-20-2
Vol. 5 Thermal Desorption, 1997, ISBN 1-883767-21-0
Vol. 6 Thermal Destruction, 1997, ISBN 1-883767-22-9
Vol. 7 Vapor Extraction and Air Sparging, 1997, 1-883767-23-7

Balshaw-Biddle, K., C. L. Oubre, and C. H. Ward. 2000. Steam and electro-heating remediation of tight soils. Boca Raton, FL: Lewis Publishers.

Beeman, R. E., J. E. Howel, S. H. Shoemaker, E. A. Salazar, and J. R. Buttram. 1994. A field evaluation of in situ microbial reductive dehalogenation by the biotransformation of chlorinated ethenes. Pp. 14 In: Bioremediation of chlorinated and polycyclic aromatic compounds. R.E. Hinchee et al. (eds.). Ann Arbor, MI: Lewis Publishers.

Boulicault, K. J., R. E. Hinchee, T. H. Wiedemeier, S. W. Hoxworth, T. P. Swingle, E. Carver, and P. E. Haas. 2000. Vegoil: a novel approach for stimulating reductive dechlorination. Pp. 1–7 In: Bioremediation and phytoremediation of chlorinated and recalcitrant compounds. G. B. Wickrama-nayake et al. (eds.). Columbus, OH: Battelle Press.

Boyle, S. L., V. B. Dick, M. N. Ramsdell, and T. M. Caffoe. 2000. Enhanced closure of a TCE site using injectable HRC®. Pp. 255–262 In: Bioremedia-tion and phytoremediation of chlorinated and recalcitrant compounds. G. B. Wickramanayake et al. (eds.). Columbus, OH: Battelle Press.

Bradley, S. P., A. C. Hax, and T. L. Magnanti. 1977. Applied mathematical programming. Boston, MA: Addison Wesley.

Brown, R. A., W. Mahaffey, and R. D. Norris. 1993. In situ bioremediation: the state of the practice. In: In situ bioremediation: when does it work? Washington, DC: National Academy Press.

Carr, C. S., S. Garg, and J. B. Hughes. 2000. Effect of dechlorinating bacteria on the longevity and composition of PCE-containing nonaqueous phase liq-uids under equilibrium dissolution conditions. Environ. Sci. Technol. 34:1088–1094.

Cherry, J. A., S. Feenstra, and D. M. Mackay. 1997. Developing rational goals for in situ remedial technologies. Pp. 75–98 In: Subsurface Restoration. C. H. Ward, J. A. Cherry, and M. R. Scalf (ed.). Chelsea, MI: Ann Arbor Press, Inc.

Davis, E. L. 1997. How heat can enhance in-situ soil and aquifer remediation: important chemical properties and guidance on choosing the appropriate technique. Ground Water Issue, April. EPA/540/S-97/502. Washington, DC: EPA.

Davis, E. L. 1998. Steam injection for soil and aquifer remediation. Ground Water Issue, January. EPA/540/S-97/505. Washington, DC: EPA.

Dept. of the Navy. 1998. Technical guidelines for evaluating monitored natural attenuation of petroleum hydrocarbons and chlorinated solvents in ground-water at Naval and Marine Corps facilities. Prepared by T. H. Wiedemeier and F. H. Chapelle.

Durham, L., D. Conboy, R. Johnson, and T. Sydelko. 1999. Precise excavation: an alternative approach to soil remediation. Pp. 93–98 In: Proceedings of National Defense Industrial Association meeting, Denver, Colorado, March 19–April 1.

Electric Power Research Institute (EPRI). 1999. Review of sediment removal and remediation technologies at MGP and other contaminated sites. TR-

113106. Palo Alto, CA: EPRI.

Environmental Protection Agency (EPA). 1994. Remediation guidance document, assessment and remediation of contaminated sediments program. EPA 905-R94-003. Washington, DC: EPA.

EPA. 1995. In situ remediation technology status report: electrokinetics. EPA 542-94-007. Washington, DC: EPA

EPA. 1997a. Remediation technologies screening matrix and reference guide (3rd Edition). EPA 542-B-93-005. Washington, DC: EPA.

EPA. 1997b. Recent developments for in situ treatment of metal contaminated soils. Contract Number 68-W5-0055. Prepared by PRC Environmental Management, Inc.

EPA. 1997c. Technology alternatives for the remediation of soils contaminated with As, Cd, Cr, Hg, and Pb. Engineering Bulletin EPA/540/S-97/500. Washington, DC: EPA.

EPA. 1998a. Field application of in situ remediation technologies: chemical oxidation. EPA 542-B-98-008. Washington, DC: EPA.

EPA. 1998b. Permeable reactive barrier technologies for contaminant remediation. EPA/600/R-98/125. Washington, DC: EPA.

EPA. 1999a. Hydraulic optimization demonstration for groundwater pump-and-treat systems. Volume I: pre-optimization screening (method and demonstration). EPA/542/R-99/011A. Washington, DC: EPA Office of Solid Waste and Emergency Response.

EPA. 1999b. Hydraulic optimization demonstration for groundwater pump-and-treat systems. Volume II: application of hydraulic optimization. EPA/542/R-99/011B. Washington, DC: EPA Office of Solid Waste and Emergency Response.

EPA. 1999c. Field applications of *in situ* remediation technologies: permeable reactive barriers. EPA 542-R-99-002. Washington, DC: EPA.

EPA. 1999d. Solidification/stabilization resource guide. EPA/542-B-99-002. Washington, DC: EPA.

EPA. 1999e. Phytoremediation resource guide. EPA 542-B-99-003. Washington, DC: EPA Office of Solid Waste and Emergency Response.

EPA. 2000a. Transmittal of final FY '00–FY'01 Superfund reforms strategy. OSWER Directive No. 9200.0-33. Washington, DC: EPA Office of Solid Waste and Emergency Response.

EPA. 2000b. Superfund reform strategy pump and treat optimization implementation plan. OSWER Directive No. 9283.1-13. Washington, DC: EPA Office of Solid Waste and Emergency Response.

EPA. 2000c. FRTR cost and performance remediation case studies and related information. EPA-542-C-00-001. Washington, DC: Federal Remediation Technologies Roundtable.

EPA. 2000d. Test methods for evaluating solid waste, physical/chemical methods. SW-846 Draft Update IVB. Washington DC: EPA Office of Solid Waste and Emergency Response.

EPA. 2000e. Innovative remediation technologies: field-scale demonstration projects in North America (2nd edition). Washington, DC: EPA Office of Solid Waste and Emergency Response.

EPA. 2000f. Solidification/stabilization use at Superfund sites. EPA-542-R-00-010. Washington, DC: EPA Office of Solid Waste and Emergency Response.

EPA. 2001a. Remediation technology cost compendium—year 2000. EPA-542-R-01-001. Washington, DC: EPA Office of Solid Waste and Emergency Response.

EPA. 2001b. Treatment technologies for site cleanup annual status report (10th edition). Washington, DC: EPA Office of Solid Waste and Emergency Response.

EPA. 2001c. Monitored natural attenuation: USEPA research program—an EPA Science Advisory Board review. EPA-SAB-EEC-01-004. Washington, DC: Science Advisory Board (1400A).

ESTCP. 1999. Technology status report: in situ oxidation. Washington, DC: EPA Office of Solid Waste and Emergency Response.

Evanko, C. R., and D. A. Dzombak. 1997. Remediation of metals-contaminated soils and groundwater. Technology Evaluation Report TE-97-01. Pittsburgh, PA: Ground-Water Remediation Technologies Analysis Center (GWRTAC).

Federal Remediation Technology Roundtable (FRTR). 1997. Remediation technologies screening matrix and reference guide (3rd edition). Washington, DC: FRTR.

Federal Remediation Technologies Roundtable (FRTR). Year. Remediation Case Studies.

Vol. 1 Bioremediation, 1995, EPA-542-R-95-002

Vol. 2 Groundwater treatment, 1995, EPA-542-R-95-003

Vol. 3 Soil vapor extraction, 1995, EPA-542-R-95-004

Vol. 4 Thermal desorption, soil washing, and *in situ* vitrification, 1995, EPA-542-R-95-005

Vol. 5 Bioremediation and vitrification, 1997, EPA-542-R-97-008

Vol. 6 Soil vapor extraction and other *in situ* technologies, 1997, EPA-542-R-97-009

Vol. 7 Ex situ soil treatment technologies (bioremediation, solvent extraction, thermal desorption), 1998, EPA-542-R-98-011

Vol. 8 *In situ* soil treatment technologies (soil vapor extraction, thermal processes), 1998, EPA-542-R-98-012

Vol. 9 Groundwater pump and treat (chlorinated solvents), 1998, EPA-542-R-98-013

Vol. 10 Groundwater pump and treat (nonchlorinated solvents), 1998, EPA-542-R-98-014

Vol. 11 Innovative groundwater treatment technologies, 1998, EPA-542-R-98-015

Vol. 12 On-site incineration, 1998, EPA-542-R-98-016

Fennell, D. E., and J. M. Gossett. 1998. Modeling the production of and competition for hydrogen in a dechlorinating culture. Environ. Sci. Technol. 32:2450–2460.

Ferrey, M. L., J. R. Lundy, and P. Estuesta. 2001. The effect of groundwater aeration on PCE natural attenuation patterns. Bioremediation Journal 5(3):211–224.

Fiorenza, S., C. L. Oubre, and C. H. Ward. 2000. Phytoremediation of hydrocarbon contaminated soil. Boca Raton, FL: Lewis Publishers. 164 pp.

Fitzpatrick W. 1994. Personal communication in the ARCS Remediation Guidance Document, EPA 905-R94-003. Washington, DC: EPA.

Foster Wheeler, Corp. 1999. New Bedford Harbor cleanup dredge technology review, final report.

Fountain, J. C. 1998. Technologies for dense nonaqueous phase liquid source zone remediation. GWRTAC Technology Evaluation Report TE-98-02. Pittsburgh, PA: GWRTAC.

Gallinati, J. D., S. D. Warner, C. L. Yamane, F. S. Szerdy, D. A. Hankins, and D. W. Major. 1995. Design and evaluation of an in-situ ground water treatment wall composed of zero-valent iron. Ground Water 33(5):834–835.

Gates, D. D., and R. L. Siegrist. 1995. In situ oxidation of trichloroethylene using hydrogen peroxide. Journal of Environmental Engineering 121:639–644.

Gavaskar, A. R., N. Gupta, B. M. Sass, R. J. Janosy, and D. O'Sullivan. 1998. Permeable barriers for groundwater remediation: design, construction, and monitoring. Columbus, OH: Battelle Press.

Gavaskar, A. R., N. Gupta, B. M. Sass, R. J. Janosy, and J. Hicks. 2000. Design guidance for application of permeable reactive barriers for groundwater remediation. Prepared for the Air Force Research Laboratory, Tyndall AFB, FL.

Gavaskar, A., B. Sass, N. Gupta, E. Drescher, W-S. Yoon, J. Sminchak, and J. Hicks. 2001. Evaluating the longevity and hydraulic performance of permeable reactive barriers at Department of Defense sites. Final report for NFESC, Charles Reeter, Project Officer.

Gill, P. E., W. Murray, and M. H. Wright. 1981. Practical optimization. San Diego, CA: Academic Press Limited.

Gorelick, S. M., C. I. Voss, P. E. Gill, W. Murray, M. A. Saunders, and M. H. Wright. 1984. Aquifer reclamation design: the use of transport simulation combined with nonlinear programming. Water Resources Research 20(4):415–427.

Gorelick, S. M., R. A. Freeze, D. Donohue, and J. F. Keely. 1993. Groundwater contamination: optimal capture and containment. Boca Raton, FL: Lewis Publishers.

GWRTAC. 1997. Technology overview report: electrokinetics. Pittsburgh, PA: GWRTAC.

Haas, P. E., P. Cork, and C. E. Aziz. 2000. In situ biowall containing organic mulch promotes chlorinated solvent bioremediation. Pp. 71-76 In: Bioremediation and phytoremediation of chlorinated and recalcitrant compounds. G. B. Wickramanayake et al. (eds.). Columbus, OH: Battelle Press.

Hansen, M. A., J. Burdick, F. C. Lenzo, and S. Suthersan. 2000. Enhanced reductive dechlorination: lessons learned at over twenty sites. Pp. 263–270 In: Bioremediation and phytoremediation of chlorinated and recalcitrant compounds. G. B. Wickramanayake et al. (eds.). Columbus, OH: Battelle Press.

Harkness, M. R. 2000. Economic considerations in enhanced anaerobic biodegradation. Pp. 9–14 In: Bioremediation and phytoremediation of chlorinated and recalcitrant compounds. G. B. Wickramanayake et al. (eds.). Columbus, OH: Battelle Press.

Hayes, D.F., T. R. Crockett, T. J. Ward, and D. Averett. 2000. Sediment resuspension during cutterhead dredging operations. J. Waterw. Port Coast Ocean Eng. 126(3):153–161.

Jackson, R. E. 2001. DNAPL remediation: which "new paradigm" will prevail? Ground Water Monitoring and Remediation 21(3):54–58.

Jafvert, C. T. 1996. Surfactants/cosolvents. GWRTAC Technology Evaluation Report TE-96-02. Pittsburgh, PA: GWRTAC.

Kauss, P. B., and P. C. Nettleton. 1999. Impact of 1996 Cole Drain Area contaminated sediment cleanup on St. Clair River water quality. ISBN 0-7778-8598-0.

Lasat, M. M. 2002. Phytoextraction of toxic metals. J. Environ. Qual. 31:109–120.

Lee, A. S., and J. S. Aronofsky. 1958. A linear programming model for scheduling crude oil production. Journal of Petroleum Technology 10(7):51–54.

Lee, M. D., and R. L. Raymond, Sr. 1991. Case history of the application of hydrogen peroxide as an oxygen source for in situ bioreclamation. Pp. 429–436 In: In situ bioreclamation: applications and investigations for hydrocarbon and contaminated site remediation. R. E. Hinchee and R. F. Olfenbuttel (eds.). Boston: Butterworth-Heinemann.

Leigh, D. P., C. D. Johnson, R. S. Skeen, M. G. Butcher, L. A. Bienkowski, and S. Granade. 2000. Enhanced anaerobic *in situ* bioremediation of chloroethenes at NAS Point Mugu. Pp. 229-235 In: Bioremediation and phytoremediation of chlorinated and recalcitrant compounds. G. B. Wickramanayake et al. (eds.). Columbus, OH: Battelle Press.

Liang, L., N. Korte, B. Gu, R. Puls, and C. Reeter. 2000. Geochemical and microbial reactions affecting the long-term performance of in situ 'iron barriers'. Adv. In Env. Research 4:273–286.

McCarty, P. L., M. N. Goltz, G. D. Hopkins, M. E. Dolan, J. P. Allan, B. T. Kawakami, and T. J. Carrothers. 1998. Full-scale application of *in situ* cometabolic degradation of trichloroethylene in groundwater through toluene injection. Environ. Sci. Technol. 32:88–100.

McDonald, M. G., and A. W. Harbaugh. 1996. A modular three-dimensional finite-difference ground-water flow model. Techniques of Water-Resources Investigations of the U.S. Geological Survey, Book 6, Chapter A1. Washington, DC: U.S. Government Printing Office.

McLellan, T. N., and R. J. Hopman. 2000. Innovations in dredging technology: equipment, operations, and management. ERDC TR-DOER-5. Vicksburg, MS: U.S. Army Corps of Engineers.

McLellan, T. N., R. N. Havis, D. F. Hayes, and G. L. Raymond. 1989. Field studies of sediment resuspension characteristics of selected dredges. Technical Report HL-89-9. Vicksburg, MS: US Army Engineer Waterways Experiment Station.

Miller, R. R. 1996a. Air sparging. GWRTAC Technology Overview Report TO-96-04. Pittsburgh, PA: GWRTAC.

Miller, R. R. 1996b. Bioslurping. GWRTAC Technology Overview Report TO-96-05. Pittsburgh, PA: GWRTAC.

Miller, R. R., and D. S. Roote. 1997. In-well vapor stripping. GWRTAC Technology Overview Report TO-97-01. Pittsburgh, PA: GWRTAC.

Morrison, S. J., D. R. Metzler, and C. E. Carpenter. 2001. Uranium precipitation in a permeable reactive barrier by progressive irreversible dissolution of zerovalent iron. Environ. Sci. Technol. 35:385–390.

Murtagh, B. A., and M. A. Saunders. 1983. MINOS 5.1 user's guide. Technical Report SOL 83-20R. Stanford, CA: Stanford University Systems Optimization Laboratory, Department of Operations Research.

Myers, T. E., M. R. Palermo, T. J. Olin, D. E. Averett, D. D. Reible, J. L. Martin, S. C. McCutcheon. 1996. Estimating contaminant losses from components of remediation alternatives for contaminated sediments. Final Report. Chicago, IL: EPA Great Lakes National Program Office.

National Research Council (NRC). 1993. In situ bioremediation: when does it work? Washington, DC: National Academy Press.

NRC. 1994. Alternatives to ground water cleanup. Washington, DC: National Academy Press.

NRC. 1997a. Innovations in ground water and soil cleanup: from concept to commercialization. Washington, DC: National Academy Press.

NRC. 1997b. Contaminated sediments in ports and waterways. Washington, DC: National Academy Press.

NRC. 1999a. Groundwater and soil cleanup: improving management of persistent contaminants. Washington, DC: National Academy Press.

NRC. 1999b. Environmental cleanup at Navy facilities: risk-based methods. Washington, DC: National Academy Press.

NRC. 2000. Natural attenuation for groundwater remediation. Washington, DC: National Academy Press.

NAVFAC. 2001. Guidance for optimizing remedial action operation (RAO). Special Report SR-2101-ENV. Prepared for the Naval Facilities Engineering Service Center. Research Triangle Park, NC: Radian International.

Newell, C. J., P. E. Haas, J. B. Hughes, and T. Khan. 2000. Results from two direct hydrogen delivery field tests for enhanced dechlorination. Pp. 31–37 In: Bioremediation and phytoremediation of chlorinated and recalcitrant compounds. G. B. Wickramanayake et al. (eds.). Columbus, OH: Battelle

Press.

Palermo, M. R., S. Maynord, J. Miller, and D. D. Reible. 1998. Guidance for in-situ subaqueous capping of contaminated sediments. EPA 905-B96-004. Great Lakes National Program Office, Assessment and Remediation of Contaminated Sediments Program.

Phillips, D. H., B. Gu, D. B. Watson, Y. Roh, L. Liang, L., and S. Y. Lee. 2000. Performance evaluation of a zero valent iron reactive barrier: mineralogical characteristics. Environ. Sci. Technol. 34:4169–4176.

PIANC. 2000. Innovative dredged sediment decontamination and treatment technologies. Results of a Specialty Workshop, U.S. Section of the International Navigation Association, May 2, 2000, Oakland, CA.

Rao, P. S. C., J. W. Jawitz, C. G. Enfield. R. W. Falta, Jr., M. D Annable, and A. L. Wood. 2002. Technology integration for contaminated site remediation: cleanup goals and performance criteria. Pp. 571–578 In: Groundwater Quality: Natural & Enhanced Restoration of Groundwater Pollution (Eds., S.F. Thornton & S.E. Oswald), IAHS Publication Number 275, International Association of Hydrological Sciences, Oxfordshire, United Kingdom.

Reible, D. D., R. H. Jensen; S. J. Bentley, M. B. Dannel, J. V. DePinto, J. A. Dyer, K. J. Farley, M. H. Garcia, D. Glaser; J. M. Hamrick, W. J. Lick, R. A. Pastorok, R. F. Schwer, and C. K. Ziegler. 2002. The role of modeling in managing contaminated sediments. Chapter 2 In: Environmental modeling and management: theory, practice, and future directions. Wilmington, DE: Today Media, Inc.

Rittmann, B. E., and P. L. McCarty. 2001. Environmental biotechnology: principles and applications. Boston, MA: McGraw-Hill.

Sacre, J. A. 1997. Treatment walls: a status update. TP-97-02. Pittsburgh, PA: GWRTAC.

Sale, T. 1998. Interphase mass transfer from single component DNAPLs. Ph.D. Dissertation, Department of Chemical and Bioresource Engineering, Colorado State University, Ft. Collins, CO.

Sale, T., and D. McWhorter. 2001. Steady-state mass transfer from single-component dense NAPLs in uniform flow fields. Water Resources Research 37:2:393–404.

Scherer, M. M., S. Richter, R. L. Valentine, and P. J. J. Alvarez. 2000. Chemistry and microbiology of permeable reactive barriers for *in situ* groundwater cleanup. Critical Reviews in Environmental Science and Technology 30:363–411.

Schnoor, J. L. 1998. Phytoremediation. GWRTAC Technology Evaluation Report TE-98-01. Pittsburgh, PA: GWRTAC.

Schnoor, J. L. 2002. Phytoremediation of soil and groundwater. Pittsburgh, PA: GWRTAC.

Schrage, L. 1997. Optimization modeling with LINDO (5th edition). Duxbury Press.

Semprini, L., G. D. Hopkins, P. L. McCarty, and P. V. Roberts. 1992. *In situ* transformation of carbon tetrachloride and other halogenated compounds resulting from biostimulation under anoxic conditions. Environ. Sci. Technol. 26:2454–2461.

Sun, Y.-H., and W. W.-G. Yeh. 1998. Location and schedule optimization of soil vapor extraction system design. ASCE Journal of Water Resources Planning and Management 124(1):47–58.

Till, B. A., L. J. Weathers, and P. J. Alvarez. 1998. Fe(0)-supported autotrophic denitrification. Environ. Sci. Technol. 32:634–639.

U.S. Army Corps of Engineers (USACE). 1995. Soil vapor extraction and bioventing. Engineering and Design Manual No. 1110-1-4001. Washington, DC: USACE.

Valsaraj, K. T., L. J. Thibodeaux, and D. D. Reible. 1995. Modeling air emissions from contaminated sediment dredged materials. In: Dredging, remediation, and containment of contaminated sediments. ASTM STP 1293. K. R. Demars, G. N. Richardson, R. N. Young, and R. C. Chaney, (eds.). Philadelphia, PA: American Society for Testing and Material.

Vidic, R. D., and F. G. Pohland. 1996. Treatment walls. GWRTAC Technology Evaluation Report TE-96-01. Pittsburgh, PA: GWRTAC.

Vidic, R. D. 2001. Remediation technology evaluation report: permeable reactive barriers: long-term performance (draft report). Pittsburgh, PA: GWRTAC.

Warner, S. D., C. L. Yamane, J. D. Gallinatti, and D. A. Hankins. 1998. Considerations for monitoring permeable ground-water treatment wall. Journal of Environmental Engineering 124(6):524–529.

Weathers, L. J., G. F. Parkin, and P. J. Alvarez. 1997. Utilization of cathodic hydrogen as electron donor for chloroform cometabolism by a mixed, methanogenic culture. Environ. Sci. Technol. 31:880–885.

WDNR. 2001. Draft RI/FS for Lower Fox River and Green Bay, Wisconsin. Wisconsin Department of Natural Resources, October.

Yang, Y., and P. L. McCarty. 1998. Competition for hydrogen within a chlorinated solvent dehalogenating anaerobic mixed culture. Environ. Sci. Technol. 32:3591–3597.

Yang, Y., and P. L. McCarty. 2000a. Biomass, oleate, and other possible substrates for chloroethene reductive dehalogenation. Bioremediation Journal 4:125–133.

Yang, Y., and P. L. McCarty. 2000b. Biologically enhanced dissolution of tetrachloroethene DNAPL. Environ. Sci. Technol. 34:2979–2984.

Zeman, A. J., and T. S. Patterson. 1997. Results of in-situ capping demonstration project in Hamilton Harbour, Lake Ontario. International Symposium on Engineering Geology and the Environment, Athens, Greece, June 23–27, 1997.

6

Nontechnical Issues Regarding
the Use of Adaptive Site Management

This chapter considers whether currently existing regulations and regulatory guidance from the U. S. Environmental Protection Agency (EPA) and the Department of Defense (DoD) allow for the use of adaptive site management (ASM). All the mechanisms for changing and modifying selected remedies—formal amendments to Records of Decision (ROD), RCRA permit modifications, various other documents such as the CERCLA Explanation of Significant Differences, contingency RODs, five-year reviews, impracticability waivers, and optimization studies, among others—can be encompassed by ASM. In addition to identifying significant regulatory and policy issues, the chapter also considers other relevant nontechnical issues including the role of the public and long-term stewardship (which is synonymous with long-term management in DoD guidance) in ASM. To make changes in remedial strategies, it is necessary to achieve consensus among stakeholders, including the lead regulatory agency, the responsible party, the affected public, and public or private transferees. The ASM tools described in Chapter 3 are critical to help demonstrate to diverse stakeholder groups that changes are warranted. Finally, long-term stewardship figures prominently in ASM and is an area in which federal facilities are only now gaining experience. These topics are discussed minimally in recent Navy guidance on optimization of remedies (NAVFAC, 2001). Several areas where supplemental guidance will be needed to fully adopt ASM are highlighted below.

REGULATORY AND POLICY ISSUES

Federal facility cleanups must comply with the Superfund cleanup requirements, as well as any more stringent state requirements (Section 120 of CERCLA and 10 U.S.C. § 2701). Superfund requires that each remedy be protective and attain the "applicable or relevant and appropriate requirements" (ARARs) provided in federal and state environmental laws. Protectiveness can be achieved by reducing the soil, groundwater, or surface water contaminant concentrations to below health-based levels or by preventing exposure without removing or destroying the chemicals at the site. Historically, cleanup goals for groundwater have been set at or below the drinking water standard for those contaminants that have one, or at a concentration within the risk range of 10^{-4} to 10^{-6} for carcinogens and a hazard index of less than 1.0 for noncarcinogens (EPA, 1990). A typical standard for contaminants in groundwater is in the range of 0.5 parts per billion to low parts per billion (NRC, 1994, 1999). Ecological risks or other site-specific factors may result in more stringent cleanup goals. ARARs, including drinking water standards, can be waived, for example, if among other reasons implementing the remedy would result in a greater risk to human health and the environment, compliance with the requirement is technically impracticable from an engineering perspective, or another remedial action would attain the performance equivalent of the federal or state requirement.

The lead agency for making cleanup decisions and the enforcing regulatory agency are different depending on whether the site (and associated facility) is on the National Priorities List (NPL) and on the relevant regulation used to guide cleanup. DoD acts as the lead agency for its sites regulated under CERCLA, although the agency must follow EPA cleanup guidance (DoD, 1999). EPA is the regulator only for those sites actually on the NPL. In practice, there may be DoD cleanups where EPA guidance is not as rigorously followed when EPA is not overseeing the activity (CPEO, 2002). At a federal facility, the remedy is selected by the head of the relevant department, although EPA must concur; if the federal agency and EPA are unable to reach agreement, the remedy is selected by EPA [CERCLA Section 120(e)(4)].

DoD also addresses many of its non-NPL sites under CERCLA, but in those cases, lead regulatory authority is held by the state. The enforcement role of EPA at these sites is greatly reduced, although EPA plays a role at closing bases because CERCLA Section 120(h) requires EPA to review transfers. There are a few sites, such as the Naval Ammunition Support Detachment Vieques, where DoD has invited EPA to oversee cleanup at non-NPL sites.

At DoD's RCRA sites, EPA may be the lead regulator—that is, it may issue administrative orders—if it has not delegated to the state/territory the relevant RCRA authority such as corrective action. In addition, many states assert regulatory authority under state hazardous waste laws. For example, at Rocky Mountain Arsenal, Colorado has regulatory authority throughout the process, including post-ROD remedy decisions.

Current Guidance on Optimizing and Changing Remedies and Remedial Goals

The greatest utility of using ASM lies in the ability to make changes over time as new information on site conditions and on the effectiveness of remedies becomes available. The approach identifies periods during which decisions can be made regarding the optimization of existing remedies, the changing of remedies, and the addition of new technologies to speed restoration—even if the existing remedy is maintaining protectiveness. Significant new information might include post-ROD, pre-implementation sampling concerning the extent or degree of contamination or a risk assessment that indicates the remedial action is unacceptable or overly protective. Typically, minor or insignificant adjustments do not require public comments (EPA, 1999a). For sites where contamination remains onsite following implementation of the remedy—such as NPL sites with dense nonaqueous phase liquid (DNAPL) contamination—the CERCLA five-year review process provides a long-term opportunity to make changes to the chosen remedy (although as discussed later, it is rarely used in this capacity).

Changing remedies over time is already addressed in a number of EPA regulations, policies, and guidance as well as in DoD guidance (Air Force, 2001; DoD, 2001a; NAVFAC, 2001). For example, EPA regulations (EPA, 1990) and policy (EPA, 1996a) clearly provide for modification of the remedy when new information is obtained that could affect the implementation of that remedy. The party seeking the change (e.g., the Navy) must generate the information needed to justify such a change. EPA guidance states that the final decision on whether to change the remedy (even at federal restoration sites) rests with EPA (EPA, 1996a,b).

Changing and optimizing remedies are widely acknowledged on a policy level as well. A number of guidance documents require potentially responsible parties (PRPs), including federal facilities, to seek to optimize the site remedial action. For example, DoD's Closeout Guidance (which cites Air Force, 2001, and EPA, 1996a) states that "emphasis should be placed on optimization" as early in the process as possible, and the remedy should be changed if new information supports it.

Thus, there are no apparent legal or regulatory prohibitions to using ASM for making decisions about optimizing or changing remedies. In fact, the approach appears to be consistent with current DoD trends toward experiential optimization. The ASM flowchart described in Chapter 2 includes specific questions that should be asked during critical decision-making periods, which goes beyond the most recent guidance document developed for experiential optimization of cleanup at Navy facilities (NAVFAC, 2001).

EPA regulations also allow for remedial goals to be changed. The most prominent example, in the case of contaminated groundwater, is a technical impracticability (TI) waiver under Superfund or RCRA, which can be issued at sites where remedies are not meeting cleanup goals (EPA, 1993). If granted an impracticability waiver pursuant to EPA's existing policy, the PRP must implement an alternative remedial action, which may include a new remedial goal or containment of the plume. Any alternative remedial strategy that leaves contamination onsite must remain in effect at Superfund sites so long as it is protective of human health and the environment, which has to be reassessed every five years. If a new non-health-based remedial action goal is set, then no further action would be required once this new goal is attained. Box 6-1 describes the elements that must accompany an application for a technical impracticability waiver.

To fully embrace an ASM approach, DoD should adopt a policy of applying new technologies that might attain the original cleanup goals at Superfund sites that have received technical impracticability waivers or where cleanup is considered impracticable. This could serve to stimulate research, to minimize future operation and maintenance costs, and/or to reduce risks such that additional land uses would be permitted. It is possible that at many sites the economic benefits of site redevelopment may exceed the cost of additional cleanup that would allow for a broader range of land uses.

Although the above recommendation represents an opportunity to update EPA's technical impracticability guidance to be consistent with ASM, its utility may be limited because there is little evidence that DoD intends to apply for TI waivers on a regular basis. In fact, recent NAVFAC guidance (NAVFAC, 2001) clearly favors the consideration of

BOX 6-1
Technical Impracticability for Contaminated Groundwater Sites

Technical impracticability refers to a situation "where achieving groundwater cleanup objectives is not possible from an engineering perspective" (EPA, 1993). Although there is no specific timeframe that defines impracticability, the guidance has been interpreted to mean very long timeframes (e.g., longer than 100 years) that are indicative of hydrogeologic or contaminant-related constraints to remediation. Technical impracticability (TI) waivers consider the feasibility, reliability, scale, and safety of the remedial option. Some cleanup approaches may be technically possible, but the scale of the operation might be of such magnitude that it is not technically practicable.

Requests for technical impracticability waivers are encouraged early during corrective action (e.g., during facility characterization) if a site has hydrogeologic or chemical-related features that are known to present cleanup limitations. EPA has made it clear that poor cleanup performance due to inadequate remedial design is not sufficient justification for a technical impracticability waiver. Rather, the waiver is usually based upon the presence of nonaqueous phase liquids (NAPL) and their persistence and location, as well as upon the technologies that are available to clean them up. Although the amount of characterization needed for a TI waiver will vary on a site-by-site basis, the waiver application should include (EPA, 1993):

- the spatial area over which the TI decision would apply,
- the specific groundwater cleanup objectives that are considered technically impracticable to achieve,
- the conceptual site model that describes geology, hydrology, groundwater contamination sources, transport, and fate,
- an evaluation of the "restoration potential" of the TI zone,
- cost estimates,
- any additional information EPA or the state program deems necessary (e.g., the difference in the timeframe for cleanup with and without the TI waiver), and
- an alternative remedial strategy.

The alternative remedial strategy should be technically practicable, control the sources of contamination, and prevent migration of contamination beyond the zone associated with the technical impracticability determination. It must be capable of achieving the groundwater cleanup objectives outside the zone associated with the technical impracticability determination, and it must be consistent with the overall cleanup goals for the facility.

The obligations for monitoring and containment within the TI zone continue as long as necessary to protect human health and the environment, or in the case of RCRA sites, until such time that cleanup within the TI zone becomes technically practicable and the cleanup levels are achieved throughout the entire plume.

new remedies rather than TI waivers if the original remedy reaches the point of diminishing returns. Indeed, in the last eight years, EPA has issued only 29 technical impracticability waivers to private and public PRPs (only 30 have been sought). These low numbers may reflect the high transaction cost involved in obtaining such waivers, the likely potential public backlash, and the shift since 1990 toward selecting containment, natural attenuation, and other remedies perceived to be less expensive. For example, changing to a containment remedy achieves the same cost reduction goal as a TI waiver but without permanently changing the ultimate cleanup goal for the site. Thus, many regulators may prefer requiring reasonable source control measures coupled with long-term containment of the residual over providing TI waivers.

Policy Barriers to Adaptive Site Management

Despite the fact that the current DoD and EPA guidance encourages optimization and that ASM is inherently consistent with the CERCLA and RCRA frameworks, there are potential policy barriers to adopting ASM on a widespread basis. First, as discussed in Chapter 2, there is no specific requirement under CERCLA to reconsider remedies over time that are ineffective in reaching cleanup goals as long as they are protective of human health and the environment (EPA, 1993, 2001a). Thus, there is relatively little incentive to optimize remedies once they are in place. The same is not true of RCRA sites, where EPA may revisit the remedy not just for protectiveness and reliability, but also if subsequent advances in remediation technology make attainment of the original cleanup standards technically practicable (EPA, 1993). Because the legal obligation to initiate a cleanup at RCRA sites (called a corrective action) is implemented through a hazardous waste permit or administrative order, EPA's authority to require periodic updating of the remedial action at RCRA sites may be stronger from a legal point of view. (It should be noted, however, that there is little evidence to date that EPA has utilized this authority to revisit the remedy at RCRA sites.) At a CERCLA site, a remedial project manager (RPM) would not by law be prevented from reconsidering an ineffective remedy. However, pressure to close out sites and rely on containment and institutional controls generally preclude this type of activity on any measurable scale. If ASM is to be adopted, such reconsideration must not just be allowed but should be required.

EPA's policy of not requiring additional remedial actions at CERCLA sites is based on statutory provisions and policy judgments that do not apply to federal facilities. For example, CERCLA's covenant-not-to-sue provision provides an explicit legal release of future liability if a private PRP successfully implements the remedy selected by EPA. Historically, judicial actions have generally favored finality rather than open-ended obligations. There is a long-standing policy of encouraging private parties to implement CERCLA remedies by providing this type of finality. However, because EPA cannot bring a judicial enforcement action against a federal facility, the covenant-not-to-sue does not apply to military sites. In fact, a strong public policy argument can be made that federal facilities should take the lead in encouraging the development and application of innovative technologies to hazardous waste sites where the remedy is not reaching cleanup goals. Clearly, because there is a different policy for RCRA sites than for CERCLA private sector sites, there could be a different policy for federal restoration sites that would better embrace the principles of ASM. This approach would be similar to that taken in a number of environmental statutes. For example, the Clean Water Act sets nonenforceable goals (e.g., zero discharge), but requires the step-by-step implementation of technologies to attain industry-specific discharge limits that are periodically made more stringent if new technology is developed.

Second, it has been argued that amending the ROD to change the remedy to reflect new data and advances in technologies is a "cumbersome process" (NRC, 1997), and for this reason, approaches such as ASM that encourage reconsideration of remedies over time may be less likely to succeed. However, there has been improvement over the last several years, such that EPA has updated a total of 300 remedy decisions through the end of the 1999 fiscal year, thereby saving an estimated $1.4 billion, although the costs at some sites have increased (EPA, 2001b). There were 156 updates to soil remedies and 129 updates to groundwater remedies; federal facilities updated 18 remedial actions. Over 62 percent of the changed remedial actions still involved treatment, and 17 percent were changed from groundwater treatment to monitored natural attenuation. These changes generally occurred in the remedial design stage. Most remedy changes were modifications to the original remedy, not installation of a completely new remedy.

Finally, there is little guidance available to Navy RPMs to assist them in evaluating whether remedies are operating optimally or whether remedies are unlikely to attain site-specific cleanup goals and need to be modified to ensure protectiveness—a key decision period in ASM (MDP3). For example, none of the existing guidance on changing the remedy (EPA, 1996a), on technical impracticability (EPA, 1993), on the five-year review (EPA, 2001a), and on site closeout (DoD, 1999) provides a systematic scientific approach to assessing optimization or to determining when groundwater contaminant concentration reduction has leveled off at a concentration significantly higher than the cleanup goal. The existing guidance is also inadequate to address monitoring needs after remedy implementation (MDP2). The same documents mentioned above explicitly state that additional data will be needed, but they do not provide concrete information on the types of data that are useful and when data gathering should be initiated. NAVFAC (2001) goes a long way toward providing some of this guidance and, as recommended in Chapter 2, should be considered for formal adoption by the Navy. However, this report does not discuss the research track of ASM (discussed in Chapter 4) or the reconsideration of remedies during long-term stewardship (MDP4), and both of these issues are absent from other guidance documents as well (such as the EPA reports on optimization).

PUBLIC PARTICIPATION

Since at least the early 1980s, the federal government and most other stakeholder groups have recognized public participation as an essential part of the process for cleaning up contaminated sites (EPA, 2000a). The mechanisms and timing of public involvement, however, have evolved over time. The ASM model suggests the need to update public participation methodologies once again.

CERCLA and other statutes that govern the cleanup of contaminated sites emphasize the public's right to influence the selection of remedies. The general process at federal facilities is for the lead agency to list a series of remedial options at each operable unit and propose a preferred alternative. Members of the public are then given the opportunity to comment on the proposal in writing during a brief public comment period or in person at one or more public meetings conducted during that period. This approach, however, proved inadequate at many contaminated sites, particularly large, complex federal facilities such as those owned by the Departments of Energy and Defense. Thus, in the early 1990s, the Federal Facilities Environmental Restoration Dialogue Committee (FFERDC) brought together federal agencies, state, tribal, and local government representatives, and community activists to explore ways to improve public participation in the federal cleanup process. FFERDC found that "where a public involvement process is mandated by law, the public often perceives that the process is used to defend decisions already made without meaningful dialogue with the affected public" (FFERDC, 1993). FFERDC participants labeled this model "Decide, Announce, Defend."

FFERDC laid the groundwork for a major expansion in the role of the affected public in cleanup decision making. It set a new standard—"early and often"—for public participation, going beyond the public comment opportunities required just before remedy selection (FFERDC, 1996). It recommended "regular, early, and effective public participation in federal cleanup programs" (FFERDC, 1993). This led to the creation of site-specific advisory boards at federal facilities across the United States. DoD established more than 292 Restoration Advisory Boards (RABs) to oversee environmental response at more than 356 present and former facilities. The Navy supports at least 91 such boards (Navy, 1999, RAB Supplement, p. 7). RABs provide opportunities for the public to learn about and comment on cleanup activities well beyond the minimal requirements of CERCLA and other hazardous waste laws. Although the implementation of RABs has been uneven across the military, the Navy has been vigilant and consistent in its overall efforts to involve the public in decision making. Like citizen advisory boards in other domains, the success of a RAB depends on a combination of factors, including the composition of the board and the commitment of its members, the formal and practical extent of the committee's role and influence, and the social and interpersonal environment created by the agency, facilitators, and members (Renn et al., 1995; DOE, 1997a; Chess and Purcell, 1999; Lynn et al., 2000; Murdock and Sexton, 2002).

Changing Role of Public Participation

As site cleanup has progressed in the United States, more sites are being remediated with containment and institutional controls such that significant levels of contamination remain onsite (see Figure 1-8). This trend in hazardous waste cleanup calls for another shift in the mode of public participation. Just as regulatory oversight and technological review are necessary until a site is closed out, at properties where the selected remedy is designed to leave contamination in place, public participation should not only occur early and often, but as long as contamination remains onsite at levels above cleanup goals. The rationale is simple: if the public is required to be involved in selecting the remedy because it may affect their health and well-being, then the public must similarly be involved in any significant decision to change that remedy or land use because these decisions also may affect their health and well-being.

The adoption of ASM is expected to make the public's role in cleanup more essential over time because new decisions that require their interaction will arise periodically as cleanup progresses. For example, the public may play a role in the evaluation and experimentation element of ASM, as discussed in Chapter 4. The Moffett Field Restoration Advisory Board has shown ongoing interest in and support for the Permeable Reactive Barrier (PRB) demonstration being conducted at Moffett. In fact, RAB members recently urged that monitoring at the demonstration site continue even after depletion of dedicated research funding. Furthermore, the Technical Assistance Grant consultant employed by the Silicon Valley Toxics Coalition—a local community group that has participated in Moffett oversight since 1989—took part in a national task force on permeable reactive barriers organized by the Interstate Technology Regulatory Council Working Group (P. Strauss, P. M. Strauss & Associates, personal communication, 2002).

Public participation, which is particularly critical during MDP4 of ASM, is expected to occur regularly and over the long term at sites where contamination remains in place. Personnel and contractors representing both responsible parties and regulatory agencies tend to change every few years, and in fact responsibility for cleanup is often transferred to new organizations. This can lead to a loss of institutional memory that often only public participants can fill. Continuity will require that the collaborative decision-making process involving responsible parties, regulators, and stakeholders established before remedy selection continue as long as significant contamination remains onsite. This requires that regulators approve and the public oversee cleanup decisions made after the signing of RODs, which is sometimes not the modus operandi at military cleanups. Achieving a high level of public participation years after the initial studies and the signing of the ROD may prove difficult, but it is essential to the long-term success of cleanup.

Current Trends in Public Participation During Long-Term Stewardship

Existing guidance on the latter stages of site cleanup states that there should be public involvement in updates to remedial actions, five-year reviews, technical impracticability determinations, and the site closeout decision. The degree of public involvement in changing a remedy depends upon whether the change is minor (in which case virtually no prior public involvement is required) or is a modification of the existing remedy (in which case some public involvement is necessary, but not as much as for a complete change in remedy) (EPA, 1999a). The five-year review guidance states that when no contaminants remain onsite above levels that allow for unrestricted use and unlimited exposure, a determination of closeout must be subjected to public comment (EPA, 2001a). However, it is not clear whether a public meeting is required, and the extent to which this requirement applies to non-NPL federal sites. EPA's technical impracticability guidance states that any alternative remedies must be selected using the existing CERCLA and RCRA remedy selection processes, which include public comments.

Despite these specific calls for public involvement, public interest in the cleanup process tends to peak at certain times, such as when threats to public health are discovered or disclosed, or when facilities are scheduled for closure and transfer. When remediation becomes routine, community interest tends to decline. Some RABs, such as at Moffett Field—the original model used by the Federal Facilities Environmental Restoration Dialogue Committee—have started to meet less frequently. Others, such as the RAB at the Philadelphia Navy complex, have lost members, particularly those who attended as volunteers. Thus, the RAB model must evolve to accommodate operation and maintenance activities occurring long after the signing of the ROD.

As mentioned above, it is standard to involve the public in long-term site management by inviting public comment on certain proposed changes, such as Explanations of Significant Difference, or on recurring review documents (i.e., five-year reviews). Depending upon the legal status of the cleanup, the latter usually occurs about four years after remedial construction starts, which may be several years after the remedy selection process—the initial focus of public involvement. Not surprisingly, the committee's review of over 30 recent five-year review reports found public involvement to be limited, although there are exceptions. For example, in 2000, at the Shattuck Chemical Company site in Denver, community-based critics, supported by high-level elected officials, used the five-year review to overturn the original remedy (SC&A, Inc., 1999). In 1999, the process of writing a five-year review spurred the residents of San Diego's Tierrasanta neighborhood—a former defense site contaminated with unexploded ordnance—to identify shortcomings in the educational risk management activities carried out in support of the U.S. Army Corps of Engineers cleanup of the site (Spehn, 2001). And at Hamilton Field, California, the San Francisco Bay Region Water Quality Control

Board received comments from an adjacent developer, the city of Novato, and a regional environmental organization regarding the Corps' proposed plan to reopen a landfill remedy (California RWQCB, 2001).

There are sites where public oversight on long-term review has not been encouraged. The Department of Energy's (DOE) second five-year review report for the Weldon Spring Site Remedial Action Project contained "no evidence of community involvement" (Missouri DNR, 2002). At a site in Palo Alto, CA, no five-year review has been initiated for extraction systems (California EPA, 2001), which has been attributed to the layoff of key people within the company as well as other expense-cutting measures (R. Moss, Barron Park Association Foundation, personal communication, 2002). At the MEW Study Area in Mountain View, CA, the responsible parties undertook an effective, comprehensive two-year review of source control and regional extraction remedies in 2000, but without notifying neighboring communities, leading to substantial controversy in 2001 when the neighbors became aware of the activity (Siegel, 2001).

EPA's new guidance for the five-year review (EPA, 2001a) offers detailed suggestions for involving the public in the review process, but overall it discourages the reopening of remedial decisions unless a remedy is shown to not be protective of human health and the environment. The guidance does not adequately address the challenge of engaging public participants who have become less involved in ongoing cleanup because of the amount of the time that has passed.

Strategies for Long-Term Public Involvement

The generally successful RAB model can be adapted to give the public a longer-lasting role in both regular review and any unscheduled reconsideration of remediation activity. Currently there is no DoD-wide guidance outlining how to involve the public following the selection of remedies at contaminated sites. The DoD's late-2001 promise to promulgate a rule, by mid-2003, to govern the operation of RABs (Defense Environmental Alert, 2001) provides an excellent opportunity to update long-term community relations policies. Indeed, the Army has developed guidance underscoring the importance of engaging the public after the signing of the ROD (USAEC, 1998). This guidance specifically identifies five-year reviews, remedy performance evaluations, monitoring to evaluate natural attenuation, decisions to discontinue or decrease treatment systems, technical impracticability waivers, maintenance and enforcement of institutional controls, demonstrations that the remedy is operating properly and successfully, and site close-out reports as benefiting from greater public involvement. The Army's guidance stresses that "if a RAB adjourns because there is no longer sufficient, sustained community interest, the installation must ensure that its overall community involvement programs provide for continued stakeholder input, and the installation must continue to monitor for any subsequent changes in community interest to revive the RAB." Without question, this implies much more than simply publishing a newspaper notice when site managers have a new plan or report available for public review.

Three approaches represent potential mechanisms for ensuring long-term public involvement; they may be used individually or in combination. First, once RABs determine that their remedy-selection work is done, they could schedule, with the support of both responsible parties and regulators, annual "reunions." Former board members and other members of the public could arrange to receive presentations on the status of long-term stewardship activities. Such reunions would be an excellent time to solicit public comment on any decisions that may be up for reconsideration. If changes are proposed in the middle of the year, the "reunion" participants would be invited to a special meeting.

Second, community oversight could be turned over to other government agencies. In locations where remaining contamination represents a visible health threat, the local health department might be best situated to assume such oversight. Where property has been transferred, local planning jurisdictions or recipient federal agencies could provide oversight for the contamination and its remedies. Grassroots involvement could be incorporated into other community relations activities conducted by such agencies. If local health or planning departments are given these new duties, federal funds need to be available to ensure that the departments have the appropriate expertise.

Third, at active federal facilities, RABs could be transformed into broader environmental advisory boards whose scope would include environmental compliance, pollution prevention, conservation, and other environmental issues. In many communities, residents actually care more about ongoing environmental issues than cleanup, but rarely has the military been willing to involve the public in resolving those problems. Should members of the public be given the opportunity to advise on base environmental affairs in general, they would be well situated to provide advice should cleanup decisions be reopened. Groups that are monitoring the compliance of effluent and emissions standards could easily monitor land use controls that are part of a cleanup remedy. DoD and the armed services have been considering this third path, but it raises internal organizational and financial obstacles. It may take legislative intervention to authorize the improved integration of cleanup activity with other environmental programs.

Whatever mechanism is utilized to encourage continuing public involvement, lead agencies should tailor their public notification activities to the level of proposed activities. For example, no special notification should be necessary for minor modifications to optimize a remedy, unless the physical location of a component is moved such that it will raise concerns among the public. On the other hand, if remedies in operation reach a point of diminishing returns without reaching cleanup goals, then the public should have the opportunity to review proposals to shut down those remedies and to recommend new strategies designed to achieve the original cleanup goals. Where remedies include long-term containment or treatment operations, the public should be provided with quantitative data that will allow them to evaluate remedial decisions being proposed by the responsible parties and regulatory agencies. Utilizing some of the tools described elsewhere in this report, lead agencies should publish data describing treatment results (such as trends in contaminant concentration versus time, mass removed versus time, or cost of mass removed versus time) and the specific monitoring values utilized to determine the effectiveness of the remedial action. The public is unlikely to comment constructively—in fact, they may not even take part in the process—if other decision makers are not providing a complete and comprehensible picture of the state of the cleanup.

The goal of cleanup is to protect public health and the environment, and the public's role continues as long as contamination remains in place at levels that pose a potential risk. The concerned, affected public should be made aware of the progress of remedies, they should have access to comprehensible summaries of innovative alternative technologies, and they should have the opportunity to present concerns and offer advice early enough to influence decisions.

LONG-TERM STEWARDSHIP:
AN INTEGRAL PART OF ASM

As demonstrated in Chapter 1, more remedies today are being selected that utilize containment and institutional controls rather than treatment of the contaminant source. Institutional controls include covenants, zoning restrictions, well drilling bans, and public advisories that limit public access to residual contamination. Along with physical controls such as fences and buffer zones, institutional controls and containment are referred to as land use controls. Residual contamination is expected to remain at these sites such that unrestricted use of soil, groundwater, and surface water will not be permitted. As a consequence, containment technologies, institutional controls, and physical controls must be maintained as long as the potential risk remains in order to protect human health and the environment. The activities needed to maintain such remedies collectively are called long-term stewardship.

There has been growing awareness of this long-term stewardship responsibility by the federal government, particularly within DOE. In 1997, DOE published the first comprehensive analysis of contamination generated by the production of nuclear weapons, in which it acknowledged that it will not be possible to remediate all sites for unrestricted use (DOE, 1997b). DOE then started planning for implementation of long-term stewardship by addressing information needs (ICF Kaiser, 1998), by identifying implementation issues (Probst and McGovern, 1998; NRC, 2000; DOE, 2001a), by describing the scope and cost (DOE, 1999; DOE, 2001b), by evaluating funding mechanisms (Bauer and Probst, 2000), by evaluating the role of local governments (Pendergrass and Kirshenberg, 2001), and by analyzing how long-term stewardship considerations have been factored into remedial decisions (DOE, 2001c). DOE has initiated efforts to develop a long-term stewardship strategic plan, to identify the ultimate responsibility for long-term stewardship, to engage the public in a dialog on long-term stewardship, and to participate in intra-agency discussions on long-term stewardship.

Long-term stewardship is an integral part of ASM. As shown in Figure 2-7, if residual contamination remains in place following an attainment of "response complete," then the site is subject to long-term stewardship. Long-term stewardship starts when remediation, disposal, or stabilization is complete, or, in the case of long-term remedial actions such as groundwater treatment, when the remedy is shown to be functioning properly. Long-term stewardship ensures that remediation remains effective for an extended, or possibly indefinite, period of time until residual hazards are reduced sufficiently to permit unrestricted use and unlimited access. ASM specifically requires that during long-term stewardship, the existing remedy be reconsidered periodically to determine if it could be optimized or if it should be replaced by a new technology that could lead to unrestricted use of the site. This might lead to the replacement of containment or institutional controls with a more active remedial system. The motivation for periodically reconsidering the remedy is to be able to reach site closeout, which is not possible unless contamination is permanently reduced to levels below that which pose an unacceptable human health or environmental risk. This reconsideration represents a significant departure from the way PRPs usually conduct long-term stewardship.

Basic Elements of Long-Term Stewardship

Long-term stewardship requires stewards, operations, information systems, research, public participation, and public education—all of which should be laid out in advance in a long-term stewardship plan (Oak Ridge, 1998, 1999; Probst and McGovern, 1998; Bauer and Probst, 2000; NRC, 2000). Stewards—those responsible for developing, implementing, and overseeing the activities necessary to maintain the remedy—should be selected based on the following criteria:

• appropriate technical expertise so that the remedy can be properly operated, maintained, monitored, evaluated, and modified to ensure protectiveness,
• knowledge of developing technologies so that a change to the remedy can be evaluated,
• ability to enforce land use controls,
• institutional longevity in order to be in existence as long as the remedy is needed,
• property ownership (e.g., federal government, local government, or private sector),
• longevity of the funding source,
• ability to oversee multiple sites for economies of scale,
• experience in public participation and public education and thus an ability to obtain public trust and confidence,
• ability to adapt to changing land use,
• institutional memory, and
• ability and authority to make decisions.

It is likely that not just one steward, but rather a consortium of stewards working through a coordinating group will be the most effective and efficient approach. Examples of potential stewards include the party responsible for the contamination, a new federal long-term stewardship agency, an existing federal agency assigned with long-term stewardship responsibility, a host state or a multi-state consortium, an insurance company, and a nonprofit organization. The goals in assigning responsibility for long-term stewardship to one or more entities are to ensure attentiveness to the long-term stewardship tasks, to achieve economies of scale, to utilize experienced personnel, to create an incentive to implement innovative technologies, and to increase public trust and confidence.

The "operations" element of long-term stewardship refers to those activities necessary to ensure the integrity of the engineering technologies, institutional controls, and physical controls, and it includes inspection, monitoring, maintenance, surveillance, modification, replacement, enforcement, and evaluation. The "information systems" element, which includes the maintenance of records of residual contamination, associated risks, and required long-term stewardship activities, must be maintained as long as the residual contamination poses a risk to human health and the environment. The "research" element is needed to understand such issues as the long-term performance of stabilization and containment technologies and the long-term migration of contaminants in order to reduce the cost of long-term stewardship and the risk of residual contamination.

Public participation is integral to the selection, implementation, and performance review of the remedy and to long-term stewardship activities. As discussed previously, engaging the public during long-term stewardship can be a challenge. Indeed, only engineering technologies, institutional controls, and physical controls (and not long-term stewardship operations) are described in a decision document, which is the major opportunity for public involvement (see Figure 1-1). Members of the public who live around restoration sites need assurance that the remedial actions are operated in a manner that maintains effectiveness over a very long time period. Along with public participation, public education is necessary to ensure that the nature and risk of the residual contamination and the resultant types of land use controls are understood. This understanding will facilitate the enforcement of land use controls.

One of the greatest obstacles to long-term stewardship is the lack of a stable source of funding, particularly one that is independent of budget cycles. EPA and the state regulatory agencies do not have the authority to consistently fund long-term stewardship activities because such money must be appropriated by Congress every year. Lump sum payments and long-term contracts can be entered into, but federal entities are also subject to Congress appropriating money for the project. EPA and state agencies often do not have the administrative resources or, at times, the willingness to require long-term stewardship. This problem tends to increase with the passage of time as competing issues arise that require funding and attention. Fortunately, EPA, the Navy, and other federal agencies are exploring the use of trusts and other lump sum payment devices. Box 6-2 contains a discussion of funding options for long-term stewardship.

In order to ensure the long-term institutional management of contaminated sites, the Navy should perform all of the basic elements of long-term stewardship as a matter of policy. Additionally, long-term stewardship should be integrated into the remedial decision-making process such that site characterization, remedial alternative assessments,

---

BOX 6-2
Funding Options for Long-Term Stewardship

The uncertainty of the length and scope of long-term stewardship presents a challenge for identifying sustainable funding mechanisms. Currently, the federal budget process provides funding for long-term stewardship for which the federal government is responsible. However, the annual budget process does not guarantee funding for long-term stewardship, which is a concern to local governments and stakeholders.

The following four funding options (English et al., 1997; Probst and McGovern, 1998; Bauer and Probst, 2000; Defense Environmental Alert, 2000; Department of Energy, 2001b) are representative of those being considered as sustainable funding sources for long-term stewardship:

Annual congressional appropriations. The federal agency requests funds for long-term stewardship on an annual basis, and Congress appropriates what it considers necessary. This is not a guaranteed funding mechanism and can be affected by changing national priorities.

Trust funds. A long-term stewardship trust fund produces a predictable source of funds. A trust fund can be created at the national, state, or local level. New legislation may be necessary to create a trust fund. The initial funding source can be congressional appropriations, fees, or sales of assets.

Fees/sales of assets. Government agencies might create revenue by selling assets or by providing services for which a fee is charged. The income from sales or services can be collected in a fund to support long-term stewardship. Because this income now goes to the general Treasury, new legislation will be needed.

Public–private partnerships. Private entities can lease government assets at below-market rates in return for assuming responsibility for long-term stewardship. This option also may require new legislation.

and decision documents evaluate long-term stewardship as part of the remedy (Pendergrass and Kirshenberg, 2001). **Because all federal agencies with environmental restoration programs face this issue, ideally the Administration should convene an interagency task force to develop a government-wide policy and mission for long-term stewardship at federal sites.** This group, which should include independent experts and representatives of major stakeholder groups, could recommend how to integrate the costs and the challenges of long-term stewardship into the decision-making and budgeting processes and into any new legislation (Probst and McGovern, 1998). This policy would help develop a clear model for how to pay for long-term stewardship activities.

Limitations of Land Use Controls

The rationale for MDP4 is to focus PRPs on eventual site closeout rather than on the indefinite maintenance of land use controls. In the case of contaminants such as recalcitrant organic compounds, heavy metals, and radionuclides, land use controls may be required for hundreds or thousands of years. Over this timeframe, the cost and viability of land use controls are highly uncertain. Cleanup to unrestricted use removes the uncertainty surrounding the long-term effectiveness of land use controls.

Many documents have noted the limitations of land use controls, particularly institutional controls, for a variety of reasons (NRC, 1999, 2000; English et al., 1997; Pendergrass, 2000; Pendergrass and Kirshenberg, 2001). For example, local governments often are responsible for implementing institutional controls but usually are not consulted in evaluating and selecting a remedy; thus, they may not have the resources or authority to implement the controls. In other cases, RODs include only general descriptions of institutional controls, which makes implementation difficult. Monitoring of institutional controls is poorly understood and thus may not be done frequently enough to identify weaknesses before failure. And very often the public does not understand the nature of the hazard or the required maintenance of institutional controls, which adversely affects the rigor with which the institutional controls can be enforced.

Nonetheless, for the present time, land use controls will be part of many site remedies. Better information is needed on the number of public and private sites that rely on or will rely on land use controls so that

DoD can develop a consistent approach to estimating the annual and life-cycle cost of maintaining such controls and to evaluating their performance (Probst and McGovern, 1998; NEPI, 1999). Research should be conducted on where and under what conditions land use controls are successful or unsuccessful. This information will be helpful in determining the national infrastructure and information needs for long-term stewardship, in defining the local and federal government roles in long-term stewardship, and in determining how to fund long-term stewardship and how to design future facilities with long-term stewardship in mind. As described in Box 6-3, DoD has established an overall framework for implementing, documenting, and managing land use controls for both closing and active facilities (DoD 1997a,b, 2001b) that should help to overcome many of the limitations of these controls.

BOX 6-3
DoD Policy on Land Use Controls
Source: DoD (2001b).

The Department of Defense policy for land use controls for active facilities and those being transferred out of federal control requires:

• using multiple, overlapping land use controls,
• modifying or terminating land use controls after going through the same process used to set the land use controls in the first place,
• considering the costs of implementing and maintaining the land use controls in the remedy determination,
• maintaining a central database of properties restricted by land use controls and using state registries where they exist,
• using existing processes and mechanisms in the development, implementation, and management of land use controls,
• managing and maintaining land use controls at the local level where possible,
• reviewing the maintenance of land use controls and notifying the installation officials immediately if a land use control is being violated,
• identifying in the proposed plan, record of decision, or other decision documents the future land use assumption that was used to develop the remedy, the specific land use restrictions necessitated by the selected remedy, and the possible mechanisms for implementing and enforcing those use restrictions, and
• developing enforceable land use controls based on state property and environmental law.

Revisiting the Remedy During the Five-Year Review

MDP4 of ASM provides an opportunity for the remedy to be re-evaluated to see if it still represents the optimum solution. In many cases, at the time of initial remedy selection, no technology may be available to clean up the site to unrestricted use at a reasonable cost. However, in ten or 20 years, such a technology may exist. Because of changing conditions, there may be opportunities to achieve the remedial goals for less money or in less time, or there may be an opportunity to achieve more aggressive remedial goals for the same money and time. Or the contaminated site may become sufficiently valuable to stakeholders in the future that they would be willing to support more cleanup than they were previously. Indeed, a study at DOE sites (Pendergrass and Kirshenberg, 2001) concluded that local governments prefer to remediate to levels that permit unrestricted use and to avoid long-term stewardship costs because land use restrictions may have long-term detrimental effects on economic development potential.

Five-year reviews are required by CERCLA at sites where contaminants remain above levels allowing unrestricted use. The purpose of these reviews is to determine if the selected remedy remains protective of human health and the environment. The three basic questions the five-year review is intended to answer are (1) is the remedy functioning as intended, (2) are the exposure assumptions, toxicity data, cleanup levels, and remedial action objectives still valid, and (3) has any other information (such as the discovery of new contamination) come to light that could call into question the protectiveness of the remedy (EPA, 2001a). The five-year review must specifically evaluate whether there has been a change in land use or exposure pathways. A remedy that was protective when it was adopted may not be protective in the future because of changes in land use or other site conditions. Five-year reviews generally include document reviews, reviews of cleanup standards, interviews, inspections, technology reviews, and preparation of a report summarizing the findings and recommendations. The five-year review is not considered a vehicle for adopting new technology.

EPA guidance (EPA, 2001a) provides useful tables describing situations where remedies are protective or not protective and provides several case examples. [Protectiveness is defined by the acceptable risk range of 10^{-6} to 10^{-4} for carcinogens and a hazard index of less than 1 for noncarcinogens (EPA, 2001a)]. Although the guidance is an improvement over prior guidance, it is still a general framework document. It does not suggest analytical methods that can be used to make decisions.

At federal facilities, the responsible federal agency performs the five-year review (even for properties already transferred to nonfederal ownership), although EPA has final authority at NPL sites over whether the five-year review is protective (EPA, 2001a). For non-NPL federal sites, EPA has no statutorily defined role, although EPA may comment on non-NPL sites on a case-by-case basis.

Once the five-year reviews are begun, they may be discontinued only if levels of contaminants allow for unrestricted use and unlimited exposure and if appropriate documentation and notification are given (EPA, 2001a). As a result, the five-year review may become a virtually perpetual requirement for sites where containment is the remedy or where the soil and groundwater cleanup goals are not met by the original remedial action. MDP4 of ASM provides an opportunity to use the five-year review as a mechanism for achieving site closeout. That is, in addition to asking whether the remedy remains protective during the five-year review, it should be asked whether there are newly available technologies that could expeditiously lead to site closeout—even if the current remedy is protective. If there were a more effective remedy available, the user would cycle back through the previous parts of ASM (see Figure 2-7). This consideration of new technologies that might optimize remedial performance and/or reduce lifecycle costs has been explicitly endorsed in new DOE guidance on the five-year review process (DOE, 2002).

MDP4 expands the scope of the five-year review process to include the basic elements of long-term stewardship—stewards, operations, information systems, research, public participation, and public education. First, the five-year reviews should evaluate operations with greater emphasis placed on enforcing and monitoring institutional controls, as there is little information available on their long-term effectiveness. EPA and DoD have initiated efforts to ensure that institutional controls are being properly implemented (EPA, 2000b; DoD, 2001b), but the detailed results are not yet available. During five-year reviews, the information system should be evaluated to see if the proper documents are being maintained in a manner accessible to the public. Typical documents that should be reviewed include RODs, state and federal environmental laws and regulations, remedial action reports, as-built drawings, monitoring data, operations and maintenance manuals and reports, institutional controls (e.g., deed notices, easements, and covenants), and community involvement plans. As discussed previously, there is little evidence of public involvement in long-term stewardship gleaned from previous five-year review reports. The involvement of the public in post-remediation decision making and activities should be evaluated as regularly as the remedy. The performance and capability of the stewards to fulfill the criteria outlined earlier should also be evaluated. Finally, the five-year review should evaluate the adequacy of funding for long-term stewardship. A lack of funding may be the cause for some of the deficiencies identified in other areas.

Expanding the role of the five-year review program to encompass remedy reconsideration should accompany general improvements to the program. Several EPA and independent studies have concluded that EPA's five-year reviews have inadequately supported the determination of "protective" (EPA, 1999b; Nakamura and Church, 2000; Probst and Konisky, 2001). Indeed, Resources for the Future (Probst and Konisky, 2001) reviewed 99 completed nonfederal remedial actions and found that at 48 percent of the sites, statements concerning the protectiveness of the remedy were insufficiently substantiated or were questionable because the remedies were not fully in place, were not functioning as intended, or were not likely to achieve remedial objectives. The committee's limited review of five-year review reports was consistent with these observations. Interestingly, although institutional controls were part of the remedial action at 61 percent of the sites, the institutional controls required were not fully implemented or had an "unknown" status at 28 percent of these sites (Probst and Konisky, 2001). Resources for the Future and others have recommended that EPA improve the quality of its five-year reviews.

Assessing Life-Cycle Costs

As discussed in Chapters 2 and 4, ASM may result in short-term cost increases at sites, partly because of the need for evaluation and experimentation activities that occur in parallel with remedy implementation. An important task is to determine whether the costs associated with ASM will be balanced by the savings that result from switching to a more efficient and effective technology or by overall life-cycle savings. There should be no debate that if a net savings (considering both implementation and life-cycle costs) can be achieved by changing to a remedy that is equally or more effective in meeting cleanup requirements, then the new remedial action should be implemented. For example, in some cases equivalent or superior long-term cleanup performance with lower life-cycle costs could be realized for groundwater if the remedy is converted from a pump-and-treat system to a passive, *in situ* system. However, making these cost assessments can be complicated. To date, few efforts have been made to determine whether remedies are cost-effective over the life span of a project, including design, construction, operations and maintenance, and closeout (EPA, 2000c). Furthermore, a selected remedy may initially be cost-effective, but over time new technologies may have been developed that could be implemented at decreased costs.

Ideally, the use of ASM necessitates that the full range of costs over the life of a chosen technology (e.g., those associated with materials and energy use and indirect pollutant emissions) be considered when determining whether and what additional site management is necessary [1]. The current practice at most sites (once the magnitude of site contamination, exposure, and potential risk have been characterized, and forecasts have been provided for how these might change under alternative technologies and management strategies) is to determine what the short-term costs of various different remedies will be to achieve the site cleanup goal. This is most effective when done for alternatives that are "comparably effective," (i.e., they accomplish the same end) (EPA, 1990). Factors other than immediate cost that may impact remedial choices, like stakeholder preferences and values, are generally addressed with group deliberation and participatory processes (Webler et al., 1995; NRC, 1996; Renn, 1999). These exercises could be improved by bringing more quantitative tools for valuation, cost-benefit analysis, and life-cycle analysis to bear on site management issues (Arrow et al., 1996; NRC, 1996; Farrow and Toman, 1999).

Although cost-benefit analysis is based on the well-established procedures of engineering economics, long-term costs from various operations (such as the management of treatment residuals, site monitoring, site security, and component depreciation) can be difficult to forecast. Evaluating the benefits associated with improvement in health, environmental quality, and community welfare likely to occur following implementation of different remedial options is even more difficult (Hull, 1993; Matthews and Lave, 2000). Some believe that cost-benefit analysis cannot capture the full range of social, political, and ethical factors that individuals and society consider when making environmental choices (e.g., Sagoff, 1988, 2000; Dower, 1990). It is not the intent of this report to delve into these complications, but rather to suggest that cost-benefit tools and life-cycle assessment have potential value for im-

---

[1] Although such full environmental "life-cycle assessment" has not to our knowledge been adopted in the evaluation of site cleanup alternatives, as it has for alternative product designs and processes (e.g., Curran, 1996; Graedel, 1998; Joshi, 1999), evaluations of the broader regional implications of alternative remediation strategies have been conducted (Schwarzenbach et al., 1999).

proving site management (for further discussion of this debate, see EPA, 1987; Freeman and Portney, 1989; Stroup, 1991; Sexton and Zimmerman, 1999; Fischhoff, 2000; Spash, 2000). The committee recognizes that such full-cost accounting may be too complex and costly to be incorporated into practical applications of ASM on a regular basis.

One important point is highlighted because it is a factor regardless of the complexity of the cost analysis that is undertaken—the issue of discount rates. Typically, feasibility studies use a 30-year net present value cost estimate that only includes direct costs of the remedy in decision making (EPA, 1988). The net present value methodology and the 30-year time frame may not be appropriate for comparing alternatives with long-term stewardship requirements that extend indefinitely (Portney and Weyant, 1999; Okrent and Pidgeon, 2000; DOE, 2001a; EPA, 2000c). This is because when usual discount rates and factors are used, the present value of future costs is essentially zero after several decades, such that an alternative with a lower initial construction cost almost always will have a lower life-cycle cost than an alternative with a higher initial cost. This is true even if the former alternative requires long-term stewardship costs indefinitely, and the latter only requires long-term stewardship costs for a short period of time. At a minimum, the sensitivity of cost analyses or predicted cost-benefit ratios to the selected discount rate should be evaluated.

To ensure that the full set of economic impacts is considered, the evaluation of cost effectiveness needs to be expanded to reflect indirect or opportunity costs that arise when a site's use is restricted (Pendergrass, 2000, 2001). These costs include lower property values, lower taxes, and lower social benefit to the community than if no land use restrictions existed. There also is the economic benefit in preventing or significantly minimizing potential future legal liability. The Navy, in conjunction with other federal agencies, should develop a life-cycle cost estimating technique that reflects the timeframes for which long-term stewardship will be needed, the indirect costs, and methods and procedures for appropriate discounting in computations of net present value for these applications.

Regulatory Oversight

Any changes made in remedies as a result of MDP4 during long-term stewardship should involve EPA and the state regulatory agency. This is necessary to preserve the checks and balances provided by the federal regulatory system and to ensure public confidence in the safety of the remedy. The existing regulatory programs provide shared authority for initial remedy selection between federal PRPs and regulatory authorities (EPA). The federal government has not developed a generally applicable, consistent position on the role of the regulatory agencies (federal or state) versus DoD in making post-ROD remedy modifications. There is no logical or policy rationale for using a different process for changing or terminating the remedy than for initial remedy selection. The continuing conflict and/or ambiguity over whether regulators may review decisions to change remedies should be resolved expeditiously. Without both public and regulatory review of DoD's remedial decisions, these decisions are unlikely to garner public support.

MAJOR CONCLUSIONS AND RECOMMENDATIONS

The underlying statutes on hazardous waste management are consistent with adaptive site management, and existing regulatory guidance could be modified to be more so. EPA's policy rationale for not requiring the implementation of additional technologies at CERCLA private sites is not applicable to federal facilities and should not be used as justification for not implementing ASM. The Navy and other federal agencies responsible for restoring sites should adopt ASM and develop agency-specific risk management policies and detailed guidance requiring that it be utilized. Many recent efforts (such as NAVFAC, 2001) are an attempt to provide some of the guidance that would be needed, although such documents must be strengthened to mention the research track of ASM and the reconsideration of remedies over the long term. The Navy may wish to issue its own technical impracticability guidance, either alone or as part of its ASM risk management policy, so that the consistency of technical impracticability waivers with ASM is clear.

The responsible federal agency should solicit public involvement during each of the four management decision periods of ASM. Changes to the remedy, the remedial goals, and future land use should be issued only after consideration of public comments, particularly the proposed easing of remedial objectives or suggestions that remedies be "turned off" before reaching established objectives. Although many individual guidance documents mention public involvement, there is no coherent public involvement process described in existing guidance or practiced in the field after remedy selection. As part of the RAB rule development process, DoD should work with regulators, public representatives, and other stakeholders to develop a menu of options for involving the public in the long-term oversight of cleanup programs at facilities where remedies or long-term stewardship activities are continuing.

During long-term stewardship, the remedy should be reconsidered as part of the five-year review, even if it is currently protective of human health and the environment. Because of changing conditions or the development of new technologies, there may be opportunities to achieve remedial goals for less money or in less time or achieve more aggressive remedial goals for the same money and time. Thus, it may be possible to replace land use controls with treatment remedies that will achieve unrestricted use and lead to site closeout. Only if unrestricted use levels are attained can the military and other agencies permanently remove sites from federal stewardship. The benefits of achieving site closeout include not only cost savings from reduced long-term operation and maintenance costs, but also increased taxes and minimization of potential future legal liability.

A government-wide policy for long-term stewardship (also known as long-term management) at federal sites is needed. This activity is needed to legitimize the basic elements of long-term stewardship and the expenditure of resources on these elements. As part of this effort, it will be important to develop a life-cycle cost estimating technique and appropriate discounting methods that reflect the timeframes for which long-term stewardship will be needed.

REFERENCES

Air Force. 2001. Final remedial process optimization handbook. Prepared for the Air Force Center for Environmental Excellence, Technology Transfer Division, Brooks Air Force Base, San Antonio, Texas, and Defense Logistics Agency, Environmental Safety Office, Fort Belvoir, VA.

Arrow, J. K., M. L. Cropper, G. C. Eads, R. W. Hahn, L. B. Lave, R. G. Noll, P. R. Portney, M. Russell, R. Schmalensee, V. K. Smith, and R. N. Stavins. 1996. Is there a role for benefit-cost analysis in environmental, health and safety regulation? Science 272:221–222.

Bauer, C., and K. N. Probst. 2000. Long-term stewardship of contaminated sites trust funds as mechanisms for financing and oversight. Washington, DC: Resources for the Future.

California EPA. 2001. Hillview-Porter regional program. Fact Sheet No. 19. Sacramento, CA: CA EPA Department of Toxic Substances Control.

California RWQCB. 2001. Cleanup and abatement order No. 01-139. CA RWQCB San Francisco Bay Region, November 28.

Center for Public Environmental Oversight (CPEO). 2002. Military list archive. Available at www.cpeo.org/lists/military/2002/msg00055.html.

Chess, C., and K. Purcell. 1999. Public participation and the environment: do we know what works? Environ. Sci. Technol. 33(16):2685–2692.

Curran, M.A. 1996. Environmental life-cycle assessment. New York: McGraw Hill.

Defense Environmental Alert. 2000. Federal trusts inappropriate for institutional controls. December 19, 2000.

Defense Environmental Alert. 2001. Pentagon agrees to develop Restoration Advisory Board Rule. December 18, 2001. Pp. 15–16.

Department of Defense (DoD). 1997a. Institutional controls: what they are and how they are used. BRAC Environmental Program Fact Sheet. Washington, DC: DoD Office of the Deputy Undersecretary of Defense (Environmental Security).

DoD. 1997b. Guidance responsibility for additional environmental cleanup after transfer of real property. Washington, DC: DoD Office of the Deputy Under Secretary of Defense (Acquisition and Technology).

DoD. 1999. The environmental site close out process guide. Washington, DC: EPA and DoD.

DoD. 2001a. Management guidance for the Defense Environmental Restoration Program. Washington, DC: DoD Office of the Deputy Under Secretary of Defense (Installations and Environment).

DoD. 2001b. Guidance on land use controls association with environmental restoration activities for active installations. 2001. Washington, DC: Department of Defense.

Department of Energy (DOE). 1997a. Site-specific advisory board initiative 1997, evaluation survey results. Washington, DC: DOE Office of Environmental Management,

DOE. 1997b. Linking legacies: connecting the cold war nuclear weapons production processes to their environmental consequences. Washington, DC: DOE Office of Environmental Management.

DOE. 1999. From cleanup to stewardship: a companion report to accelerating cleanup: paths to closure. Washington, DC: DOE Office of Environmental Management.

DOE. 2001a. Long-term stewardship study. Washington, DC: DOE Office of Environmental Management.

DOE. 2001b. A report to Congress on long-term stewardship. Washington, DC: DOE Office of Environmental Management.

DOE. 2001c. Long-term stewardship case study report, final draft. Washington, DC: DOE Office of Environmental Management.

DOE. 2002. Comprehensive Environmental Response, Compensation and Liability Act (CERCLA) five-year review guide. Washington, DC: DOE Office of Environmental Management.

Dower, R. 1990. Hazardous wastes in public policies for environmental protection 151, 180. Washington, DC: Resources for the Future.

English, M., D. Feldman, R. Inerfedl, and J. Lumley. 1997. Institutional controls at Superfund sites: a preliminary assessment of their efficacy and public acceptability. Knoxville, TN: University of Tennessee Energy, Environment, and Resources Center.

Environmental Protection Agency (EPA). 1987. Unfinished business: a comparative assessment of environmental problems. Washington, DC: EPA.

EPA. 1988. Guidance for conducting remedial investigations and feasibility studies under CERCLA, Interim Final. EPA/540/G-89/004. Washington, DC: EPA.

EPA. 1990. National oil pollution and hazardous substances contingency plan. Federal Register 56:8666, 8728.

EPA. 1993. Guidance for evaluating the technical impracticability of groundwater restoration. OSWER Dir. No. 9234.2-25. Washington, DC: EPA.

EPA. 1996a. Superfund reforms: updating remedy decisions. EPA 540-F-96-026, OSWER 9200.0-22. Washington, DC: EPA.

EPA. 1996b. Guidance for evaluation of federal agency demonstrations that remedial actions are operating properly and successfully under CERCLA Section 120(h)(3). Washington, DC: EPA.

EPA. 1999a. A guide to preparing Superfund proposed plans, record of decision, and other remedy selection decision documents. EPA 540-R-98-031. Washington, DC: EPA.

EPA. 1999b. Backlog of five-year reviews increased nearly three-fold. Washington, DC: EPA Office of Inspector General.

EPA. 2000a. Draft public involvement policy. Federal Register 65(250):82335-82345.

EPA. 2000b. Institutional controls: a site manager's guide to identifying, evaluating and selecting institutional controls at Superfund and RCRA corrective action cleanups. EPA 540-F-00-005, OSWER 9355.0-74FS-P. Washington, DC: EPA.

EPA. 2000c. A guide to developing and documenting cost estimates during the feasibility study. EPA 540-R-00-002, OSWER 9355.0-75. Washington, DC: EPA.

EPA. 2001b. Updating remedy decisions at select Superfund sites: cumulative summary report FY 1996 through FY 1999. EPA 540-R-01-002. Washington, DC: EPA.

EPA. 2001a. Comprehensive five-year review guidance. EPA 540-R-01-007. Washington, DC: EPA.

Farrow, S., and M. Toman. 1999. Using environmental benefit-cost analysis to improve government performance. Environment 41(2):12–15, 33–37.

FFERDC. 1993. Interim report of the Federal Facilities Environmental Dialogue Committee, The Keystone Center, February.

FFERDC. 1996. Final report of the Federal Facilities Environmental Dialogue Committee, The Keystone Center, April.

Fischhoff, B. 2000. Informed consent for eliciting environmental values. Environ. Sci. Technol. 34:1439–1444.

Freeman, M., and P. R. Portney. 1989. Economics and the rational management of risk. Discussion Paper CRM-89-5. Washington, DC: Resources for the Future.

Graedel, T. E. 1998. Streamlined life-cycle assessment. Upper Saddle River, NJ: Prentice Hall.

Hull, B. 1993. Valuing the environment: full-cost pricing-an inquiry and a goal 2-3. Report 103-93 to the Conf. Bd. of Canada, 1993.

ICF Kaiser Consulting Group. 1998. Managing data for long-term stewardship (working draft). Prepared for EG& G Technical Services of West Virginia under DOE Prime Contract No. DE-AC-95MC31346. Available at http://lts.apps.em.doe.gov/center/reports/doc1.html.

Joshi, S. 1999. Product environmental life cycle assessment using input-output techniques. Journal of Industrial Ecology 3(2/3):95–120.

Lynn, F. M., G. Busenberg, N. Cohen and C. Chess. 2000. Chemical industry's Community Advisory Panels: What has been their impact? Environ. Sci. Technol. 34(10):1881–1886.

Matthews, H. S., and L. B. Lave. 2000. Applications of environmental valuation for determining externality costs. Environ. Sci. Technol. 34(8):1390–1395.

Missouri Department of Natural Resources (DNR). 2002. Comments on the Weldon Spring site remedial action project second five-year review, August 2001.

Murdock, B. S., and K. Sexton. 2002. Promoting pollution prevention through community-industry dialogues: the good neighbor model in Minnesota. Environ. Sci. Technol. 36(10):2130–2137.

Nakamura, R., and T. Church. 2000. Reiventing Superfund: an assessment of EPA's administrative reforms. Washington, DC: National Academy of Public Administration.

National Environmental Policy Institute (NEPI). 1999. Rolling stewardship: beyond institutional controls. Washington, DC: NEPI.

National Research Council. 1994. Alternatives for ground water cleanup. Washington, DC: National Academy Press.

NRC. 1996. Understanding risk: informing decisions in a democratic society. Washington, DC: National Academy Press.

NRC. 1997. Innovations in ground water and soil cleanup: from concept to commercialization. Washington, DC: National Academy Press.

NRC. 1999. Environmental cleanup at navy facilities: risk-based methods. National Academy Press. Washington, D.C.

NRC. 2000. Long-term institutional management of U.S. Department of Energy legacy waste sites. Washington, DC: National Academy Press.

Navy. 1999. Defense Environmental Restoration Program Annual Report to Congress, Fiscal Year 1999.

NAVFAC. 2001. Guidance for optimizing remedial action operation (RAO). Special Report SR-2101-ENV. Prepared for the Naval Facilities Engineering Service Center. Research Triangle Park, NC: Radian International.

Oak Ridge Reservation End Use Working Group Stewardship Committee. July 1998. Stakeholder report on stewardship, Volume I. Washington, DC: DOE Environmental Management.

Oak Ridge Reservation End Use Working Group Stewardship Committee. December 1999. Stakeholder report on stewardship, Volume II. Washington, DC: DOE Environmental Management.

Okrent, D., and N. Pidgeon. 2000. Introduction: dilemmas in intergenerational versus intragenerational equity and risk policy. Risk Analysis 20(6):759–762.

Pendergrass, J. A. 2000. Protecting public health at Superfund sites: can institutional controls meet the challenge? Washington, DC: Environmental Law Institute.

Pendergrass, J. A., and S. Kirshenberg. 2001. Role of local government in long-term stewardship of DOE facilities. Washington, DC: Environmental Law Institute.

Portney, P. R., and J. P. Weyant (Eds.) 1999. Discounting and intergenerational equity. Washington, DC: Resources for the Future.

Probst, K. N., and M. H. McGovern. 1998. Long-term stewardship and the nuclear weapons complex: the challenge ahead. Washington, DC: Resources for the Future.

Probst, K. N., and D. M. Konisky. 2001. Superfund's future: what will it cost? Washington, DC: Resources for the Future.

Renn, O., T. Webler, and P. Wiedermann. 1995. Fairness and competence in citizen participation: evaluating models for environmental discourse. Boston, MA: Kluwer Academic Publishers.

Renn, O. 1999. A model for an analytic-deliberative process in risk management. Environ. Sci. Technol. 33(18):3049–3055.

Sagoff, M. 1988. The economy of the Earth: philosophy, law and the environment. Cambridge, UK, Cambridge University Press.

Sagoff, M. 2000. Environmental economics and the conflation of value and benefit. Environ. Sci. Technol. 34(8):1426–1432.

SC&A, Inc. 1999. Special five-year review report for Denver Radium Site, S.W. Shattuck Chemical Operable Unite #8, November 12, 1999.

Sexton, K., and R. Zimmerman. 1999. The emerging role of environmental justice in decision making. Pp. 419–443 In: Better environmental decisions: strategies for governments, businesses, and communities. K. Sexton, A. A. Marcus, K. W. Easter, and T. D. Burkhardt (eds.). Washington, DC: Island Press.

Schwarzenbach, R. C., R. W. Scholz, A. Heitzer, B. Staubli, and B. Grossmann. 1999. A regional perspective on contaminated site remediation: fate of materials and pollutants. Environ. Sci. Technol. 33(14):2305–2310.

Siegel, L. 2001. Emissions questioned at redeveloped MEW Superfund site. Center for Public Environmental Oversight. Available at http://www.cpeo.org/lists/brownfields/2001/msg00122.html.

Spash, C. L. 2000. Multiple value expression in contingent valuation: economics and ethics. Environ. Sci. Technol. 34(8):1433–1438.

Spehn, D. 2001. Presentation to the NRC Committee on Environmental Remediation at Naval Facilities. San Diego, CA. February 28.

Stroup, R. 1991. Chemophobia and activist environmental antidotes: is the cure more deadly than the disease? Economics and the Environment 193.

USAEC. 1998. U.S. Army restoration advisory board and technical assistance for public participation guidance. SFIM-AEC-ERP (200). Aberdeen Proving Ground, MD: Army Environmental Center.

Webler, T., H. Kastenholz, and O. Renn. 1995. Public participation in impact assessment: a social learning perspective. Environmental Impact Assessment Review 15:443–463.

Appendixes

Appendix A

ACRONYMS

AFCEE	Air Force Center for Environmental Excellence
APG	Aberdeen Proving Ground
ARAR	Applicable or relevant and appropriate requirement
ASM	Adaptive Site Management
ASTD	Accelerated Site Technology Deployment
ASTM	American Society for Testing and Materials
BGS	Below ground surface
BTEX	Benzene, toluene, ethylbenzene, and xylene
BRAC	Base Realignment and Closure
CAD	Confined aquatic disposal
CDF	Confined disposal facility
CERCLA	Comprehensive Environmental Response, Compensation, and Liability Act
COC	Constituent of concern
CVOC	Chlorinated volatile organic compound
DCE	Dichloroethylene
DNAPL	Dense nonaqueous phase liquid
DOD	Department of Defense
DOE	Department of Energy
EPA	Environmental Protection Agency
ESTCP	Environmental Security Technology Certification Program
ETV	Environmental Technology Verification Program

FFERDC	Federal Facilities Environmental Restoration Dialogue Committee
FRTR	Federal Remediation Technologies Roundtable
GAO	General Accounting Office
GIS	Geographic information systems
GWETER	Groundwater Extraction and Treatment Effectiveness Review
GWRTAC	Ground-Water Remediation Technologies Analysis Center
HSRC	Hazardous Substance Research Center
ISTD	*In situ* thermal desorption
LIBS	Laser-induced breakdown spectroscopy
LIF	Laser induced fluorescence
LNAPL	Light non-aqueous phase liquid
MAROS	Monitoring and Remediation Optimization System
MCL	Maximum contaminant level
MDP	Management decision period
MIP	Membrane interface probe
MNA	Monitored natural attenuation
MTBE	Methyltertbutylether
NAPL	Nonaqueous phase liquid
NETTS	National Environmental Technology Test Site
NFESC	Naval Facilities Engineering Service Center
NCP	National Contingency Plan
NPL	National Priorities List
NRC	National Research Council
ISO	*In situ* chemical oxidation
ONR	Office of Naval Research
PAH	Polyaromatic hydrocarbon
PCB	Polychlorinated biphenyl
PCE	Perchloroethylene
PRB	Permeable reactive barrier
PRP	Potentially responsible party
PVC	Polyvinyl chloride
RAB	Restoration Advisory Board
RCRA	Resource Conservation and Recovery Act
RPM	Remedial project manager
ROD	Record of Decision
RTDF	Remediation Technology Development Forum
SARA	Superfund Amendments and Reauthorization Act
SBRP	Superfund Basic Research Program

SCAPS	Site Characterization and Analysis Penetrometer System
SCM	Site conceptual model
SEAR	Surfactant enhanced aquifer remediation
SERDP	Strategic Environmental Research and Development Program
SITE	Superfund Innovative Technology Evaluation
SVE	Soil vapor extraction
SVOC	Semi-volatile organic compound
SWAC	Surface area weighted average concentration
TCE	Trichloroethylene
TI	Technical impracticability
TIO	Technology Innovation Office
UAD	Unconfined aquatic disposal
UPL	Upland disposal
UPS	Upland secure disposal
USAFRL	U.S. Air Force Research Laboratory
VC	Vinyl chloride
VOC	Volatile organic compound
XRF	X-ray fluorescence spectrometry

Appendix B

REMEDIATION CASE STUDIES

Many methods for remediation of contaminated soil and groundwater are characterized by an initial phase of relatively high effectiveness, followed by a prolonged period of much lower effectiveness. This appendix includes case studies that document this system performance over time. These studies can be found in the general scientific literature, as well as in various reports issued by the U.S. EPA, DOE, and other government organizations. The case studies summarized here are all taken from a set of volumes published by the U.S. EPA under the auspices of the Federal Remediation Technologies Roundtable (FRTR). This information is available in hard copy, on CD-ROM, or via the web at http://www.frtr.gov. An attempt was made to balance the quality of the available data with the goal of presenting a variety of remediation technologies and contaminants. Only a very small portion of the information contained in these case studies is presented here.

Pump-and-Treat Systems

City Industries Superfund Site, Orlando, Florida (EPA 542-R-98-014)

This is a former hazardous waste Treatment, Storage, and Disposal Facility where the contaminants of concern include chlorinated solvents and BTEX. Maximum contaminant concentrations detected prior to remediation operation include 1,1-DCE (6,000 µg/l), acetone (146,000 µg/l), methylene chloride (165,000 µg/l), vinyl chloride (2,400 µg/l), and

333

MIB78,000 µg/l). The pump-and-treat system consists of 13 wells installed across the width of the initial contaminant plume.

The following figure (Figure 3 in the report) shows the average of the total VOCs in all the monitoring wells at the site. There is a trend of decreasing concentration, but levels of all VOCs remain above cleanup goals. The "tailing" effect is shown more dramatically in the time history of the total VOCs at monitoring wells located in the most heavily contaminated portion of the plume. This figure is also shown below (Figure 6 in the report); note the logarithmic scale for concentration.

Finally shown is the overall mass removal (daily rate and cumulative removal—Figure 7 in the report). Although the initial mass removal rate is much larger than its value at later times, the rate does not seem to continually decline as observed at many other pump-and-treat sites; that is, the cumulative mass removal seems to continually increase. This is attributed to the relatively homogeneous and high hydraulic conductivity at the site.

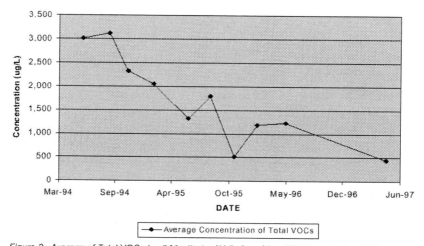

Figure 3. Average of Total VOCs in all Monitoring Wells from May 1994 through May 1997

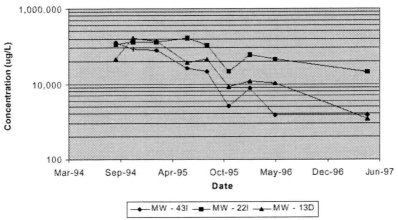

Figure 6. Total VOCs Concentrations in Highly Contaminated Wells, May 1994 through May 1997

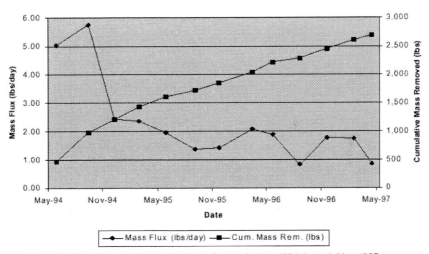

Figure 7. Mass Flux and Cumulative Mass Removal, June 1994 through May 1997

United Chrome Superfund Site, Corvallis Oregon (EPA 542-R-98-014)

This is an example where the contaminant of concern is a heavy metal rather than an organic compound. The site is a former chrome plating facility that operated from 1956 until early 1985. The site hydrogeology consists of a shallow surficial aquifer about 15 to 18 feet thick, a relatively thin aquitard, and a deep aquifer about 15 to 25 feet thick. Testing in 1983 to 1984 revealed chromium concentrations up to 3,619 mg/l in the shallow aquifer and 30 mg/l in the deep aquifer. The pump-and-treat system currently consists of nine wells in the upper aquifer and one well in the deep aquifer. This is a subset of the original system which consisted of 23 upper aquifer wells and seven deep aquifer wells; wells were retired from operation as remediation progressed and chromium levels decreased.

The following figures (Figures 2, 3, and 4 from the report) show (1) the time history of the average chromium concentration measured in the extracted groundwater and the mass removal rate and cumulative mass removed for the (2) upper and (3) lower aquifers. The figures show that the system removal rate has decreased over time. Cleanup goals for the site are 10 mg/l for the upper aquifer and 0.1 mg/l (the MCL) for the lower aquifer. These goals have been met at 11 out of 23 wells in the upper aquifer and at six out of seven deep aquifer wells.

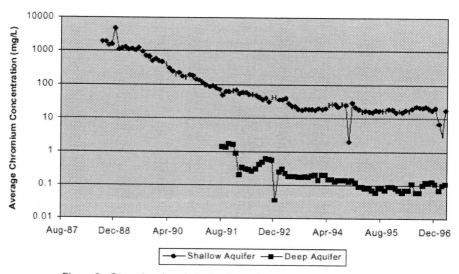

Figure 2. Chromium Levels in the Groundwater as a Function of Time

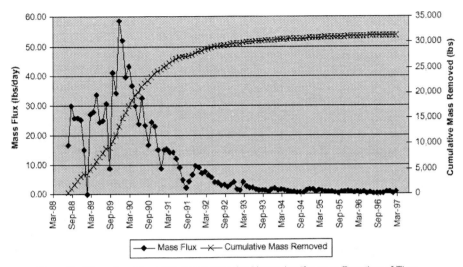

Figure 3. *Chromium Mass Removed from the Upper Aquifer as a Function of Time*

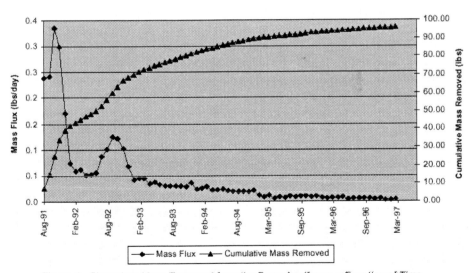

Figure 4. *Chromium Mass Removed from the Deep Aquifer as a Function of Time*

Groundwater Containment at OU1 Area A, Shaw AFB, SC (EPA 542-R-98-012)

This case study is similar to the Pope Air Force Base case study described in Chapter 3, and it includes a more detailed economic analysis of remediation system cost performance. At this site there is an estimated 400,000 gallons of JP-4 free product, with an extensive dissolved-phase groundwater plume. As part of an Interim Response at Area A, a groundwater containment and free product recovery system were installed. The groundwater containment system consisted of nine recovery wells and a groundwater treatment system.

The performance of the containment and free product removal systems are shown below (Figures 17 and 18, respectively, from the report). It can be seen that both systems eventually reach a point where removal rates are negligible. The free product system also exhibits an early period of low removal rate; the reasons for this are not explained in the report.

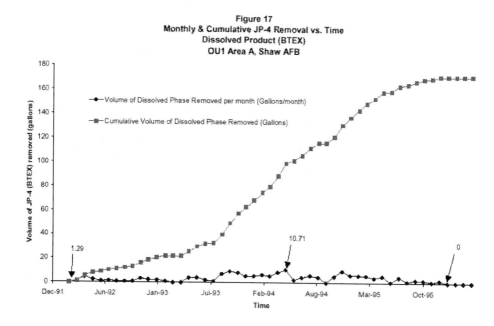

Figure 17
Monthly & Cumulative JP-4 Removal vs. Time
Dissolved Product (BTEX)
OU1 Area A, Shaw AFB

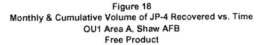

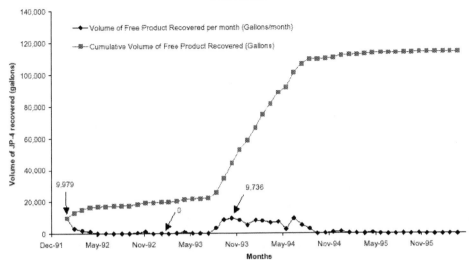

Figure 18
Monthly & Cumulative Volume of JP-4 Recovered vs. Time
OU1 Area A, Shaw AFB
Free Product

Data were also available on monthly operations and maintenance costs at this site. Combining this data with the removal information gives the graph shown next (Figure 19 from the report). This figure shows the relationship between pollutant removal and cost and again is a dramatic display of how the marginal cost of removal (slope of the curve) increases over time. (The large marginal costs at the start are due to the low effectiveness of the free product recovery system at early time, as previously mentioned in relation to Figure 18). In November 1996 the containment and recovery system was shut down because the operating objectives were no longer being met.

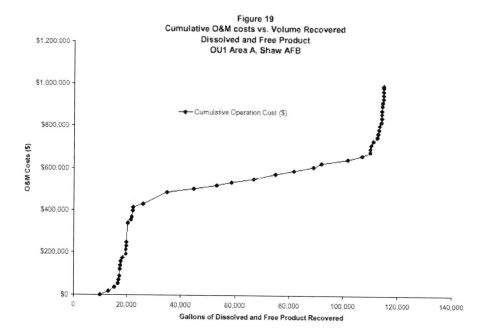

Figure 19
Cumulative O&M costs vs. Volume Recovered
Dissolved and Free Product
OU1 Area A, Shaw AFB

Soil Vapor Extraction (SVE) Systems

Sacramento Army Depot Superfund Site, Burn Pits OU (EPA 542-R-98-013)

As part of regular operations at the Sacramento Army Depot, a variety of wastes were generated. The Burn Pits Operable Unit was the location of two rectangular trenches constructed in the late 1950s and used intermittently as incineration pits until 1966. Materials reportedly buried and/or burned in the pits included plating shop wastes, oil and grease, batteries, and construction debris. Studies in 1981 revealed that groundwater under a portion of the Depot was contaminated, and the most likely source of contamination was identified as the burn pits.

A Record of Decision (ROD) for the Burn Pits OU was signed in 1993. In addition to soil excavation and institutional controls, the ROD required SVE for all soils in the area containing VOCs. The SVE system used was a special patented fluid injection/vapor extraction system, that included both injection and extraction wells to produce relatively larger subsurface pressure gradients and higher flow rates of extracted vapors than would be achieved using conventional vapor extraction technology. The SVE system was operated from May 1994 to January 1995 and again from March 1995 to September 1995, for a total of 347 days of run time.

The mass removal rate and cumulative mass removal of TCE, PCE, and 1,2-DCE for the first six months of operation are shown in the figures below (Figures 7 and 8, respectively, from the report). Again, these figures demonstrate typical "tailing" behavior; for example, the mass extraction rate decreased from an average of 4 lbs/day during the first 20 days of operation to less than 1 lb/day after 40 days of operation. More than 80 percent of the total mass was removed during the first 42 days of operation. Soil borings collected in September 1995 after the system was shut down confirm that the average concentrations of each of the three target compounds were less than the cleanup standards set in the ROD.

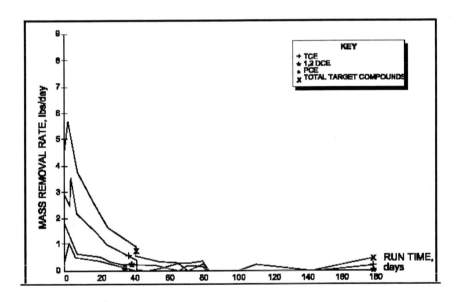

Figure 7. Mass Removal Rates of TCE, PCE, and 1,2-DCE

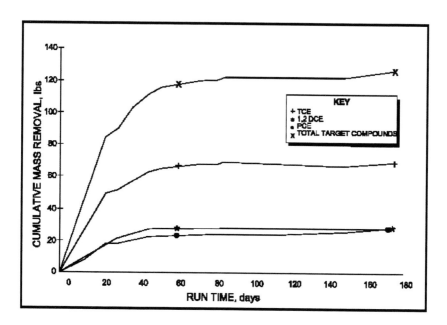

Figure 8. Cumulative Mass Extracted of TCE, PCE, and 1,2-DCE

Intersil/Siemens Superfund Site, Cupertino, CA (EPA 542-R-98-012)

At this site, TCE contamination of soil and groundwater was caused by leaks and spills of solvents used in semiconductor manufacturing. In order to address soil contamination, an interim SVE system consisting of four vertical extraction wells began operating in May 1988, and as part of the final remedy specified in the ROD this system was expanded in May 1991 to include three additional wells.

The mass removal rate over time is shown below (Figure 2 in the report). It can be seen that the removal rate for TCE decreased from approximately 15 lbs/day to less than 0.5 lbs/day from May 1988 to December 1992. The system was shut down in August 1993, after determination that the remedial goal of 1 mg/kg total VOCs had been satisfied.

Figure 2. SVE Total System Removal Rate and Cumulative Removal Mass of TCE (May 1998 - Dec 1992)

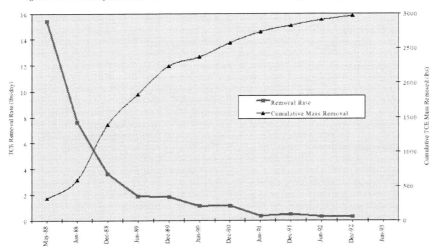

Thermal Processes

Broadhead Creek Superfund Site (EPA 542-R-98-013)

This is the site of a former coal gasification plant that operated from 1888 until 1944. Coal tar from these operations was disposed of in open pits. Free phase coal tar was identified at the site, in addition to soil and groundwater contamination with PAHs, other SVOCs, VOCs, and metals. An enhanced thermal recovery process was selected for removal of free-phase coal tar from the subsurface soils. The Contained Recovery of Oily Waste (CROW)^{TM} process was used; this involves injection of hot water to decrease coal tar viscosity and recovery of the water and coal tar via extraction wells.

It is much more difficult to estimate recovery for this process than for conventional pump-and-treat and SVE processes. Nevertheless, an estimate of the cumulative recovery as a function of pore volume is given in the graph below (Figure 4 from the report). It can be seen that the majority of the coal tar recovered occurred in the during the early stage of operation; for example, approximately half the coal tar was recovered in

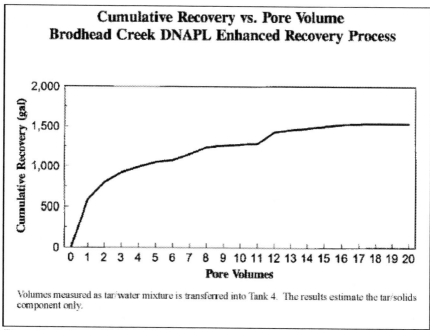

Figure 4. Estimated Cumulative Recovery of Tar Over Life of Project

the first three pore volumes, while an additional 17 pore volumes were required to recover the second half. The CROW [TM] system was in operation from m December 1994 until June 1996, when the EPA determined that the performance standard had been met.

Innovative Technologies

There are a number of case studies that considered technologies such as permeable reactive barriers (PRBs), air sparging, and *in situ* bioremediation, usually in conjunction with pump-and-treat systems. In general, the data for these newer technologies is less developed than for the case studies presented above.

Pump-and-treat and Permeable Reactive Barriers to Treat Contaminated Groundwater at the Former Intersil Site, Sunnyvale, CA (EPA 542-R-98-015)

This is a chlorinated solvent site at the location of a semi-conductor manufacturing facility. A pump-and-treat system operated at the site from 1987 to 1995. The original system was expanded in 1989 and 1991. The mass removal by the pump-and-treat system was steadily declining over time, as would be expected.

Due to the declining efficiency of the pump-and-treat system, in 1993 Intersil examined alternative technologies. The selected technology was an *in situ* granular iron treatment wall system. The added benefit of this technology is that it would allow Intersil to dismantle the pump-and-treat system and return the property to leasable/sellable conditions.

The following figures (Figures 4 and 5 from the report) show the average concentration in monitoring wells across the plume over time, and the mass removal rate. Although it appears that the treatment wall has a lower mass removal rate than the pump-and-treat system, the former technology is passive and should be less costly over the long term. Also, use of the treatment wall had the added benefit of allowing sale/lease of the property. (No information was given whether or not such a transaction has occurred.)

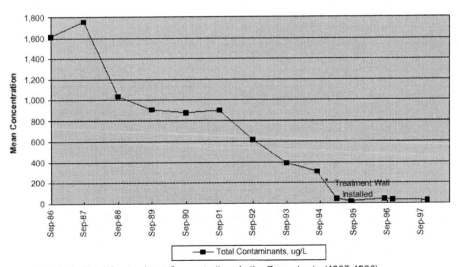

Figure 4. Total Contaminant Concentrations in the Groundwater(1987-1996)

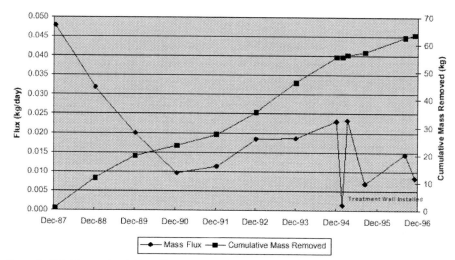

Figure 5. Total Contaminant Mass Flux and Mass Removed as a Function of Time (1987-1996)

Pump-and-treat and In Situ Bioremediation at the French Ltd. Superfund Site, Crosby TX (EPA 542-R-98-015)

This was a former sand mine site that was permitted by the State of Texas to accept industrial waste from 1966 to 1971. The facility's permit was revoked in 1973 and it was placed on the NPL in 1981. Contaminants of concern in groundwater include benzene, toluene, chloroform, 1,2-DCA, and vinyl chloride.

Active remediation was conducted at the site from January 1992 through December 1995. Initially groundwater extraction and above-ground treatment was used. Then an enhanced *in situ* bioremediation strategy was instituted. Nitrate and diammonium phosphate was mixed with clean water and injected for 90 days, followed by injection of oxygenated water for 44 months.

The figures below (Figures 3 and 5 from the report) show the time history of average contaminant concentrations in monitoring wells and the mass removal rate. It can be seen that the contaminant levels are declining. There was no documentation provided regarding what portion of the removal could be attributed to *in situ* bioremediation.

A modeling study conducted in 1995 demonstrated that natural attenuation would reduce contaminant concentrations below the remedial goals within ten years after shut down of the system. As a result, EPA allowed active systems to be shut down in December 1995.

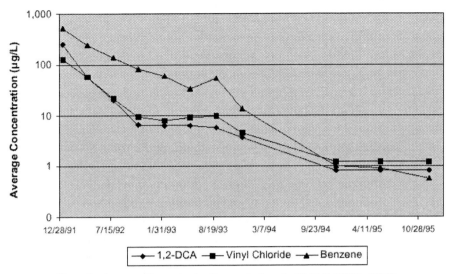

Figure 3. Average Groundwater Concentrations in S1 Unit (1992 - 1995)

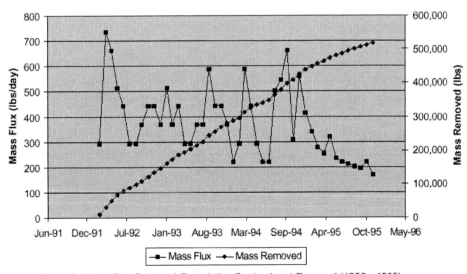

Figure 5. Mass Flux Rate and Cumulative Contaminant Removal (1992 - 1995)

Appendix C

Tables of Innovative Technology Demonstration Projects

TABLE C-1 Summary of In Situ Oxidation (Phase II) Field Sites

Site Location	Area of Concern	Contaminants of Concern	Regulatory Driver	Oxidant	Scale	Remedial Objectives	Ability to Meet Objectives	Follow-up Actions
Anniston Army Depot Calhoun County, AL	SWMU 12 soils in the former industrial lagoon area	VOCs in soil, primarily TCE	RCRA Corrective Action –Emergency Removal Action	Hydrogen peroxide	Pilot and full scale	Reduce chemical contamination that may be contributing to exceedances of health-based concentration limits in onsite and offsite groundwater	Reports claim up to 90% removal of total VOCs. Post-treatment sampling data show several areas above the 41-ppm TCE soil cleanup criteria.	Additional polishing treatment in selected locations.
Cherry Point UST Bogue, Carteret County, NC	Vadose zone soils and groundwater in former UST area.	Gasoline and diesel range organics in soil. VOCs in groundwater, primarily benzene	NCDENR environmental regulations and site cleanup criteria	Hydrogen peroxide	Pilot scale	Demonstration project to remediate soil and groundwater within the 1,000 µg/L benzene contour interval to levels acceptable to the NCDENR.	Project caused pavement upheaval, underground explosions, and fire. Post-incident sampling indicates that significant contamination still remains that will most likely require further site characterization and remediation by other means.	Additional site characterization and remediation alternatives will have to be evaluated.
CRREL Grafton County, NH	AOC 2 (Former TCE UST) & 9 (Ice Well) Vadose Zone Soils	TCE and MEK in soil.	Voluntary cleanup. NHDES is the primary oversight agency.	Potassium perman-ganate	Pilot and full scale	Reduce soil concentrations of TCE in vadose zone source area.	Pilot-scale testing is currently inconclusive. Increases in chloride concentrations in pore water during injection provide evidence that the dechlorination reaction is occurring, although pre- and post-injection soil samples indicate limited effect.	Full-scale trial is currently in operation.

Site Location	Area of Concern	Contaminants of Concern	Regulatory Driver	Oxidant	Scale	Remedial Objectives	Ability to Meet Objectives	Follow-up Actions
Kings Bay NSB Camden County, GA	Site 11- Former sanitary landfill along the western boundary of the NSB with contaminant plume moving toward residential area	VOCs in soil, primarily PCE	RCRA cleanup under a GDEP consent order	Hydrogen peroxide	Full scale	Aggressive source reduction with chemical oxidation to 100 ppb VOCs in source area	In situ oxidation was able to reduce total VOCs in the primary treatment zone to below 100 ppb. The success of this project may be linked to the sandy soil with high hydraulic conductivities (30 ft/day). GDEP rescinded the consent order and allowed the shutdown of the pump and treat system.	Natural attenuation to polish residuals outside the source area that are less than 100 ppb.
LEAD Franklin County, PA	OBP vadose zone Soils	VOCs in soil, primarily TCA	The OBP is located within an NPL site at LEAD; working under an IAG with USEPA/PADEP.	Hydrogen peroxide	Full scale	Reduce soil concentrations of select VOCs below Act 2 Soil Cleanup Criteria for groundwater.	In situ oxidation provided significant removals of contaminants. However, in situ oxidation did not meet cleanup objectives, and additional alternatives will have to be evaluated to achieve greater reductions.	LEAD evaluating enhanced biological treatment; soil vapor extraction; hot spot excavation; and natural attenuation
LEAD Franklin County, PA	SE DA bedrock aquifer	VOCs in groundwater, primarily TCE	The entire SE Area including the DA is an NPL site.	Hydrogen peroxide	Pilot scale	Evaluate pilot test results to determine whether ISO alone or combined with other technologies can be used for full-scale remediation	Not available	Not available
Nellis AFB Northeast of Las Vegas, NV	Site ST-44 along the flight line with a plume of TCE-impacted groundwater	TCE in saturated soils and groundwater	Environmental investigations undertaken through the IRP and overseen by the Nevada DEP; must comply with Nevada ARARs	Ozone sparging	Pilot scale	Determine the feasibility of using in situ ozone sparging to reduce TCE contamination at ST-44	While in situ ozone sparging appeared to be able to reduce TCE contamination at ST-44 by varying amounts in some wells and spargers, there were increases in other wells and rebound was seen in 4 out of 5 wells and 1 out of 3 spargers.	Full-scale treatment with system modifications

Site Location	Area of Concern	Contaminants of Concern	Regulatory Driver	Oxidant	Scale	Remedial Objectives	Ability to Meet Objectives	Follow-up Actions
Pensacola NAS Pensacola, FL	Former sludge drying beds were open surface impoundments that accepted paint wastes and listed hazardous waste	VOCs in groundwater, primarily TCE	State order based on monitoring data showing impact to groundwater. RCRA-regulated soils were removed and a P&T system installed.	Hydrogen peroxide	Full scale	To significantly reduce contamination in the aquifer	Groundwater results after Phase 2 indicate that in situ oxidation was successful in remediating chlorinated organics found in the treatment zone. However, the site experienced a rebound after Phase I, and the RPM expects it again after Phase 2.	Performing a phased evaluation of natural attenuation as a polishing technology
Shaw AFB Sumter, SC	OU 4 – Former Fire Training Area No. 1. Soil and groundwater contaminated from the use of combustible liquids in the fire training exercises	VOCs in soil and groundwater, primarily TCA and DCA	Site under an ACO. Pilot test performed under TERC.	Hydrogen peroxide	Pilot scale	1. Determine if groundwater contamination at OU-4 can be treated and significantly reduced using hydrogen peroxide solution. 2. Define the radius of influence of an injection well. 3. Gather sufficient data to support the design of a full-scale remediation system.	1. Pilot test showed that groundwater can be partially treated with in situ oxidation; it remains to be seen whether significant reductions achievable. 2. Pilot test further defined the radius of influence of an injection well at the site based on the interpolation of several different field measurements. 3. Pilot test data allowed contractors to adjust chemical requirements, further define radius of influence, and estimate an approximate cost for full-scale remediation. More information is required for remediating contaminants in vadose zone and the lower portion of the aquifer, for selecting appropriate injection rates for large-scale areas, and controlling releases of VOCs to the air.	The draft Pilot Test Report recommended performing an air sparging pilot test using the existing injector and monitoring well system

Site Location	Area of Concern	Contaminants of Concern	Regulatory Driver	Oxidant	Scale	Remedial Objectives	Ability to Meet Objectives	Follow-up Actions
DOE Kansas City Plant Kansas City, MO	Former Ponds Site with VOC contamination in vadose and saturated zone soils	TCE and DCE in soil	Not available	Potassium permanganate	Field demonstration	Evaluate feasibility of degrading VOCs in situ by addition of $KMnO_4$ using a DSM process and evaluate impact of $KMnO_4$ addition to chemical, physical, and biological properties of the soil being treated	Oxidant introduction through the DSM process resulted in significant reductions of TCE within the vertical profile of the soil columns and homogenization of the treatment region. Average reduction of TCE levels by 67% in the test cells compared favorably with the 70% treatment goal.	Not discussed
DOE PORTS Piketon, OH	Former area X-701B holding pond used for the neutralization and settling of metal-bearing acidic wastewater and solvent-contaminated solutions	TCE in groundwater	Pond was closed under RCRA closure action; site agreed to collaborate with ORNL and support ISO field test at Area X-701B	Potassium permanganate	Full scale demonstration	Field-scale treatability study of ISCOR to evaluate effectiveness in reducing sources of groundwater plumes and minimizing time pump-and-treat facilities are required to be operational	In situ oxidation via injection of $KMnO_4$ solution resulted in dramatic removal of TCE from the Gallia aquifer. However, the persistence of TCE in surrounding units will result in recontamination of the Gallia aquifer over time. Groundwater samples collected 12 weeks after the conclusion of the test suggest that the rate of contaminant rebound will be slow, and that the ISCOR test was successful in reducing the overall mass of TCE within the aquifer unit.	Monitor TCE groundwater levels to determine if further action is needed.

Site Location	Area of Concern	Contaminants of Concern	Regulatory Driver	Oxidant	Scale	Remedial Objectives	Ability to Meet Objectives	Follow-up Actions
DOE Savannah River Site Aiken, SC	A/M Area	PCE and TCE in vadose zone soils and soils below the water table	A/M Area RCRA Groundwater Corrective Action; Integrated Demonstration R&D Activities	Hydrogen peroxide	Pilot scale	Pilot-scale demonstration to evaluate the ability of Fenton's Reagent to destroy DNAPL (TCE & PCE) at a field site and assess the efficiency of Fenton's Reagent deployed at depth (150 ft).	In situ oxidation provided significant removals of DNAPL found below the water table in the test zone. Although in situ oxidation met the pilot test objectives (verify an alternative DNAPL destruction technology), additional remediation technologies would have to be used in conjunction with in situ oxidation to meet typical aquifer protection standards.	Not discussed
BMC Olen Irvine, CA	Site includes an operating plant that covers much of the source area	TCE, PCE, and MC in vadose zone soils and below the water table	Regional Water Quality Control Board	Potassium permanganate	Pilot and full scale	Phased objectives: 1. Reduce dissolved TCE and MC levels to asymptote. 2. Turn off active remediation (vapor recovery and groundwater extraction). 3. Obtain site closure. 4. Achieve dissolved TVOC levels below 500 µg/L site-wide.	The treatment met the most critical goals of: 1. reducing dissolved-phase CVOC levels (estimated at 97% reduction, to low ppb levels in the injection zone), and 2. terminating active remediation by vapor and groundwater recovery. The treatment did not meet the further objectives of: 1. site closure without long-term monitoring (partly due to the unexpected presence of MC), and 2. a site-wide average dissolved TCE concentration of < 500 µg/L.	Long-term monitoring will be required as part of the monitored natural attenuation strategy for residual contaminants, particularly MC

SOURCE: ESTCP. 1999. Technology Status Report: In Situ Oxidation. http://www.cstep.org/documents/techdocs/index.cfm (November).

TABLE C-2 Summary of Selected Steam Injection Projects for Subsurface Remediation

Site	Contaminant Concentration/Volume	Geology	Treatment System Design	Removal Efficiency or Volume	Comments
Solvent Services, Inc. San Jose, CA	VOCs and nonvolatile organic contaminants at concentrations > 1,000 ppm	Silts and clays, continuous poorly sorted sand layer at bottom, 0.61 to 1.5 m	7.3-m² area treated by 6 injection wells and 1 extraction well, 1.5 m between wells, 111.6 kg/hr of steam injected for 120 hrs, then 67 kg/hr for 20 hrs	Vacuum extraction: 99 kg in 40 hrs. Steam extraction: 146 kg in 140 hrs. Followed by intermittent operations.	Pilot-scale demonstration
Annex Terminal Port of Los Angeles, San Pedro, CA	Major contaminants were TCE, PCE, chloro-benzene. Initial average concentration of 466 ppm VOCs			33-m² area treated to 1.5-m depth by steam (200°C) and compressed air (135°C)	84.7% VOCs, 55% SVOCs. Treatment time was 1 month	Detoxifier system developed by NovaTerra, Inc.
AT&T New York	Chlorinated solvents, TCE & 1,1,1-TCA, DNAPL (separate and dissolved aqueous phase)	Tight, heterogeneous		4,500 kg of hydrocarbons recovered in 2 years	Nutrients injected at 40°C, vacuum extraction of 7.8 m³/min, maximum vacuum of 41 cm Hg
Yorktown Naval Shipyards	Naval Special Fuel Oil, estimated 8,000 L [a]	Upper 6 m was homogeneous fine to coarse sand, below that was interbedded sands and clays. Water table 3.8 to 4.1 m below ground. Hydraulic conduct. 2.0 to 5.2 x 10⁻³ cm/s	83.6 m² area treated with five spot pattern of four injection wells and one extraction well, 9.1 m between injection wells; injected at 6.1- to 7.6-m depth, extracted from 3- to 9.1-m depth; Injection rate of 272 kg/hr	617 L recovered. Steam injected over 2-month period	Hot water may be more appropriate for this nonvolatile, viscous oil
Naval Air Station Lemoore, CA	JP-5, estimated 757,000 L	Sands and silts with hydraulic conduct. 3.9 x 10⁻³ to 1.4 x 10⁻² cm/s. Water table at 4.9 m	12,140-m² area treated with two injection wells at the center, and eight vapor/groundwater extraction wells; Injection depth of 6 m	Approximately 976,000 L recovered in 3 months of operation. Final vadose zone concentration of 20 to 50 ppm TPH; 20,000 ppm remains at water table	Demonstration project
Pinellas Plant Northeast Site, Largo, FL	Volatile organic compounds including BTEX and chlorinated solvents	Silty sands, water table at 1 m below ground surface	1,566 m³ treated by 48 holes to a depth of 9.8 m	Approximately 544 kg recovered	Dual auger rotary steam stripping demonstration project

SOURCE: Davis, E. L. 1998. Steam Injection for Soil and Aquifer Remediation, Ground Water Issue, EPA/540/S-97/505, January.

[a] The report does not contain an estimate of the amount of oil contained in the area treated by the pilot study. This estimate was made using the same assumptions made in the report to estimate the oil contained within the entire contaminated region, with an estimate of the contamination extending to a depth of 1.5 m.

TABLE C-3 Compilation of In Situ Thermal Treatment Projects

Project	Contaminant(s)	Technology	Status
LLNL Gas Pad	Gasoline	Steam Injection	Complete (115,000 lbs recovered)
Visalia Pole Yard NPL Site	Creosote, pentachlorophenol	Steam Injection	Ongoing full-scale (1 million lbs recovered in 18 months)
Skokie, IL	TCE	6-phase heating	Full-scale cleanup completed
Seattle, WA	PCE	6-phase heading	Full-scale cleanup completed
Ft. Richardson, AK	PCE	6-phase heading	Full-scale cleanup completed
Atlanta, GA	Diesel fuel	6-phase heading	Full-scale cleanup completed
Portland, IN	PCE, TCE	In situ thermal desorption	Full-scale cleanup completed
Tanapag, Saipan, NMI	PCBs	In situ thermal desorption	Full-scale cleanup completed
Fuel Terminal, Eugene, OR	Diesel fuel	In situ thermal desorption	Full-scale cleanup completed
Naval Facility, Ferndale, CA	PCBs	In situ thermal desorption	Full-scale cleanup completed
Dragstrip, Glen Falls, NY	PCBs	In situ thermal desorption	Demonstration project completed
Mew, Cape Girardeau, MO	PCBs	In situ thermal desorption	Demonstration project completed
Navy BADCAT, Vallejo, CA	PCBs	In situ thermal desorption	Demonstration project completed
Lemoore NAS, CA	JP-5	Steam injection	Full-scale cleanup completed
Petrochemical, TX	Solvents	3-Phase electrical heating	Sequential full-scale cleanup of hot spots
NAS North Island, San Diego, CA	TCE, JP-5	In situ thermal treatment	Full-scale cleanup underway following successful pilot
Yorktown Navy Facility, VA	Fuel oil	In situ steam heating	Full-scale project underway. Steam in pipes used to reduce viscosity, facilitate recovery in trenches
Rainbow Disposal, Huntington Beach, CA	Diesel fuel	Steam injection	EPA SITE demonstration
DESC, Whittier, AL	JP-5	Steam injection	Full-scale design and construction
Bulk oil plant, Jacksonville, FL	Motor oil	Steam injection	Full-scale design/startup Fall 2000
Metal recycling facility, Boston, MA	Heavy machine oil	In situ thermal treatment	Procurement and fabrication underway
Aircraft engine plant, Lynn, MA	PCBs	Steam injection	Design completed/implementation 2001
Safety Kleen Breslau, Ontario	PCBs	In situ thermal treatment	Pilot test

Project	Contaminant(s)	Technology	Status
DESC, San Pedro, CA	Diesel fuel	In situ thermal treatment	Pilot test
PSNS, Bremerton, WA	Fuel oil	In situ thermal treatment	Pilot test
Ft. Hood, TX	JP-8	Steam injection/3-phase heat	Demonstration
Panama City, FL	Diesel fuel	Steam injection	Full-scale
Plating facility, Danbury, CT	CVOCs	Steam injection	Full-scale designed and constructed
DOE Savannah River, SC	PCE, TCE	Steam injection	Contract awarded
A.G. Communications, North Lake, IL	Solvents	Steam injection	Ongoing
Waukegon, IL	Methylene chloride	6-phase heating	Full-scale cleanup in progress
Long Beach, CA	PCE	6-phase heating	Pilot project (awarded)
Portland, OR	TCE	6-phase heating	Full-scale cleanup (awarded)
Newark, CA	EDB	6-phase heating	Pilot project (awarded)
Air Force Plant 4, Ft. Worth, TX	Solvents	6-phase heating	Pilot project underway
Holyoke, MA	Styrene	Steam injection	Ongoing
Alameda NAS	TCE, diesel, motor oil	Steam injection	Pilot project completed
DOE Portsmouth, OH	TCE	Steam injection	Pilot project completed
Solvent Services, San Jose, CA	Chlorinated solvents	Steam injection	Pilot project completed
Port of Ridgefield, WA	PAHs, PCP	Steam injection	Contract awarded
Cape Canaveral, FL	TCE	Steam injection/oxidation	Joint DOD/DOE/EPA/NASA "treat-off"
Mobil Oil	Petroleum	RF Heating	Full-scale completed
Ashland Refinery, St. Paul, MN	Petroleum	Microwave Heating	Full-scale completed
Wyckoff Wood Treater NPL Site	Creosote Pentachlorophenol	In situ thermal treatment	Signed ROD, conceptual design underway
Rocky Mt. Arsenal Hex Pit Commerce City, CO	Pesticides	In situ thermal desorption	Full-scale design/EPA SITE demonstration
Pole Yard, Alhambra, CA	Creosote, PAHs	In situ thermal desorption	Under contract
N. Ryan St. Site, Lake Charles, LA	PAHs, PCBs	In situ thermal desorption	Administrative Order on Consent Action Memorandum issued

SOURCES: (1) GWRTAC. 2000. Advances on Innovative Ground-Water Remediation Technologies and In Situ Thermal Treatment, Conference Proceedings, Boston, MA, June and (2) J. Cummings, 2001, EPA Technology Innovation Office, personal communication.

Appendix D

Committee Member And Staff Biographies

EDWARD J. BOUWER, *Chair,* is a professor of environmental engineering at the Johns Hopkins University. His research interests include biodegradation of hazardous organic chemicals in the subsurface, biofilm kinetics, water and waste treatment processes, and transport and fate of bacteria in porous media. Dr. Bouwer has served on several NRC committees, including the U.S. National Committee for SCOPE, the Steering Committee on Building Environmental Management Science Programs, and the Committee on Groundwater Cleanup Alternatives. He received a Ph.D. in environmental engineering from Stanford University. Dr. Bouwer is currently director of an EPA Hazardous Substance Research Center at Johns Hopkins.

SIDNEY B. GARLAND is the manager for strategic planning at Bechtel Jacobs Co. in Oak Ridge, Tennessee. He was previously employed as a group leader, project manager, and program manager for Lockheed Martin Energy Systems, Inc. He has extensive professional experience in the development of multi-million-dollar contaminant remediation strategies, integration of related planning activities, and management of remediation projects. In particular, he is involved in long-term stewardship of contaminated sites at Department of Energy facilities, which includes remedial operations, implementation and maintenance of institutional controls, long-term monitoring, and site closure. Mr. Garland received a B.E. in civil engineering from Vanderbilt University, an M.S. in environmental health engineering from the University of Texas, and an M.P.A. from the University of Oklahoma.

PATRICK E. HAAS is employed by Mitretek Systems, a nonprofit, public interest corporation that works with federal, state, and local governments as well as with other nonprofit, public interest organizations in areas of information and scientific systems. He was previously an environmental engineer in the Technology Transfer Division of the Air Force Center for Environmental Excellence located at Brooks Air Force Base, Texas. He represented the Air Force in domestic and international efforts to improve the regulatory acceptance of innovative environmental cleanup technologies and strategies and served as project manager on several nationwide environmental remediation technology application initiatives. Mr. Haas received his B.S. in chemistry and biology from Southwestern University and his M.S. in environmental sciences from the University of Texas, San Antonio.

ROBERT JOHNSON is an engineer in the Geosciences and Information Technology Section of the Environmental Assessment Division at the Argonne National Laboratory. His current areas of research include design of precision excavation programs for soil remediation, application of geostatistical methods to the design of environmental sampling programs, numerical modeling of groundwater flow and contaminant transport, and optimization techniques applied to subsurface remedial action design. He is also a lead developer of Adaptive Sampling and Analysis Program (ASAP) characterization and remedial technologies to support cleanup activities at hazardous waste sites. Dr. Johnson received his B.S. in mathematics from Calvin College, his M.S. in environmental engineering systems from Johns Hopkins University, and his Ph.D. in soil and water resources from Cornell University.

MICHELLE M. LORAH is a research hydrologist with the U.S. Geological Survey in Baltimore, Maryland. Her research and project management expertise includes biodegradation of organic contaminants, hydrology and biochemistry in wetlands and other groundwater/surface water interfaces, and subsurface remediation at contaminated Army sites, particularly Aberdeen Proving Ground. Dr. Lorah received her B.S. in geosciences from Pennsylvania State University, her M.S. in aqueous geochemistry from the University of Virginia, and her Ph.D. in environmental chemistry from the University of Maryland.

GENE F. PARKIN is a professor of civil and environmental engineering at the University of Iowa, where he served as chair from 1990 to 1995. His teaching interests include biological wastewater treatment, environmental chemistry, and remediation of hazardous wastes. He has conducted research in bioremediation, the fate and effects of toxic chemicals (including metals) in the subsurface and aboveground treatment systems, anaerobic biological treatment, and biological nitrogen removal. He serves as the director of the Center for Health Effects of Environmental Contamination at the University of Iowa. Dr. Parkin received his Ph.D. in environmental engineering from Stanford University.

FREDERICK G. POHLAND is professor and Edward R. Weidlein chair of environmental engineering at the University of Pittsburgh. His research focuses on environmental engineering operations and processes, solid and hazardous waste management, and environmental impact monitoring and assessment. He has been a visiting scholar at the University of Michigan and a guest professor at the Delft University of Technology in the Netherlands. Dr. Pohland has previously served on many NRC committees, most recently on the Committee on Technologies for Cleanup of Subsurface Contaminants in the DOE Weapons Complex. He is a member of the National Academy of Engineering. Dr. Pohland received his B.S. in civil engineering from Valparaiso University and his M.S. and Ph.D. in environmental engineering from Purdue University.

DAN D. REIBLE is a professor of chemical engineering and director of the EPA Hazardous Substance Research Center at Louisiana State University. His research focuses on transport phenomena and their applications to environmental mechanics, especially as related to contaminated sediments and dredged materials. Dr. Reible also directs projects on the remediation of contaminated soils and a program on the transport of airborne contaminants released from sediments along coastlines. He is currently serving on the NRC Committee on Remediation of PCB-Contaminated Sediments. Dr. Reible received his Ph.D. in chemical engineering from the California Institute of Technology.

LENNY M. SIEGEL is the director of Center for Public Environmental Oversight, a project of San Francisco State University's Urban Institute. He is one of the environmental movement's leading experts on military base contamination and has worked as a consultant to a wide range of organizations. Mr. Siegel is or was recently a member of several government advisory committees including the Defense Science Board Task Force on Unexploded Ordnance, the Federal Facilities Environmental Restoration Dialogue Committee, the Subcommittee on Waste and Facility Siting of the National Environmental Justice Advisory Committee, and the Moffett Field Restoration Advisory Board. Mr. Siegel edits the Citizens Report on the Military and the Environment, and his organization runs Internet forums both on military environmental issues and Brownfields.

MITCHELL J. SMALL is a professor in the departments of civil and environmental engineering and engineering and public policy at Carnegie Mellon University. His research interests include mathematical modeling of environmental quality, statistical methods and uncertainty analysis, human exposure modeling, human risk perception and decision making, and groundwater and soil pollution monitoring. He has previously served on several NRC committees, including the Committee on USGS Water Resources Research. Dr. Small received his B.S. in civil engineering/engineering and public affairs from Carnegie Mellon University and his M.S. and Ph.D. in environmental and water resources engineering from the University of Michigan.

RALPH G. STAHL, JR., is a senior consulting associate with DuPont Engineering Corporate Remediation. His research primarily focuses on evaluating the effects of chemical stressors on aquatic and ecological ecosystems. Most recently, Dr. Stahl has been responsible for leading DuPont's corporate efforts in ecological risk assessment and natural resource damage assessments for site remediation. He is also currently the chair of the Chemical Manufacturers Association Ecological Risk Assessment Task Group. Dr. Stahl received his B.S. in marine biology, his M.S. in biology from Texas A&M University, and his Ph.D. in environmental science and technology from the University of Texas School of Public Health.

ALICE D. STARK is director of the Bureau of Environmental and Occupational Epidemiology at the New York State Department of Health. She has conducted health assessments among populations exposed to toxic substances from hazardous waste sites and other sources of environmental exposure. Dr. Stark is also an adjunct professor of anthropology at SUNY, Albany, where she was formerly an associate professor of environmental health and toxicology and epidemiology. She is currently conducting a follow-up health study at Love Canal and a Farm Family Health and Hazard Survey. Dr. Stark recently served on two National Cancer Institute committees, one of which investigated the role of xenobiotic chemicals in causing cancer.

ALBERT J. VALOCCHI is a professor of civil and environmental engineering at the University of Illinois. His research focuses upon mathematical modeling of pollutant fate and transport in porous media, with applications to groundwater contamination and remediation. Dr. Valocchi specializes in the development and application of models that couple physical, geochemical, and microbiological processes over scales ranging from the pores (micrometers) to the field (kilometers). Some current research projects include investigation of pore-scale processes controlling pollutant fate, impact of spatial variability upon the modeling of in situ biodegradation, and application of advanced scientific computers to reactive solute transport. He received his Ph.D. in civil engineering from Stanford University.

WILLIAM J. WALSH is an attorney and a partner in the Washington, D.C., office of Pepper, Hamilton LLP. He has also served as a section chief in the EPA Office of Enforcement and was the lead attorney for EPA in the Love Canal litigation, which involved four large hazardous waste landfills in Niagara Falls, New York. His legal experience encompasses environmental litigation on a broad spectrum of issues pursuant to a variety of environmental statutes, including the Resources Conservation and Recovery Act (RCRA) and the Toxic Substances Control Act (TSCA). He represents trade associations in rulemakings and other public policy advocacy, represents individual companies in environmental actions, and advises technology developers and users concerning ways to take advantage of the incentives for and eliminate the regulatory barriers to the use of innovative environmental technologies. He has previously served on several NRC committees, most recently on the Committee to Review the Army Non-Stockpile Chemical Material Disposal. Mr. Walsh received his B.S. in physics from Manhattan College and holds a J.D. from George Washington University Law School.

CLAIRE WELTY is an associate professor of civil engineering at Drexel University. Her current research entails stochastic analyses of dispersive transport in porous media and virus transport through aquifers. She has also concentrated on providing improved computer code for numerical contaminant transport models. She previously worked as an environmental scientist in the Hazardous and Industrial Waste Division at EPA. Dr. Welty received her Ph.D. in civil engineering from the Massachusetts Institute of Technology.

STAFF

LAURA J. EHLERS is a senior staff officer for the Water Science and Technology Board of the National Research Council. Since joining the NRC in 1997, she has served as study director for nine committees, including the Committee to Review the New York City Watershed Management Strategy, the Committee on Riparian Zone Functioning and Strategies for Management, and the Committee on Bioavailability of Contaminants in Soils and Sediment. She received her B.S. from the California Institute of Technology, majoring in biology and engineering and applied science. She earned both an M.S.E. and a Ph.D. in environmental engineering at the Johns Hopkins University.